Handbook on Freshwater Aquaculture

Handbook on Freshwater Aquaculture

Editor

Abilio Cuevas

scitus academics

Handbook on Freshwater Aquaculture

Edited by **Abilio Cuevas**

ISBN: 978-1-68117-540-9
Library of Congress Control Number: 2016942940

www.scitusacademics.com
Box No. 4766, 616 Corporate Way,
Suite 2, Valley Cottage,
NY 10989

Printed in United States of America on Acid Free Paper

Preface

Aquaculture is the practice of cultivating aquatic animals and plants in managed aquatic environments. Aquaculture in salt-water or marine environments is called mariculture. Fish culture, or pisciculture, refers to the husbandry of finfish. The most popular aquaculture species are finfish grown in fresh waters, accounting for over 40 percent of total aquaculture production. Aquaculture has a long history, but for much of the world it remains somewhat of a novelty, being practiced less than agriculture or capture fisheries. During the last 30 years of the twentieth century, aquaculture grew at an average annual rate of 10 percent, and emerged as the only growth sector of the fisheries industry. At the beginning of the twenty-first century, aquaculture's share of total fish production worldwide was 25 percent, and that proportion is projected to increase. Even though the production of fish from capture fisheries has not substantially increased over the past decade, capture fisheries nevertheless account for a far greater percentage than aquaculture. Aquaculture is practiced for a number of reasons, chief among them being food production and income generation. Most fresh-water aquaculture production (over 70 percent) comes from low-income, food-deficit countries. Even in the poorest countries, fish farming is seldom solely a subsistence activity. So while farmers may consume some of their product, typically fish are sold, thereby enabling farmers to earn income to purchase other goods and services. Aquaculture has two types, freshwater aquaculture and salt-water aquaculture. With the ever increasing demand of fish and increased fish catching activities, sea is facing shortage of fish and cannot fulfil this much demand. Freshwater aquaculture has also improved economies of many areas by providing new job opportunities. The fish produced there is mostly used by industries for processing which is then made available as canned food item.

Handbook on Freshwater Aquaculture is the ultimate guide to freshwater aquaculture, an essential resource for both professional aquaculturists and backyard fish growers.

Table of Contents

Chapter 1

Evaluation of the impacts of aquaculture and freshwater habitat on the status of Atlantic salmon stocks in Ireland

David Jackson[*], Tom McDermott, Frank Kane, Pauline O'Donohoe, Suzanne Kelly

Marine Institute, Galway, Ireland

ABSTRACT

Atlantic salmon (Salmo salar) returns to rivers in Ireland have fallen in the last decade resulting in the dramatic closure or curtailment of traditional fisheries. Concerns that stocks were slow to recover prompted further investigation of all factors believed to impact on salmon. In examining geographic and temporal trends in Atlantic salmon stock abundance at a River Basin District level, the effect of salmon aquaculture sites and freshwater habitat quality as potential drivers of stock abundance are evaluated. This study found no correlation between the presence of aquaculture and the performance of adjacent wild salmon stocks. Freshwater habitat quality was found to have a highly significant correlation with stock status, suggesting that it may be a key driver, implicated in the survival of individual stocks.

KEYWORDS

Aquaculture; Atlantic Salmon; Marine Survival; Habitat Quality; Sea Lice; Escapees

1. INTRODUCTION

Stock levels of Atlantic salmon in the Northeast Atlantic had been declining since the 1970s prompting studies into possible factors affecting survival, including climate change [1-4], overexploitation [5,6], by-catch in commercial marine fisheries [7,8], predation [9], freshwater habitat quality [10-12] and aquaculture [13-16]. In Ireland marine survival, as presented by the Standing Scientific Committee on the Status of Irish Salmon Stocks [17] shows a general downward trend from the mid 1980s to 2008 (**Figure 1**) which is attributed to "very poor marine conditions leading to poor survival". Other potential factors that may contribute to the observed reduction are Aquaculture and Freshwater Habitat.

Sea lice (Lepeophtheirus salmonis) levels on returning wild Atlantic salmon off the coast of Ireland regularly approach 100% prevalence, with a mean abundance of 10.9 L. salmonis per fish. Significant numbers of returning wild salmon remain in coastal and estuarine waters off Ireland's west coast for sufficient periods to be both the source of, and subject to, infestation with L. salmonis [18]. A number of investigations in the last decade have focussed on the impacts of sea lice on outwardly migrating salmon along Ireland's coastline [19-22]. To determine what impact such levels of lice have on salmon migrating to sea as post smolts, a large scale metaanalysis involving 352,142 salmon was undertaken from 2001 to 2009 at 8 locations along Ireland's south and west coasts [22]. The results suggest the observed level of marine mortality attributable to sea lice is small in absolute terms (approximately 1%), and as a proportion of the overall marine mortality [22]. At the levels found it is unlikely to influence the conservation status of stocks and is neither a significant driver of marine mortality nor has been implicated in the observed decline in marine survival in the stocks studies

[19]. These results are consistent with a Norwegian research that generated broadly similar findings [23].

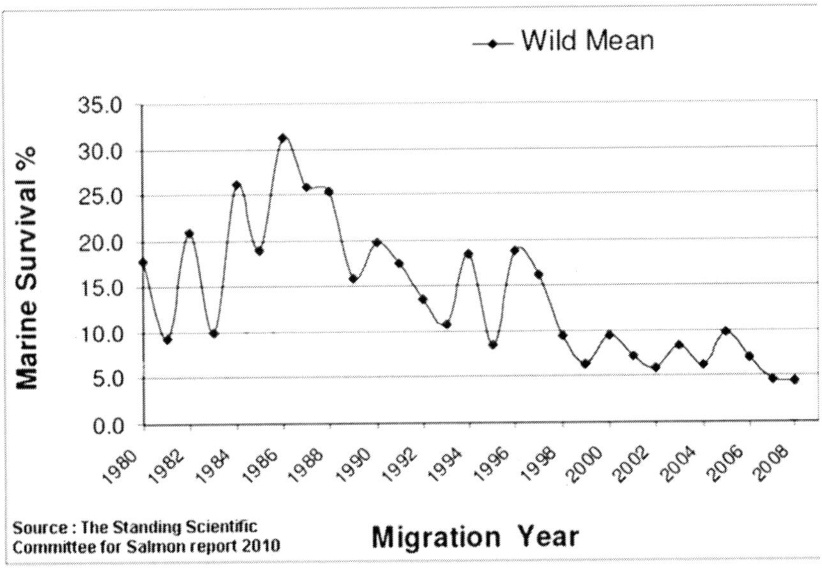

Figure 1. Marine survival (from smolt release to adult returns to the coast of Ireland) for wild salmon [17].

Rivers in Ireland are divided into seven River Basin Districts for administrative purposes with Conservation Limits established for each salmon river from 2007 [17]. Conservation Limits for stocks are based on reaching escapement targets of spawning fish into rivers that allow for a sustainable harvest of salmon into the future. Conservation limits for salmon stock complexes may therefore be defined as “the level of stock (number of spawners) that will achieve long-term average maximum sustainable yield (MSY)”. The status of salmon stocks in each river is assessed based on meeting their Conservation Limit, and is classified as Open, Catch & Release (C&R) or Closed. There is significant variability in stock status between River Basin Districts. Such geographical variability points to near field factors are implicated in influencing stock status.

In examining geographic and temporal trends in stock abundance of Atlantic salmon at a River Basin District level, the authors are seeking to evaluate the effect of salmon aquaculture sites and fresh water habitat quality as potential drivers of stock abundance.

2. MATERIALS AND METHODS

Published national statistics were examined to evaluate stock status on a catchment and River Basin District basis. Conservation limits and stock status, including surplus above Conservation Limits available for exploitation for each salmon river, was sourced from the reports of the Standing Scientific Committee to the Department of Communications, Energy and Natural resources [24-27]. Where rivers are meeting, or exceeding, their Conservation Limits, this surplus is available for exploitation by angling and commercial fisheries (i.e. the rivers are open).

The locations and active salmon aquaculture sites was obtained from the series of annual reports on the results of the National Survey of sea lice Lepeophtheirus salmonis Kroyer and Caligus elongatus Nordmann on fish farms in Ireland [28-30].

Water quality data was obtained from the Environmental Protection Agency (EPA). River water quality was classified by percentage of channel length meeting Class A unpolluted standard [31].

River Basin Districts are the administrative areas into which inland waters are divided for the purpose of management under the European Union's Water Framework Directive (WFD: 2000/60/EC). The proportion of these rivers meeting their Conservation Limits was established for each year. Salmon rivers with a population of less than ten salmon were excluded from the analysis.

3. RESULTS

Table 1 presents the data on rivers open for exploitation, including the taking of fish and on a catch and release basis. There is an increase in the rivers open nationally from 42 in 2007 to 48 in 2011. When broken down by River Basin District the trend is similar with numbers open either remaining static or increasing. The total number of rivers assessed varies slightly from year to year. This is a feature of the reports and arises due to sub-catchments being classified differently from time to time. **Figure 2** shows the proportion of rivers either Open, C&R or Closed in each River Basin District by year, including locations of active fish farm sites.

The same data is also presented as the proportions of fluvial area accessible to salmon (**Figure 3**) with commercial draft net fishing on open rivers highlighted. The West and Southwest River Basin Districts have consistently the highest proportion of rivers open throughout the period; the results are similar when expressed in terms of fluvial area accessible to salmon. The next highest proportion of rivers open is in the Northwest River Basin District.

There is no geographic correlation between the presence of salmon farms and failure of rivers to meet their Conservation Limits at a River Basin District level. In fact, the rivers in the River Basin Districts with salmon farms have performed best in terms of meeting their Conservation Limits and also in terms of ability to support a commercial catch by way of a commercial draft net fishery (**Figure 3**).

The percentage of channel length in each River Basin District meeting Class A unpolluted status is plotted for each River Basin District against the percentage of rivers meeting their conservation objective in each district (**Figure 4**). A significant geographical correlation exists between water quality in the catchment as measured by percentage Class A channel length and percentage of rivers meeting Conservation Limit (R^2 = 89.1%, p = 0.001).

4. DISCUSSION

The data presented here shows that the Conservation Limits are being met in many Irish salmon rivers and that there is a steady and sustained improvement in the overall status of Irish salmon stocks. The rivers on the West coast are in general better in this regard, and rivers in the Northwest, West and Southwest River Basin Districts support both angling and commercial draft net fisheries. There is no relationship between the presence of salmon farms and difficulties with rivers meeting their Conservation Limits. Numerous studies point to the fact of sea lice not being a factor in declining marine survival of wild salmon. The absence of any evidence for a negative correlation between salmon stock status and aquaculture operations would suggest that there is also no discernible negative impact from escapees on the stocks studied. This conclusion is supported by the finding of O'Maoleidigh et al. [32] that fish farm escapees in Ireland are at a low level and contribute little to spawning stocks.

There is a significant correlation between unpolluted water quality in catchments and the numbers of rivers meeting their Conservation Limits (**Figure 4**). This correlation would suggest that the quality of freshwater habitat may be a key driver, implicated in the survival of individual wild salmon stocks.

Since the introduction of Conservation Limits in 2007 exploitation of salmon stocks has been reduced. The commercial offshore mixed-stock drift net fishery was closed since 2007. In addition angling pressure has been curtailed by closure of certain rivers, the introduction of Bye-Laws limiting catches and restrictions to angling for Spring salmon. The improvement of stock status over the study period would suggest that these conservation measures are contributing to stock recovery in a period of low marine survival.

Table 1. Salmon rivers open for exploitation, including catch and release, by River Basin District. Where rivers are open they are above their Conservation Limit, where they are C&R (catch and release of fish caught) they are close to their Conservation Limit [24,27] (Source: Conservation of salmon and sea trout bye-laws as published annually by the Department of Communications, Marine and Natural Resources in Ireland).

River Basin District		Number of Rivers				Percentage of Rivers (%)		
	Year	Open	C & R	Closed	Total	Open	C & R	Closed
Northwest	2007	9	0	6	15	60	0	40
	2008	7	0	8	15	47	0	53
	2009	8	0	7	15	53	0	47
	2010	8	2	5	15	53	13	33
	2011	9	2	3	14	64	14	21
West	2007	16	4	4	24	67	17	17
	2008	17	5	1	23	74	22	4
	2009	15	5	2	22	68	23	9
	2010	18	4	1	23	78	17	4
	2011	17	5	1	23	74	22	4
Shannon	2007	2	0	4	6	33	0	67
	2008	2	0	3	5	40	0	60
	2009	2	0	3	5	40	0	60
	2010	2	0	3	5	40	0	60
	2011	2	0	3	5	40	0	60
Southwest	2007	14	0	6	20	70	0	30
	2008	15	4	2	21	71	19	10
	2009	19	3	4	26	73	12	15
	2010	19	0	5	24	79	0	21
	2011	16	5	1	22	73	23	5
Southwest	2007	14	0	6	20	70	0	30
	2008	15	4	2	21	71	19	10
	2009	19	3	4	26	73	12	15
	2010	19	0	5	24	79	0	21
	2011	16	5	1	22	73	23	5
Southeast	2007	0	2	2	4	0	50	50
	2008	0	3	3	6	0	50	50
	2009	0	3	3	6	0	50	50
	2010	1	2	2	5	20	40	40
	2011	2	1	2	5	40	20	40
East	2007	0	0	2	2	0	0	100
	2008	0	1	2	3	0	33	67
	2009	0	1	2	3	0	33	67
	2010	0	1	2	3	0	33	67
	2011	0	1	2	3	0	33	67
Neagh-Bann	2007	1	1	2	4	25	25	50
	2008	1	1	2	4	25	25	50
	2009	1	1	2	4	25	25	50
	2010	1	1	2	4	25	25	50
	2011	2	1	1	4	50	25	25
National	2007	42	7	26	75	56	9	35
	2008	42	14	21	77	55	18	27
	2009	45	13	23	81	56	16	28
	2010	48	10	20	78	62	13	26
	2011	48	15	13	76	63	20	17

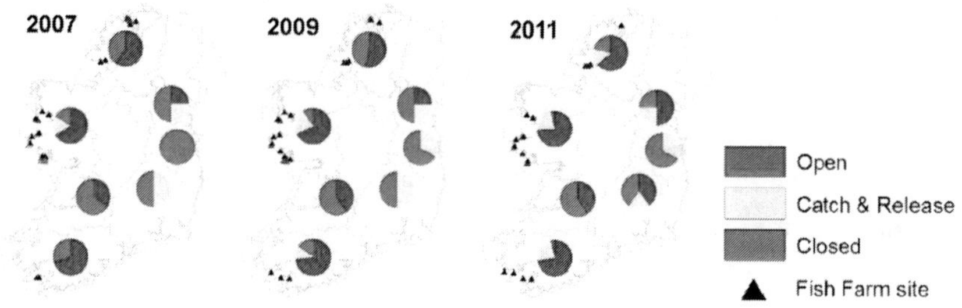

Figure 2. Map showing the proportion of rivers in each River Basin District which are open for fishing, being fished on a Catch & Release basis or closed to exploitation [24-27]. Locations of salmon farm sites operational in each year [29-31] are indicated. (Source: Conservation of salmon and sea trout bye-laws as published annually by the Department of Communications, Marine and Natural Resources in Ireland).

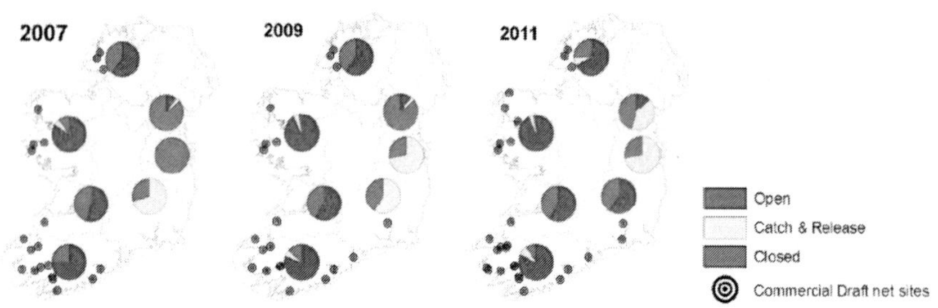

Figure 3. Map showing the proportion of rivers, measured as fluvial area accessible to salmon (m^2), in each River Basin District which are open for fishing, being fished on a Catch & Release basis, or closed to angling [24-27]. Locations of commercial salmon draft net fishing sites are indicated. (Source: Conservation of salmon and sea trout bye-laws as published annually by the Department of Communications, Marine and Natural Resources in Ireland).

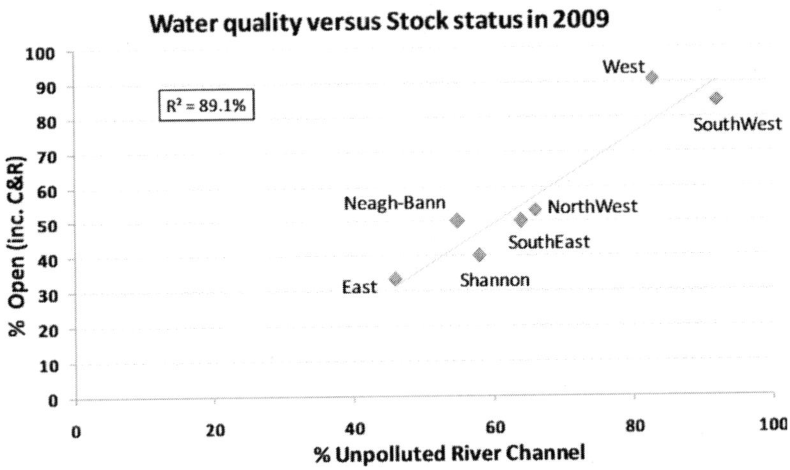

Figure 4. Relationship between salmon stock status as measured by percentage of rivers open for exploitation [24-27], and water quality of river channels [31] grouped by River Basin Districts. A regression line is fitted with R^2 value indicated.

5. CONCLUSION

This study found no correlation between the presence of aquaculture and the performance of adjacent wild salmon stocks. This finding supports previous research on farm escapees and sea lice which found little influence of escaped farmed salmon on spawning stocks [32], and sea lice were a minor and irregular component in marine mortality [22]. Freshwater habitat quality was found to have a highly significant correlation with stock status.

REFERENCES

1. Friedland, K.D., Chaput, G. and MacLean, J.C. (2005) The emerging role of climate in post-smolt growth of Atlantic salmon. ICES Journal of Marine Science, 62, 1338- 1349.

2. Friedland, K.D., MacLean, J.C., Hansen, L.P., Peyronnet, A.J., Karlsson, L., Reddin, D.G., Maoiléidigh, N.Ó. and McCarthy, J.L. (2009) The recruitment of Atlantic salmon in Europe. ICES Journal of Marine Science, 66, 289-304.

3. Peyronnet, L.A., Friedland, K.D., Maoileidigh, N.Ó., Manning, M. and Poole, W.R. (2007) Links between patterns of marine growth and survival of Atlantic salmon Salmosalar L. Journal of Fish Biology, 71, 684-700.

4. Todd, C.D., Hughes, S.L., Marshall, C.T., MacLean, J.C., Lonergan, M.E. and Biuw, E.M. (2008) Detrimental effects of recent ocean surface warming on growth condition of Atlantic salmon. Global Change Biology, 14, 958- 970.

5. Rago, P.J., Reddin, D.G., Porter, T.R., Meerburg, D.J., Friedland, K.D. and Potter, E.C.E. (1993) A continental run reconstruction model for the non-maturing component of North American Atlantic salmon: Analysis of fisheries in Greenland and Newfoundland-Labrador, 1974- 1991. ICES Document, CM 1993/M: 25.

6. Potter, E.C.E. and Dunkley, D.A. (1993) Evaluation of marine exploitation of salmon in Europe. In: Mills, D., Ed., Salmon in the Sea, and New Enhancement Strategies, Fishing News Books, Oxford, 203-219.

7. Hansen, L.P., Holm, M., Hoist, J.C. and Jacobsen, J.A., (2003) The ecology of post-smolts of Atlantic Salmon Chapter 4. In: Salmon at the Edge, Blackwell Science Ltd., 25-39.

8. Mork, K.A., Gilbey, J., Hansen, L.P., Jensen, A.J., Jacobsen, J.A., Holm, M., Holst, J.C., Maoiléidigh, N.Ó., Vikebø, F., McGinnity, P., Melle, W., Thomas, K., Verspoor, E. and Wennevik, V. (2012) Modelling the migration of postsmolt Atlantic salmon (Salmosalar) in the Northeast Atlantic. ICES Journal of Marine Science, fss108v1-fss108.

9. Butler, J.R.A., Middlemas, S.J., Graham, I.M., Thompson, P.M. and Armstrong, J.D. (2006) Modelling the impacts of removing seal predation from Atlantic salmon, Salmosalar, rivers in Scotland: A tool for targeting conflict resolution. Fisheries Management and Ecology, 13, 285- 291.

10. Gibson, R.J. (1993) The Atlantic salmon in fresh water: Spawning, rearing and production. Reviews in Fish Biology and Fisheries, 3, 39-73.

11. Mäki-Petäys, A., Huusko, A., Erkinaro, J. and Muotka, T. (2002) Transferability of habitat suitability criteria of juvenile Atlantic salmon (Salmosalar). Canadian Journal of Fisheries and Aquatic Sciences, 59, 218-228.

12. Einum, S. and Nislow, K.H. (2011) Variation in population size through time and space: Theory and recent empirical advances from Atlantic salmon. In: Aas, Ø., Einum, S., Klemetsen, A. and Skurdal, J., Eds., Atlantic Salmon Ecology, Wiley-Blackwell, New York, 277-298.

13. Crozier, W.W. (2000) Escaped farmed salmon, Salmosalar L., in the Glenarm River, Northern Ireland: Genetic status of the wild population 7 years on. Fisheries Management and Ecology, 7, 437-446.

14. Skilbrei, O.T. and Wennevik, V. (2006) Survival and growth of sea-ranched Atlantic salmon, Salmosalar L., treated against sea lice before release. ICES Journal of Marine Science, 63, 1317-1325.

15. Karlsson, S., Moen, T., Lien, S., Glover, K. and Hindar, K. (2011) Generic genetic differences between farmed and wild Atlantic salmon identified from a 7K SNP-chip. Molecular Ecology Resources, 11, 247-253.

16. Glover, K.A., Quintela, M., Wennevik, V., Besnier, F., Sørvik, A.G.E., et al. (2012) Three decades of farmed escapees in the wild: A spatio-temporal analysis of Atlantic salmon population genetic structure throughout Norway. PLoS ONE, 7, e43129.

17. Anon (2010) Report of the standing scientific committee to the department of communications, energy and natural resources—The status of Irish salmon stocks in 2009 and precautionary catch advice for 2010. Department of Communications, Marine and Natural Resources, Dublin.

18. Jackson, D., Kane, F., O'Donohoe, P., Mc Dermott, T., Kelly, S., Drumm, A. and Newell, J. (2013) Sea lice levels on wild Atlantic salmon, Salmosalar L., returning to the coast of Ireland. Journal of Fish Diseases, 36, 293- 298.

19. Jackson, D., Cotter, D., O'Maoileidigh, N., O'Donohoe, P., White, J., Kane, F., Kelly, S., McDermott, T., McEvoy, S., Drumm, A., Cullen, A. and

Rogan, G. (2011) An evaluation of the impact of early infestation with the salmon louse Lepeophtheirus salmonis on the subsequent survival of outwardly migrating Atlantic salmon, Salmosalar L., smolts. Aquaculture, 320, 159-163.

20. Jackson, D., Cotter, D., O'Maoileidigh, N., O'Donohoe, P., White, J., Kane, F., Kelly, S., McDermott, T., McEvoy, S., Drumm, A. and Cullen, A. (2011) Impact of early infestation with the salmon louse Lepeophtheirus salmonis on the subsequent survival of outwardly migrating Atlantic salmon smolts from a number of rivers on Ireland's south and west coast. Aquaculture, 319, 37-40.

21. Gargan, P.G., Forde, G., Hazon, N., Russell, D.J.F. and Todd, C.D. (2012) Evidence for sea lice-induced marine mortality of Atlantic salmon (Salmosalar) in western Ireland from experimental releases of ranched smolts treated with emamectin benzoate. Canadian Journal of Fisheries and Aquatic Sciences, 69, 343-353.

22. Jackson, D., Cotter, D., Newell, J., McEvoy, S., O'Donohoe, P., Kane, F., McDermott, T., Kelly, S. and Drumm, A. (2013) Impact of Lepeophtheirus salmonis infestations on migrating Atlantic salmon, Salmosalar L., smolts at eight locations in Ireland with an analysis of lice—Induced marine mortality. Journal of Fish Diseases, 36, 273-281.

23. Skilbrei, O.T., Finstad, B., Urdal, K., Bakke, G., Kroglund, F. and Strand, R. (2013) Impact of early salmon louse, Lepeophtheirus salmonis, infestation and differences in survival and marine growth of sea—Ranched Atlantic salmon, Salmosalar L., smolts 1997-2009. Journal of Fish Diseases, 36, 249-260.

24. Anon (2007) Report of the standing scientific committee of the national salmon commission—The status of Irish salmon stocks in 2006 and precautionary catch advice for 2007. Department of Communications, Marine and Natural Resources, Dublin.

25. Anon (2008) Report of the standing scientific committee of the national salmon commission—The status of Irish salmon stocks in 2007 and precautionary catch advice for 2008. Department of Communications, Marine and Natural Resources, Dublin.

26. Anon (2009) Report of the standing scientific committee to the department of communications, energy and natural resources—The

status of Irish salmon stocks in 2008 and precautionary catch advice for 2009. Department of Communications, Marine and Natural Resources, Dublin.

27. Anon (2011) Report of the standing scientific committee for salmon to the department of communications, energy and natural resources—The status of Irish salmon stocks in 2010, with precautionary catch advice for 2011. Department of Communications, Marine and Natural Resources, Dublin.

28. O'Donohoe, P., Kane, F., Kelly, S., Nixon, P., Power, A., Naughton, O. and Jackson, D. (2008) National survey of sea lice (Lepeophtheirus salmonis Krøyer and Caligus elongatus Nordmann) on fish farms in Ireland—2007. Irish Fisheries Bulletin No. 31, Marine Institute.

29. O'Donohoe, P., Kane, F., Kelly, S., Nixon, P., Power, A., McDermott, T., Drumm, A. and Jackson, D. (2010) National survey of sea lice (Lepeophtheirus salmonis Krøyer and Caligus elongatus Nordmann) on fish farms in Ireland—2009. Irish Fisheries Bulletin No. 33, Marine Institute.

30. O'Donohoe, P., Kane, F., Kelly, S., McDermott, T., Drumm, A. and Jackson, D. (2012) National survey of sea lice (Lepeophtheirus salmonis Krøyer and Caligus elongatus Nordmann) on fish farms in Ireland—2011. Irish Fisheries Bulletin No. 40, Marine Institute.

31. McGarrigle, M., Lucey, J. and Ó'Cinnéide, M. (2010) Water quality in Ireland 2007-2009. EPA, Ireland, 138.

32. O'Maoileidigh, N., Cullen, A., McDermott, T., Bond, N. and McLaughlin, D. (2002) Review of Irish salmon aquaculture escapee data. International Council for the Exploration of the Sea, Copenhagen.

Chapter 2

Identification of Bacterial Community Composition in Freshwater Aquaculture System Farming of *Litopenaeus vannamei* Reveals Distinct Temperature-Driven Patterns

Yuyi Tang [1], Peiying Tao [1], Jianguo Tan [2], Haizhen Mu [2], Li Peng [2], Dandan Yang [2], Shilu Tong [3] and Lanming Chen [1,*]

[1]Key Laboratory of Quality and Safety Risk Assessment for Aquatic Products on Storage and Preservation (Shanghai), China Ministry of Agriculture, College of Food Science and Technology, Shanghai Ocean University, 999 Hu Cheng Huan Road, Shanghai 201306, China
[2]Shanghai Key Laboratory of Meteorology and Health, 951 Jinxiu Road, Shanghai 200135, China
[3]School of Public Health, Institute of Health and Biomedical Innovation, Queensland University of Technology, Brisbane, Australia Kelvin Grove, QLD 4059, Australia

ABSTRACT

Change in temperature is often a major environmental factor in triggering waterborne disease outbreaks. Previous research has revealed temporal and spatial patterns of bacterial population in several aquatic ecosystems. To date, very little information is available on aquaculture environment. Here, we assessed environmental temperature effects on bacterial community composition in freshwater aquaculture system farming of *Litopenaeus vannamei* (FASFL). Water samples were collected over a one-year period, and aquatic bacteria were characterized by polymerase chain reaction-denaturing gradient gel electrophoresis (PCR-DGGE) and 16S rDNA pyrosequencing. Resulting DGGE fingerprints revealed a specific and dynamic bacterial population structure with considerable variation over the seasonal change, suggesting that environmental temperature was a key driver of bacterial population in the FASFL. Pyrosequencing data further demonstrated substantial difference in bacterial community composition between the water at higher (WHT) and at lower (WLT) temperatures in the FASFL. *Actinobacteria*, *Proteobacteria* and *Bacteroidetes* were the highest abundant phyla in the FASFL, however, a large number of unclassified bacteria contributed the most to the observed variation in phylogenetic diversity. The WHT harbored remarkably higher diversity and richness in bacterial composition at genus and species levels when compared to the WLT. Some potential pathogenenic species were identified in both WHT and WLT, providing data in support of aquatic animal health management in the aquaculture industry.

KEYWORDS

aquaculture; *Litopenaeus vannamei*; bacterial community; temperature; pathogen

1. INTRODUCTION

The Pacific white shrimp *Litopenaeus vannamei* is the most widely cultured and productive alien crustacean worldwide [1]. It is native to the western Pacific coast of Latin America, and introduced commercially since 1996 into China and several countries in Asia [2]. In China, the yearly estimated *L. vannamei* production was over one million tons during the past several years [1]. The freshwater culture of *L. vannamei* has proven even more successful than brackish water culture conditions [2], which play very important roles in the shrimp production in the southeast littoral provinces in China. Along with the fast growing shrimp-production industry, however, aquatic animal diseases caused by waterborne pathogens have also rapidly increased, which led to huge economic losses in the past decades [3]. Previous culture-based studies have revealed that*Vibrionaceae*-related organisms, being virtually ubiquitous in aquatic environments [4], were the major pathogens causing disease outbreaks and mortality of early stage hatchery-reared shrimps including *L. vannamei* [5,6,7,8,9]. It is well known that some pathogenic *Vibrios* are also serious human foodborne pathogens causing worldwide cholera epidemics and diarrheal disease [10]. Recently, *Streptococcosis* in farmed *L. vannamei* was reported as a new emerging bacterial disease of penaeid shrimp [11]. To elucidate the mechanism underlying the emergence and resurgence of the waterborne pathogens, a complete understanding of the composition and dynamics of microbial population in aquaculture ecosystems is required.

Recent significant development of metagenomic techniques has allowed for culture-independent genomic fingerprinting of bacterial communities in the aquaculture environment [12]. Using the DGGE technique, bacterial community structures have been investigated in various aquaculture settings in different parts of the world, e.g., culturing the tropical rock lobster (*Panulirus ornatus*) in Australia [13], the Pacific white shrimp (*L. vannamei*) in the USA [14], the Asian tiger shrimp (*Litopenaeus monodon*) in Thailand [15], the Atlantic cod (*Gadus morhua* L.) in Norway [16], the grass carp (*Ctenopharyngodon idellus*) in Zhangjiang, China [17], as well as the shrimp (*Penaeus vannamei,Penacus orientalis*), abalone (*Haliotis diversicolor*) and reef cod (*Epinephelus diacanthus*) in coastal mariculture ponds in Southeast China [18]. These studies have proposed strong links between environmental variables (e.g., season, water flow rate and aeration) and bacterial population structures in the aquaculture niches. Compared to the DGGE and traditional Sanger sequencing of 16S rDNA clone libraries [19], the developed second-generation sequencing techniques significantly improved the researcher's ability to achieve a more definitive phylogenetic-based description of microbial communities in aquatic ecosystems, e.g., ocean [20], deep sea [21], river [22], spring [23], drinking water [24] and wastewater [25]. Previous studies using the approaches have revealed temporal [26,27,28] and spatial [29] patterns of bacterial communities in marine, lake and river environments. Nevertheless, very little information is available to date for microbial community dynamics in aquaculture ecosystems, despite their great significance in economy and human health. In this study, we combined PCR-DGGE and 454-pyrosequencing techniques to evaluate the influence of environmental temperature change on the structure, composition and diversity of bacterial population in the FASFL in Shanghai, one of the major shrimp production regions in China. Our data uncovered a temperature-driven substantial shift in bacterial population in the FASFL. The information will facilitate better understanding of possible molecular mechanisms underlying waterborne disease outbreaks and seasonal change.

2. RESULTS AND DISCUSSION

2.1. DGGE-Based Bacterial Population Profiles in the FASFL

To gain an insight into the possible influence of temperature on bacterial population in the FASFL, surface water samples were collected over a one-year period (see the Section 3), and aquatic bacteria were characterized by PCR amplification of the V3 variable region of the bacterial 16S rRNA gene. This analysis yielded clear PCR products from all the water samples (Figure not shown). Based on the amplicons, DGGE fingerprints were obtained (Figure 1A), which strongly suggested that the FASFL harbored a complex and specific bacterial community structure. Considerable variation in bacterial composition was observed across the water temperature change (Figure 1B). Cluster analysis of the DGGE fingerprints revealed three distinct groups, designated Group I, II and III (Figure 1C). Group I contained the samples collected in the summer and early autumn (June to October 2012), whereas Group III was comprised of the samples in the early spring (March to April 2013). The late autumn and late spring samples fell into Group II (November 2012, May 2013), showing mosaic fingerprints of the former two Groups. The similarity among the samples was observed in the range from 52% to 74% (Figure 1C), indicating significant difference of intergroup bacterial composition. The data provided the first example of a dynamic bacterial population driven by water temperature in aquaculture environment. Previous studies also revealed distinct seasonal patterns of bacterial population in other freshwater habitats (e.g., river and lake) by PCR-DGGE analysis [27,28].

Selected dominant and distinct bands on the DGGE gel were subsequently sequenced and analyzed (Figure 1A), and revealed phylogenetic diversity of bacterial population in the FASFL across different sampling times (Table 1). The DGGE fingerprints of Group I samples were dominated by three major bacterial bands: one (band 1) was affiliated closely with an uncultured*Rickettsiales* bacterium clone FWB6C1-73; one (band 2) with an uncultured *Micrococcineae* bacterium clone D7N78; the other (band 3) with an uncultured *Methylophilus* sp. clone JA127_2010-09-15 and *Sphingomonas oligophenolica* strain R2A-AUG-EA-11. Interestingly, these bacteria appeared highly sensitive to lower temperature since none of any corresponding bands was observed in Group III patterns, suggesting their thermophilic property. Along with the dropping temperature, Group III fingerprints again changed largely where several bands were dominant in the profile. Almost all of these (band 4–9) were affiliated closely with uncultured bacterium clone sequences belonging to*Bacteroidetes*, *Actinobacteria* and *Firmicutes* (Table 1). These bacteria were absent from Group I samples, the majority of which only occurred in the sample collected in March 2013 with the lowest water temperature, displaying their psychrotolerent feature. In addition, one of the dominant sequences retrieved from Group II patterns (band 10) showed 100% identity to an uncultured *Intrasporangiaceae bacterium* clone M7N57 belong to *Actinobacteria* (Table 1).

2.2. Environmental Temperature and Bacterial Population in the FASFL

To assess the relationship between water temperature and bacterial composition in the FASFL, we did redundancy analysis (RDA) of the DGGE data. As presented in Figure 2, the resulting RDA plot clearly indicated that temperature was positively and closely associated with the bacterial population retrieved from the WHT (≥30 °C). In contrast, a negative correlation was observed for the bacterial community of the WLT (≤20 °C), which located in a completely opposite direction to the temperature variable in the plot. This result was consistent with those of the cluster analysis and band sequencing. Further analysis of similarity (ANOSIM) generated the *R* value of 0.75 (p <0.05) between the WHT and WLT, strongly suggesting a significant separation of the bacterial communities. Thus, the WHT and WLT were subjected for deeper-sampling by pyrosequencing (see below).

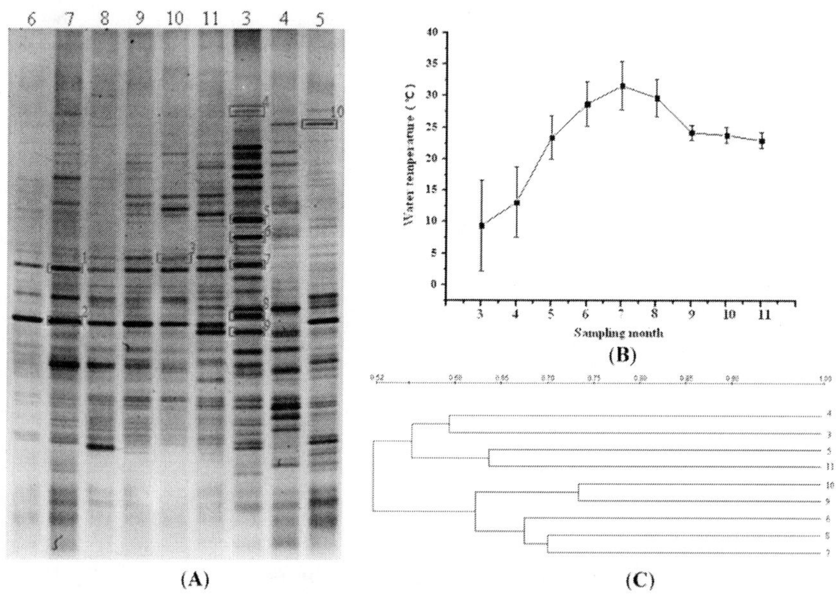

Figure 1. PCR-DGGE analysis of bacterial community structures derived from the water samples in the FASFL in different seasons. On the DGGE gel (**A**), Lane 6 to 11 and 3 to 5 represent the water samples collected from June to November in 2012, and from March to May in 2013, respectively. The DNA bands marked with red boxes on the DGGE gel (**A**) were individually excised and subjected for DNA sequencing. The numbers on the horizontal axis of the temperature curve (**B**) and clustering profile (**C**) represent the sampling months as correspondingly shown in the DGGE gel (**A**).

Table 1. Phylogenetic identity of dominant bands on the DGGE profile.

Band	Clustering Group	Length (bp)	Closest Relative and Database Accession Number	Identity (%)	Taxonomic Description
B1	III	169	Uncultured *Rickettsiales* bacterium clone FWB6C1-73, KF583165.1	100%	α-*Proteobacteria*
B2	III	174	Uncultured *Micrococcineae* bacterium clone D7N78, KC006224.1	100%	*Actinobacteria*
B3	III	194	Uncultured *Methylophilus* sp. clone JA127_2010-09-15, JN866934.1	100%	β-*Proteobacteria*
		169	*Sphingomonas oligophenolica* strain R2A-AUG-EA-11, JX237432.1, KC836618.1	100%	α-*Proteobacteria*
B4	I	194	*Pediococcus ethanolidurans* strain RU12-4	100%	*Firmicutes*
B5	I	189	Uncultured *Bacteroidetes* bacterium, FR647662.1	100%	*Bacteroidetes*
		189	Uncultured *Cryomorphaceae* bacterium clone Jab PL2W2H12, HM486317.1	100%	*Bacteroidetes*
B6	I	174	*Actinobacterium* MS-B-64, FJ460153.1	100%	*Actinobacteria*
B7	I	189	Uncultured *Bacteroidetes* bacterium clone XSLJ052, KC246401.1	100%	*Bacteroidetes*
B8	I	174	Uncultured *actinobacterium* clone FF1G3, EU117678.1	100%	*Actinobacteria*
		174	Uncultured *actinobacterium* clone B12-88, JN371245.1	100%	*Actinobacteria*
		174	*Actinobacterium* SCGC AAA043-A09, HQ663377.1	100%	*Actinobacteria*
B9	I	174	Uncultured *Micrococcineae* bacterium clone D7N78, KC006224.1	100%	*Actinobacteria*
B10	II	174	Uncultured *Intrasporangiaceae bacterium* clone M7N57, KC006381.1	100%	*Actinobacteria*

Overall, our data highlighted a dynamic bacterial population existed in the FASFL, which was directly and closely associated with the water temperature. The FASFL is a complex ecosystem where microflora and aquatic animals coexist and interact. It has been reported that live feed is not a major determinant of microbiota associated with cod larvae (*Gadus morhua*) [30]. Nevertheless, in this study, we could not rule out possible influence of some other environmental variables such as total dissolved solids (TDS) on bacterial community in the niche analyzed, since the TDS values were observed varying as seasonal change as well (data not shown).

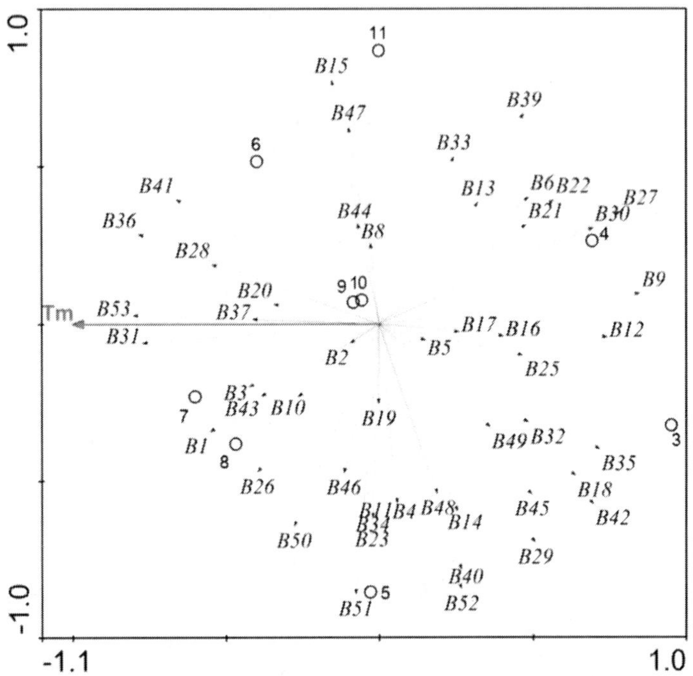

Figure 2. The relationship between environmental variable and bacterial community composition derived from different water samples in the FASFL. DGGE banding scores were plotted using the program CANOCO (Version 4.5). Tm represents the environmental variable of the water temperature. The open closed circles and numbers 3 to 11 indicate different sampling months as described in Figure 1, while the numbers B1 to B52 represent the DNA bands on the PCR-DGGE gel analyzed by the CANOCO, respectively.

2.3. Overall Bacterial Community Composition in the FASFL

2.3.1. Bacterial Richness and Diversity

Aquatic bacteria derived from the WHT and WLT were determined by high-throughput 454-pyrosequencing of the 16S rDNA to achieve a detailed phylogenetic-based description of the bacterial compositions. This analysis yielded a total of 18,427 high quality sequences with an average sequence length of 414 bp after the implementation of the quality control criteria as described in Section 3. Of these, 12,802 sequences were retrieved from the WHT, while 5625 sequences were retrieved from the WLT. A large number of operational taxonomic units (OTUs) at the species level were identified using the MOTHUR software, which underlined a highly diverse bacterial community composition in the FASFL analyzed in this study (Table S1).

On the basis of the identified OTUs, bacterial richness was examined by the rarefaction analysis. As shown in Figure S1, the rarefaction curves for the WHT and WLT extended to the horizontal at a phylogenetic distance of 0.20, indicating that almost all of the bacterial diversity at the phyla level was revealed by the pyrosequencing-based analysis, which gave the Good's coverage of 99.6% and 99.8%, respectively. This result was consistent with the predicted values of OTUs by Chao1 richness calculation (Table S1). However, at phylogenetic distances of 0.05 and 0.03, the identified OTUs derived from the WHT represented 49.0%–54.5% of those estimated by Chao1 analysis (Table S1), implying that more sequencing efforts will be put into this sample in the future to cover the full taxonomic diversity at the genus and species level. Lower bacterial richness assessed by rarefaction analysis than by Chao1 and ACE estimator has also been observed at phylogenetic distances below 5% in previous studies (e.g., [31]). For the WLT, 98.3% and 97.2% coverage of bacterial richness was achieved at the distances below 5%. Moreover, the number of identified OTUs derived from the WLT was *ca.* 6-fold lower than that encountered in the WHT. These results highlighted distinct richness of the bacterial community between the WLT and WHT.

Bacterial diversity of the samples was also assessed by Shannon diversity indices. At the three different phylogenetic distances analyzed in this study, the values of Shannon index ranged from 5.79 to 3.45 for the WHT, which were higher than those encountered in the WLT (4.64 to 3.29) (Table S1). These data revealed a higher degree of bacterial diversity in the WHT when compared to the WLT.

2.3.2. Bacterial Community Composition

Phylum level affiliations of the sequences retrieved from the samples revealed distinct differences in phylum-level community composition in the FASFL in different seasons. As presented in Figure 3, out of the twelve phyla and candidate phyla identified in this study, the dominant phyla in the bacterial community derived from the WHT were *Actinobacteria*, unclassified bacteria, *Proteobacteria* and *Bacteroidetes*, representing 53.5%, 22.4%, 18.8% and 4.32% of all classified sequences from this sample, respectively. Four phyla were present in lower relative abundance, including *Cyanobacteria/Chloroplast* (0.289%), *Firmicutes* (0.265%), *Gemma timonadetes*(0.203%) and *Acidobacteria* (0.109%), whereas *Fibrobacteres, Fusobacteria, Spirochaetes* and*Verrucomicrobia* were found in extremely low abundance at a percentage abundance of 0.0078%, 0.0055% 0.0078%, 0.0078%, respectively. For the WLT, all sequences were affiliated to eight phyla, the dominance of which was *Actinobacteria* (41.80%), *Proteobacteria* (39.52%) and *Bacteroidetes*(17.29%). The lower abundant members included *Cyanobacteria/Chloroplast* (0.195%), *Deferribacteres*(0.04%), *Firmicutes* (0.05%) and *Gemmatimonadetes* (0.017%). Distinct from the WHT, only 1.07% of the sequences retrieved from the WLH fell into the unclassified bacteria group.

Comparison of the phylum-level community composition reinforced the temperature-driven bacterial diversity variation in the aquatic niche analyzed in this study, which lead to three major phylogenetic snapshots. *Actinobacteria* was the most dominant phylum shared between the two samples, followed by *Proteobacteria* and *Bacteroidetes*, which constituted a large proportion of the microflora in the FASFL. These bacteria are also the highest abundant phyla in water samples originated from both aquaculture and freshwater ecosystems, playing important roles in the processes of nutrient cycling and mineralization of organic compounds (e.g., [22,32,33]). Nevertheless, the unclassified bacteria contributed the most to the observed phyla-level diversity variation since *ca*. 21-fold increase in the percentage abundance was observed in the WHT when compared to the WLT. Secondly, three phyla present in lower relative abundance were distributed in the two samples, including *Cyanobacteria/Chloroplast, Firmicutes, Gemmatimonadetes*, but their relative abundance diminished largely in the WLT. Finally, *Acidobacteria*, *Fibrobacteres*, *Fusobacteria*, *Spirochaetes*, and *Verrucomicrobia* were only found in the WHT, whereas *Deferribacteres* showed an opposite pattern. Due to DGGE sensitivity limitations when dealing with low target abundance samples from natural environments [18], some phyla revealed in this study were not reported in previous DGGE-based studies on aquaculture settings.

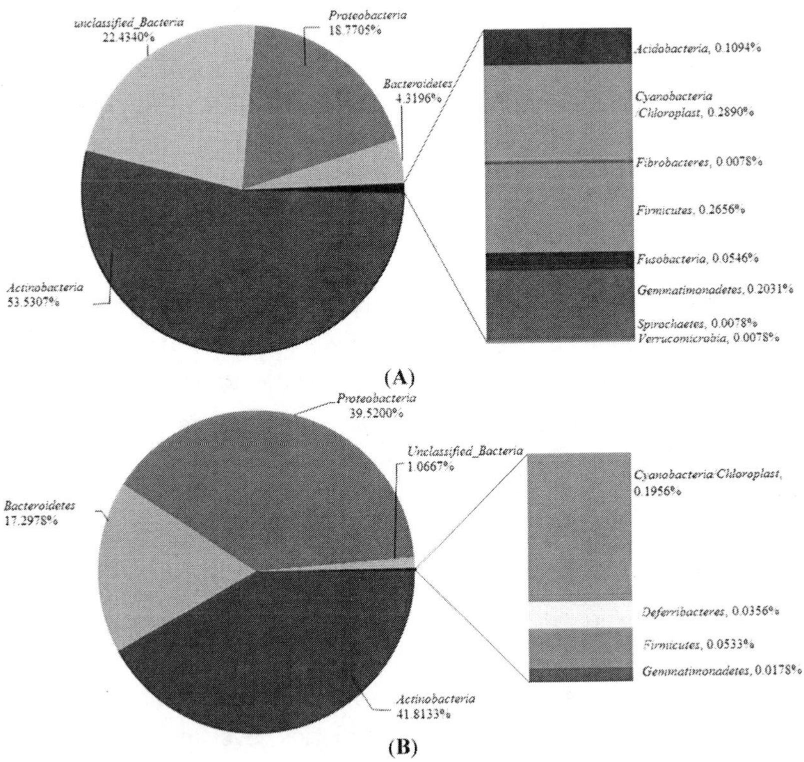

Figure 3. Change in bacterial community composition at the phylum level in the WHT (**A**) and WLT (**B**) samples from the FASFL. Phylum level affiliations of the 454-pyrosequencing sequences retrieved from the samples were performed using the MOTHUR software at a phylogenetic distance of 0.20.

At the genus level, comparison of the phylogenetic profiles provided further evidence for the temperature-shifted variation in bacterial community composition in the FASFL. As shown inFigure 4, the most abundant genus in the bacterial community derived from the WHT was*Ilumatobacter* representing 5.300% of the classified sequences, followed by *Mycobacterium* (1.523%), *Conexibacter* (1.055%), *Hydrogenophaga* (0.922%), *Rhodobacter* (0.609%) and *Polynucleobacter* (0.601%). *Ilumatobacter* was commonly isolated from the sediments of aquatic habitats (e.g., [34]). The land-based FASFL may serve as an explanation for the observed result. For the WLT, a distinct pattern was revealed, where *Limnohabitans*, *Hydrogenophaga*, *Humatobacter*, *Rhodobacter*, *Algoriphagus*,*Flavobacterium*, *Mycobacterium* and *Fluviicola* were the most responsible for the diversity difference at the genus level, displaying a relative abundance of 6.418%, 4.018%, 3.538%, 3.022%, 2.453%, 2.044%, 2.009% and 1.102%, respectively. These genera comprised *ca*. 23% of the WLT bacterial community. Among the dominant genera shown in Figure 4, *Dyadobacter* was only observed in the WLT, whereas the relative abundance of *Limnohabitans*, *Herbaspirillum*, *Pedobacter*, *Hyphomonas*,*Leucobacter*, *Flavobacterium*, *Algoriphagus* and *F luviicola* was increased by *ca*. 10–100-fold in the WLT compared to the WHT. However, lower temperature greatly reduced the total number of observed genera, and significantly diminished the relative abundance of a large number of bacteria, such as*Methylocystis*, *Conexibacter*, *Gemmatimonas* and *Sediminibacterium* (Figure 4). In addition, the dominant genera distributed in the two samples were *Ilumatobacter*, *Mycobacterium*,*Hydrogenophaga*, *Rhodobacter* and *Polynucleobacter*, which have also been reported in various aquatic ecosystems. Taken together, our data revealed an extremely diverse community composition with lower relative abundance at the species level in the WHT when contrasted to the phylogenetic profile of the WLT.

Species-level affiliation of the sequences retrieved from the WHT revealed a total of 2562 phylotypes at a phylogenetic distance of 0.03, while 406 phylotypes were identified from the WLT. Of these, a core species containing 115 phylotypes was observed, representing 4.03% of all the richness in the FASFL, regardless of the temperature change (Figure 5).

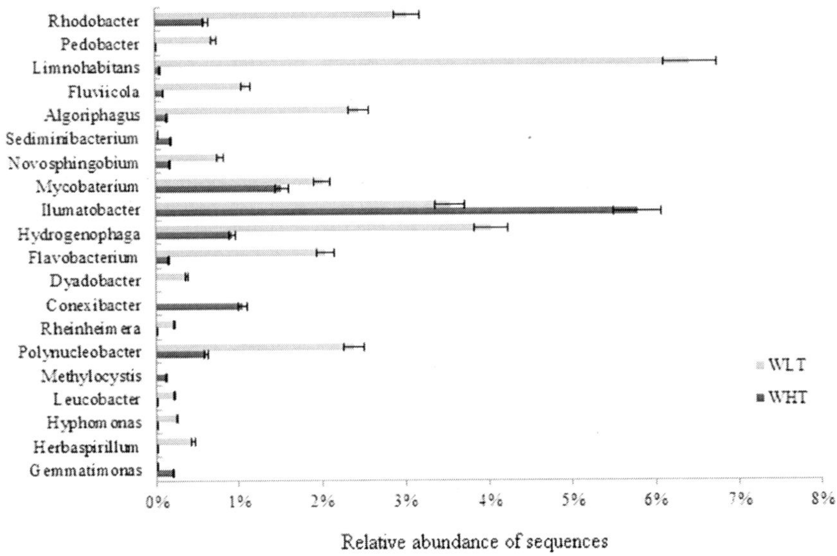

Figure 4. Change in bacterial community composition at the genus level in the WHT and WLT samples from the FASFL. Genus level affiliations of the 454-pyrosequencing sequences retrieved from the samples were performed using the MOTHUR software at a phylogenetic distance of 0.05. The resulting top twenty abundant genera were extracted and their relative abundance was compared.

2.3.3. Potential Pathogenic Bacteria

Base on the pyrosequencing datasets, a search of major bacterial pathogens to aquaculture animals and human revealed different potential risks between the WHT and WLT (Table 2). A total of 25 sequences, representing 0.1357% of all the analyzed sequences, were affiliated closely to seven genus and eleven pathogenetic species that have been reported in the literature [35,36,37]. The following potentially pathogenic species were detected in the WHT sample: *Aeromonas hydrophila*, *Aeromonas caviae*, *Bacillus anthracis*, *Mycobacterium marinum*, *Pseudomonas anguilliseptica*, *Mycobacterium avium* and *Mycobacterium fortuitum*. Among these, *A. caviae*, *M. marinum* and *M.avium* were higher temperature-associated and absent from the WLT sample. *A. caviae* is one of the most common pathogens related with aquaculture including shrimp, while *M. marinum* is also a major pathogen in fish aquaculture. *M. avium* exists in various environments including freshwater and can cause human Mycobacterium avium complex disease. Distinct from the WHT, four potential pathogens were detected from the WLT, including *Aeromonas veronii, Flavobacterium johnsoniae, Serratia marcescens* and *Vibrio cholerae*, all of which are known as human pathogens previously detected in diverse environments in nature. In addition, *A. hydrophila, B. anthracis, M.fortuitum* and *P. anguilliseptica* appeared to more adaptable to temperature change in the FASFL compared to the other pathogens, since they were detected in both WHT and WLT samples. *A.hydrophila* is one of the major pathogens in aquaculture, while the other three known as human pathogens contributed the most to the observed potential pathogen enrichment in the FASFL.

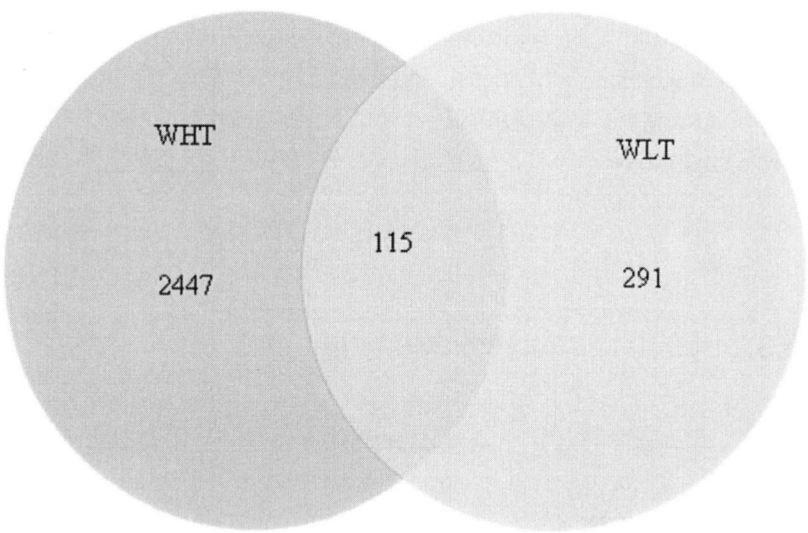

Figure 5. Venn diagram showing the change in bacterial composition at the species level in the WHT and WLT samples from the FASFL. The 454-pyrosequencing sequences retrieved from the samples were analyzed using the MOTHUR software at a phylogenetic distance of 0.03.

Table 2. Bacterial pathogens identified in the WHT and WLT samples from the FASFL.

Bacterial Pathogen [a]	Disease	Source		RAS (%) [b]	
		Environment	Main Hosts	WHT	WLT
Aeromonas hydrophila	Motile aeromonads septicaemia, cholangitis	Freshwater, brackish water, biosolid	Catfish, carp, trout, eel, sturgeon, tilapia, bass	0.0109	0.0054
Aeromonas caviae	Speticaemia, gastroenteritis, cholangitis	Freshwater, brackish water, soil, biosolid, agricultural products.	Fish, shrimp, frog, soft-shelled turtle	0.0054	0
Aeromonas veronii	Speticaemia, gastroenteritis, cholangitis	Freshwater, biosolid	Human, mosquitos, leeches	0	0.0054
Bacillus anthracis	*Anthrax*	Natural and processed water sources, sewage, biosolid	Human	0.0054	0.0054
Flavobacterium johnsoniae	False columnaris	Natural water sources, fish cultures	Barramundi	0	0.0054
Mycobacterium avium	Mycobacterium avium complex	Natural water sources, soil, biosolid	Human, farm animals, birds	0.0054	0
Mycobacterium fornitum	Osteomyelitis	River, lake, tap water, soil, dust, biosolid	Human, cattle, frog, other animals	0.0217	0.0163
Mycobacterium marinum	Mycobacteriosis	Natural water sources, fish cultures	Atlantic salmo, Seabass, turbot	0.0054	0
Serratia marcescens	Conjunctivitis, keratitis, endophthalmitis, tear duct infections	Natural water sources, soil, biosolid	Human, plants, animals	0	0.0109
Pseudomonas anguilliseptica	Pseudomonadiasis, Winter disease	Natural water sources, fish cultures	Human, seabream, eel, turbot, ayu	0.0163	0.0109
Vibrio cholerae	Vibriosis	Natural water sources, fish cultures, biosolid	Human, croaker fish, puffer fish, grouper, cod, shrimp, big-scale sand smelt, flounder, abalone, seabream, salmon, sweetfish, sheatfish, catfish	0	0.0054

[a] The bacterial pathogens were reported in literature [35–37]; [b] Relative abundance of sequence.

3. EXPERIMENTAL SECTION

3.1. Sample Collection and Bacterial Genomic DNA Extraction

Water samples were collected from a typical aquaculture farming base for *L. vannamei*, located in Fengxian district (N30°51'–E121°23'), Shanghai, China. Each closed and land-based culture pond is about 80 m long and 40 m wide with an average water depth of *ca.* 1.5 m. The ponds are equipped with aerating systems running twice daily to maintain moderate oxygen levels. Water quality was maintained approximately at pH 7.9–8.6 and 0.4‰–2.1‰ salinity during the study period. Water temperature (Wt) in the shrimp ponds is naturally adjusted by the local climate. Sampling was carried out three times monthly from June 2012 to May 2013, except December 2012 to February 2013 when the shrimp aquaculture was annually suspended due to colder weather. Wt was determined using a HACH sensION5 conductivity Meter (HACH Company, Loveland, CO, USA). Surface water was collected at three different sites in each pond using 10 L sterile plastic bottles, and immediately transferred on ice to the laboratory at Shanghai Ocean University in Shanghai, China. Bacterial cells were separated by standard sequential filtration techniques: each water sample was filtered though 8-μm qualitative filter paper to remove large suspended particles, and 1.0 L filtrate was subsequently filtered through polycarbonate membranes with 0.8- and 0.22-μm pore size (47 mm diameter, Millipore, Corcaigh, Ireland), respectively. DNA was extracted from three filters of each sample using the QIAamp DNA Stool Mini Kit (QIAGEN Biotech Co. Ltd., Hilden, Germany) according to the manufacturer's instruction. The concentration of DNA in the samples was determined using a multi-mode microplate reader BioTek Synergy™ 2 (BioTek Instruments, Inc., Winooski, VT, USA).

3.2. PCR-DGGE and Data Analysis

The V3 variable regions of bacterial 16S rRNA genes were amplified by PCR using the primer pair P3 (5'-GCCCGCCGCGCGGCGGGCGG GGCG GGGG

CACG GGGGGCCTACGGGAGGCAGCAG-3') and P2 (5'-ATTACCGC GGCT GCTGG-3') as described previously [38]. The expected length of amplified PCR products is 193 bp with a GC clamp attached to their 5'-termini [38]. Oligonucleotide primers were synthesized by Shanghai Sangon Biological Engineering Technology and Services Co., Ltd. (Shanghai, China). The PCR amplification was performed in a 20 μL reaction volume as described previously [39]. Amplification was performed in a MastercyclerW pro PCR thermal cycler (Eppendorf, Hamburg, Germany) with a touch down PCR protocol [38]. A sample (5 μL) of each PCR product was analyzed by agarose gel electrophoresis with a 1.5% agarose gel. Amplified DNA fragments were visualized and imaged by Molecular Imager® Gel Doc™ XR + System (Bio-Rad Laboratory, Hercules, CA, USA).

The vertical DGGE was performed using a DCode™ Universal Mutation Detection System (Bio-Rad, Hercules, CA, USA) according to the manufacturer's instructions. Each sample (400–500 ng) combined from three independent amplicons was loaded per well onto a 8% polyacrylamide gel (acrylamide:bis-acrylamide, 37.5:1) with a liner denaturing gradient range of 45%–55% formed with 7 M urea and 40% (*v*/*v*) deionized formamide. The denaturing gradient gel was run at 60 V for 16 h with a constant temperature of 60 °C in 1× TAE buffer (40 mM Tris base, 20 mM acetic acid, 1.0 mM Na_2-EDTA, pH 8.0). Following electrophoresis, the gel was stained three times with 6 mL of 0.01% SYBR® Green I Nucleic Acid Gel Stains (Invitrogen, Carlsbad, CA, USA) for 10 min according to the manufacturer's instructions, rinsed and imaged by Molecular Imager® Gel Doc™ XR System (Bio-Rad, Hercules, CA, USA). The DGGE profiles were analyzed using the Quantity One® 1-D Analysis software (version 4.6.2; Bio-Rad, Hercules, CA, USA). DNA banding patterns were compared using pairwise similarity matrices calculated with the Dice's coefficient [27], and cluster analysis was performed using the unweighted pair group method with arithmetic mean (UPGMA) via the software. The PRIMER (Version 5.2.8, PRIMER-E Ltd., Ivybridge, Devon, UK) was used for analysis of similarity (ANOSIM) [40]. The statistical significance of differences in microbial community composition was determined by IBM SPSS Statistics (Version 19) with one-way ANOVA (IBM, Armonk, NY, USA). The RDA analysis was performed using the CANOCO Version 4.5 [41].

3.3. Sequencing and Phylogenetic Analysis

Distinct DNA bands on the DGGE gels were individually excised and re-amplified using the primer pair 341f (5'-CCTACGGGAGGCAGCAG-3') (without GC clump) and 534r (5'-ATTACCGCGGCTGCTGG-3') according to the method described previously [38,42]. The PCR product was purified using AxyPrep DNA Gel Extraction Kit (Axygen, Union City, CA, USA), and ligated into the pGM-T cloning vector (TianGen Biotech Co., Ltd., Beijing, China) according to the method described previously [43]. Ligated DNA transformation and positive colony identification were performed as described previously [44]. The *E. coli* TOP10 (genotype: F$^-$*mcr*AΔ (*mrr-hsd*RMS-*mcr*BC) ψ80 *lac*ZΔM15Δ*lac*X74 *rec*A1 *ara*D139Δ (*ara-leu*) 7697 *gal*U *gal*K *rps*L(Str^r) *end*A1 *nup*G), (TianGen Biotech Co., Ltd., Beijing, China) was used as a host strain for DNA cloning. Plasmid DNA was prepared using the MiniBEST Plasmid DNA Extraction Kit Ver.2.0 (Japan TaKaRa BIO, Dalian Company, Dalian, China).

Automated DNA sequencing was carried out using ABI 3730xl capillary sequencer (Applied Biosystems, Foster City, CA, USA) and BigDye® terminator Version 3.1 cycle sequencing kit (Perkin-Elmer, Maltham, MA, USA) at the China Human Genome Centre (Shanghai, China). Sequencing reads were checked for chimera formation with the Ribosomal Database Project (RDP, http://rdp.cme.msu.edu/), and inferred for the closest relative using the Basic Local Alignment Search Tool (BLAST) (http://www.ncbi.nlm.nih.gov/BLAST). The sequence data was deposited in the NCBI Sequence Read Archive under SAMN02698684-97.

3.4. Pyrosequencing and Data Analysis

Pyrosequencing was carried out using the 454/Roche GS-FLX Plus System (Roche Diagnostics, Basel, Switzerland) at the Hanyu Biotech Co., Ltd., the China Human Genome Centre (Shanghai, China), according to the standard protocols of the manufacturer. The V3–V4 variable regions of bacterial 16S rRNA genes were amplified by PCR using the universal bacterial primers with an 8-bp barcode and a GS-FLX sequencing adaptor (not shown). The

MOTHUR Ver.1.32.0 software was employed for most of the sequence processing and analyses [45]. Raw sequence data were pre-processed to remove primers and barcodes, poor quality reads (>1% sequencing error rate), and the reads less than 200 bp in length. Chimeric reads were detected using the UCHIME software [46]. Silva/Greengenes database was used to remove non-targeted sequence contamination [47]. Qualified sequences were clustered into OTUs in MOTHUR at 0.03, 0.05 and 0.20 phylogenetic distance threshold, defined at species, genus and phylum level, respectively [44]. For read level taxonomic analysis, the RDP Classifier 2.2 of the RDP was used at a confidence threshold value of 80% [48]. Bacterial Chao1 richness [49], Shannon diversity index [50], Good's coverage, rarefaction curves, Venn diagram were calculated and created using the MOTHUR based on the OTUs clustering data. The sequence data was deposited in the NCBI Sequence Read Archive under the SAMN02699090-91.

4. CONCLUSIONS

This study constitutes the first investigation of temperature influence on bacterial community composition in freshwater aquaculture system farming of *Litopenaeus vannamei*. Water samples were collected over one year period, and aquatic bacteria were characterized by PCR-DGGE and pyrosequencing of the 16S rDNA. The results revealed a specific and dynamic bacterial population structure with considerable variation over the seasonal change. Phylogenetic analysis of the 18,427 sequences substantiated that the WHT harbored remarkably higher diversity and richness in bacterial composition when compared to the WLT. A large number of unclassified bacteria contributed the most to the observed variation in the system, albeit *Actinobacteria*, *Proteobacteria*and *Bacteroidetes* constituted a major proportion of the microflora in the FASFL. Three phyla present in lower relative abundance were distributed in WHT and WLT, including*Cyanobacteria/Chloroplast*, *Firmicutes* and *Gemmatimonadetes*. In addition, *Acidobacteria*, *Fibrobacteres*, *Fusobacteria*, *Spirochaetes* and *Verrucomicrobia* were only found in the WHT in extremely low abundance, whereas *Deferribacteres* showed an opposite pattern. At the genus level, phylogenetic analysis also revealed distinct profiles: *Ilumatobacter, Mycobacterium, Conexibacter, Hydrogenophaga, Rhodobacter* and *Polynucleobacter* were the highest abundant genus in the WHT, whereas*Limnohabitans, Hydrogenophaga, Humatobacter, Rhodobacter, Algoriphagus, Flavobacterium,Mycobacterium* and *Fluviicola* were the most responsible for the diversity difference. At the species level, a total of 2562 phylotypes were identified in the WHT, while 406 phylotypes were found in the WLT. A core species containing 115 phylotypes was observed, representing 4.18% of all the richness in the FASFL. Eleven potential pathogenetic species were identified with different phylogenetic patterns in the WHT and WLT. It will be interesting to characterize these bacteria in future research. Our data demonstrated a temperature-driven substantial shift in bacterial population in the FASFL. The results provide guidance for aquatic animal disease control in shrimp aquaculture, and also constitute an important step in extending our knowledge of microbial ecology in aquaculture ecosystem, particularly waterborne pathogenic bacteria mediated by climate change.

ACKNOWLEDGMENTS

This work was supported by a Grant No. QXJK201207 from the Shanghai Key Laboratory of Meteorology and Health, a Grant No. B-9500-10-0004 from the Shanghai Municipal Education Commission, and a Grant No. 31271830 from the National Natural Science Foundation of China.

AUTHOR CONTRIBUTIONS

Shilu Tong, Jianguo Tan, Haizhen Mu, Li Peng, Dandan Yang and Lanming Chen planned and coordinated the research. Shilu Tong, Jianguo Tan, Haizhen Mu and Lanming Chen participated in the design and discussion of the study. Yuyi Tang carried out the major experiments. Yuyi Tang, Peiying Tao and Lanming Chen analyzed the data. Lanming Chen wrote the manuscript. All authors read and approved the final manuscript.

REFERENCES

1. Food and Agriculture Organization of the United Nations. *FAO Yearbook 2010: Fishery and Aquaculture Statistics*; FAO: Rome, Italy, 2012.

2. Liao, C.; Chien, Y.-H. The Pacific white shrimp, *Litopenaeus vannamei*, in Asia: The world's most widely cultured alien crustacean. In *The Wrong Place-Alien Marine Crustaceans: Distribution, Biology and Impacts*; Galil, B.S., Clark, P.F., Carlton, J.T., Eds.; Invading Nature-Springer Series in Invasion Ecology: Dordrecht, The Netherlands, 2011; Volume 6, pp. 489–519.

3. Food and Agriculture Organization of the United Nations. FAO Yearbook 2009: Fishery and Aquaculture Statistics. FAO: Rome, Italy, 2011.

4. Thompson, F.L.; Iida, T.; Swings, J. Biodiversity of *Vibrios*. *Microbiol. Mol. Biol. R.* **2004**, *68*, 403–431.

5. Lavilla-Pitogo, C.R.; Leano, E.M.; Paner, M.G. Mortalities of pond-cultured juvenile shrimp,*Penaeus monodon*, associated with dominance of luminescent *Vibrios* in the rearing environment. *Aquaculture* **1998**, *164*, 337–349.

6. Sung, H.-H.; Hsu, S.-F.; Chen, C.-K.; Ting, Y.-Y.; Chao, W.-L. Relationships between disease outbreak in cultured tiger shrimp (*Penaeus monodon*) and the composition of *Vibrio*communities in pond water and shrimp hepatopancreas during cultivation. *Aquaculture***2001**, *192*, 101–110.

7. Israngkura, A.; Sae-Hae, S. A review of the economic impacts of aquatic animal disease. In*Primary Aquatic Animal Health Care in Rural, Small-Scale, Aquaculture Development*; Arthur, J.R., Phillips, M.J., Subasinghe, R.P., Reantaso, M.B., MacRae, I.H., Eds.; FAO Fish. Tech.: Rome, Italy, 2002; Volume 406, pp. 253–286.

8. Bourne, D.G.; Young, N.; Nicole Webster, N.; Payne, M.; Salmon, M.; Demel, S.; Hall, M. Microbial community dynamics in a larval aquaculture system of the tropical rock lobster,*Panulirus ornatus*. *Aquaculture* **2004**, *242*, 31–51.

9. Zhou, J.; Fang, W.; Yang, X.; Zhou, S.; Hu, L.; Li, X.; Qi, X.; Su, H.; Xie, L.A. Nonluminescent and highly virulent *Vibrio harveyi* strain is associated with "Bacterial White Tail Disease" of*Litopenaeus vannamei* shrimp. *PLoS One* **2012**, *7*, e29961.

10. Wilcox, B.A.; Colwell, R.R. Emerging and reemerging infectious diseases: Biocomplexity as an interdisciplinary paradigm. *EcoHealth* **2005**, *2*, 244–257.

11. Hasson, K.W.; Wyld, E.M.; Fan, Y.; Lingsweiller, S.W.; Wiener, S.J.; Cheng, J.; Varner, P.W.*Streptococcosis* in farmed *Litopenaeus vannamei*: A new emerging bacterial disease of penaeid shrimp. *Dis. Aquat. Organ.* **2009**, *86*, 93–106.

12. Metzker, M.L. Sequencing technologies-the next generation. *Nat. Rev. Genet.* **2010**, *11*, 31–46.

13. Payne, M.S.; Hall, M.R.; Sly, L.; Bourne, D.G. Microbial diversity within early-stage cultured*Panulirus ornatus* phyllosomas. *Appl. Environ. Microbiol.* **2007**, *73*, 1940–1951.

14. Johnson, C.N.; Barnes, S.; Ogle, J.; Grimes, D.J. Microbial community analysis of water, foregut, and hindgut during growth of Pacific white shrimp, *Litopenaeus vannamei*, in closed-system aquaculture. *J. World Aquacult. Soc.* **2008**, *39*, 251–258.

15. Sakami, T.; Fujioka, Y.; Shimoda., T. Comparison of microbial community structures in intensive and extensive shrimp culture ponds and a mangrove area in Thailand. *Fish. Sci.***2008**, *74*, 889–898.

16. Van der Meeren, T.; Brunvold, L.; Sandaa, R.-A.; Bergh, O.; Castberg, T.; Thyrhaug, R.; Mangor-Jensen, A. Water quality and microbial community structure in juvenile Atlantic cod (*Gadus morhua* L.) cultures. *Aquaculture* **2011**, *316*, 111–120.

17. Zhang, X.; Fu, L.; Deng, B.; Liang, Q.; Zheng, J.; Sun, J.; Zhu, H.; Peng, L.; Wang, Y.; Shen, W.; *et al. Bacillus subtilis* SC02 supplementation causes alterations of the microbial diversity in grass carp water. *World J. Microbiol. Biotechnol.* **2013**, *29*, 1645–1653.

18. Zeng, Y.; Ma, Y.; Wei, C.; Jiao, N.; Tang, K.; Wu, Z.; Jian, J. Bacterial diversity in various coastal mariculture ponds in Southeast China and in

diseased eels as revealed by culture and culture-independent molecular techniques. *Aquacult. Res.* **2010**, *41*, e172–e186.

19. Youssef, N.H.; Couger, M.B.; Elshahed, M.S. Fine-scale bacterial beta diversity within a complex ecosystem (Zodletone Spring, OK, USA): The role of the rare biosphere. *PLoS One***2010**, *5*, e12414.

20. Han, D.; Kang, I.; Ha, H.K.; Kim, H.C.; Kim, O.-S.; Lee, B.Y.; Cho, J.-C.; Hur, H.-G.; Lee, Y.K. Bacterial communities of surface mixed layer in the Pacific sector of the western Arctic ocean during sea-ice melting. *PLoS One* **2014**, *9*, e86887.

21. Sogin, M.L.; Morrison, H.G.; Huber, J.A.; Welch, D.M.; Huse, S.M.; Neal, P.R.; Arrieta, J.M.; Herndl, G.J. Microbial diversity in the deep sea and the underexplored "rare biosphere".*Proc. Nat. Acad. Sci. USA* **2006**, *103*, 12115–12120.

22. Vaz-Moreira, I.; Egas, C.; Nunes, O.C.; Manaia, C.M. Culture-dependent and culture-independent diversity surveys target different bacteria: A case study in a freshwater sample.*Anton. Leeuw. Int. J. G.* **2011**, *100*, 245–257.

23. Miller, S.R.; Strong, A.L.; Jones, K.L.; Ungerer, M.C. Bar-coded pyrosequencing reveals shared bacterial community properties along the temperature gradients of two alkaline hot springs in yellowstone national park. *Appl. Environ. Microbiol.* **2009**, *75*, 4565–4572.

24. Revetta, R.P.; Pemberton, A.; Lamendella, R.; Iker, B.; Santo Domingo, J.W. Identification of bacterial populations in drinking water using 16S rRNA-based sequence analyses. *Water Res.***2010**, *44*, 1353–1360.

25. Shanks, O.C.; Newton, R.J.; Kelty, C.A.; Huse, S.M.; Sogin, M.L.; McLellan, S.L. Comparison of the microbial community structures of untreated wastewaters from different geographic locales. *Appl. Environ. Microbiol.* **2013**, *79*, 2906–2913.

26. Beleneva, I.A.; Zhukova, N.V. Seasonal dynamics of cell numbers and biodiversity of marine heterotrophic bacteria inhabiting invertebrates and water ecosystems of the Peter the Great Bay, Sea of Japan. *Microbiology* **2009**, *78*, 369–375.

27. Zhang, M.; Yu, N.; Chen, L.; Jiang, C.; Tao, Y.; Zhang, T.; Chen, J.; Xue, D. Structure and seasonal dynamics of bacterial communities in three urban rivers in China. *Aquat. Sci.* **2012**,*74*, 113–120.

28. Rösel, S.; Allgaier, M.; Grossart, H.-P. Long-term characterization of free-living and particle-associated bacterial communities in lake Tiefwaren reveals distinct seasonal patterns.*Microb. Ecol.* **2012**, *64*, 571–583.

29. Jing, H.; Xia, X.; Suzuki, K.; Liu, H. Vertical profiles of bacteria in the tropical and subarctic oceans revealed by pyrosequencing. *PLoS One* **2013**, *8*, e79423.

30. Bakke, I.; Skjermo, J.; Vo, T.A.; Vadstein, O. Live feed is not a major determinant of the microbiota associated with cod larvae (*Gadus morhua*). *Environ. Microbiol. Rep.* **2013**, *5*, 537–548.

31. Nacke, H.; Thürme, A.; Wollherr, A.; Will, C.; Hodac, L.; Herold, N.; Schöning, I.; Schrumpf, M.; Daniel, R. Pyrosequencing-based assessment of bacterial community structure along different management types in German forest and grassland soils. *PLoS One* **2011**, *6*, e17000.

32. Martins, P.; Cleary, D.F.R.; Pires, A.C.C.; Rodrigues, A.M.; Quintino, V.; Calado, R.; Gomes, N.C.M. Molecular analysis of bacterial communities and detection of potential pathogens in a recirculating aquaculture system for *Scophthalmus maximus* and *Solea senegalensis*. *PLoS One* **2013**, *8*, e80847.

33. Dworkin, M., Falkow, S., Rosenberg, E., Schleifer, K.H., Stackebrandt, E., Eds.; *The Prokaryotes: A Handbook on the Biology of Bacteria*; Springer-Verlag: New York, NY, USA, 2006; pp. 3–37.

34. Matsumoto, A.; Kasai, H.; Matsuo, Y.; Ōmura, S.; Shizuri, Y.; Takahashi, Y. *Ilumatobacter fluminis* gen. nov., sp. nov., a novel actinobacterium isolated from the sediment of an estuary. *J. Gen. Appl. Microbiol* **2009**, *55*, 201–205.

35. Bibby, K.; Viau, E.; Peccia, J. Pyrosequencing of the 16S rRNA gene to reveal bacterial pathogen diversity in biosolids. *Water Res.* **2010**, *44*, 4252–4260.

36. Toranzo, A.E.; Magariños, B.; Romalde, J.L. A review of the main bacterial fish diseases in mariculture systems. *Aquaculture* **2005**, *246*, 37–61.

37. Pridgeon, J.W.; Klesius, P.H. Major bacterial diseases in aquaculture and their vaccine development. *CAB Rev.* **2012**, *7*, 048.

38. Muyer, G.; de Waal, E.C.; Uitterlinden, A.G. Profiling of complex microbial populations by denaturing gradient gel electrophoresis analysis of polymerase chain reaction-amplified genes coding for 16S Rrna. *Appl. Environ. Microbiol.* **1993**, *59*, 695–700.

39. Zhang, T.; Pan, Y.; Li, B.; Ou, J.; Zhang, J.; Chen, Y.; Peng, X.; Chen, L. Molecular cloning and antimicrobial activity of enterolysin A and helveticin J of bacteriolysins from metagenome of Chinese traditional fermented foods. *Food Control* **2013**, *31*, 499–507.

40. Clarke, K.R.; Green, R.H. Statistical design and analysis for a "biological effects" study. *Mar. Ecol. Prog. Ser.* **1988**, *46*, 213–226.

41. Ter Braak, C.J.F.; Šmilauer, P. CANOCO reference manual and CanoDraw for Windows user's guide: Software for canonical community ordination (version 4.5). In *Section on Permutation Methods. Microcomputer Power*; Ithaca: New York, NY, USA, 2002.

42. Schabereiter-Gurtner, C.; Maca, S.; Rölleke, S.; Nigl, K.; Lukas, J.; Hirschl, A.; Lubitz, W.; Barisani-Asenbauer, T. 16S rDNA-based identification of bacteria from conjunctival swabs by PCR and DGGE fingerprinting. *Investig. Ophthalmol. Vis. Sci.* **2001**, *42*, 1164–1171.

43. Shi, Y.; Pan, Y.; Li, B.; He, W.; She, Q.; Chen, L. Molecular cloning of a novel *bioH* gene from an environmental metagenome encoding a carboxylesterase with exceptional tolerance to organic solvents. *BMC Biotechnol.* **2013**, *13*, 13.

44. Song, Y.; Yu, P.; Li, B.; Pan, Y.; Zhang, X.; Cong, J.; Zhao, Y.; Wang, H.; Chen, L. The mosaic accessory gene structures of the SXT/R391-like integrative and conjugative elements derived from *Vibrio* spp. isolated from aquatic products and environment in the Yangtze River estuary, China. *BMC Microbiol* **2013**, *13*, 214.

45. Schloss, P.D.; Westcott, S.L.; Ryabin, T.; Hall, J.R.; Hartmann, M.; Hollister, E.B.; Lesniewski, R.A.; Oakley, B.B.; Parks, D.H.; Robinson,

C.J.; *et al.* ntroducing mothur: Open-source, platform-independent, community-supported software for describing and comparing microbial communities. *Appl. Environ. Microbiol.* **2009**, *75*, 7537–7541.

46. Edgar, R.C.; Haas, B.J.; Clemente, J.C.; Quince, C.; Knight, R. UCHIME improves sensitivity and speed of chimera detection. *Bioinformatics* **2011**, *27*, 2194–2200.

47. DeSantis, T.Z.; Hugenholtz, P.; Larsen, N.; Rojas, M.; Brodie, E.L.; Keller, K.; Huber, T.; Dalevi, D.; Hu, P.; Andersen, G.L. Greengenes, a chimera-checked 16S rRNA gene database and workbench compatible with ARB. *Appl. Environ. Microbiol.* **2006**, *72*, 5069–5072.

48. Maidak, B.L.; Olsen, G.J.; Larsen, N.; Overbeek, R.; McCaughey, M.J.; Woese, C.R. The Ribosomal Database Project (RDP). *Nucleic Acids Res.* **1996**, *24*, 82–85.

49. Chao, A.; Bung, J. Estimating the number of species in a stochastic abundance model.*Biometrics* **2002**, *58*, 531–539.

50. Shannon, C.E.; Weaver, W. The mathematical theory of information. *AT&T Tech. J* **1949**, *27*, 379–423.

Chapter 3

Development of Freshwater Native Species with Aquacultural Potential

Fernando Garcia-Trejo[1], Silvia Hurtado-Gonzalez[1], Genaro M. Soto-Zarazua[1] and P. J Gutierrez-Yurrita[2]

[1] División de Estudios de Posgrado Facultad de Ingeniería, Universidad Autónoma de Querétaro, C.U. Cerro de las Campanas, Qro, México

[2] Centro Interdisciplinario de Investigaciones y Estudios sobre Medio Ambiente y Desarrollo (CIIEMAD), Instituto Politécnico Nacional, México

1. INTRODUCTION

Aquaculture is an activity practiced by Chinese people since 2800 B.C. The first culture species were the carp (*Cyprinus carpio*) and mullet (*Mugil cephalus*), the first documented knowledge about fish culture in captivity belong to the carp. This activity has been supporting human demands for fish products for centuries and now is an important worldwide industry. Over the years this practice has become more technical with the objective to make the work easily but mainly in order to increase production. Nowadays this activity has grown to an entire industry that handles both supplements as the product itself. Global production from aquaculture now supplies one third of seafood consumed worldwide. With this massive increase in world production the current aquaculture industry is one of the fastest growing sectors in food production [1]. However, this rapid growth in the aquaculture industry has generated ecological damage due to a huge extractive use of water, land, and feeds. Besides, is important to consider that these requirements are associated to another impacts such as; polluting, salinization of soils, nutrient-loading, clearing of natural habitat, overexploit of ground water reserves, introduction and transmission of diseases [2]. At this respect Bailey [3] establish a new term "blue revolution", which describes the expansion of fish-farming in tropical regions, according to this idea aquaculture must provide huge quantities of fish and help to solving problems of world food security and alleviating poverty. However, production increase in aquaculture demands feeds, energy for the cultured species and almost always is obtained through catch, so the fastest growing sectors in food production threatens its ability to continue to provide increasing yields in a sustainable manner, and concerns with the resulting from fish-farming have led to calls for the "greening of the blue revolution" [4].

The situation in México is not an exception; the expansion of aquaculture has been accompanied by degradation of the natural environment, especially on marine aquaculture [5]. As it happens in the world directly impacts of fisheries and aquaculture are:

- Introduction of nonnative fish species:
- Introduction of exogenous parasites:
- Nutrient pollution:
- Habitat modification:
- Overcollection of wild seed stock:
- Changes of food webs
- Increase of interspecific competition:

It is clear that current food productions techniques in aquaculture are good just under financial point of view but always leave aside the environment aspects. The relationship between aquaculture and environment is complex specifically the biodiversity topic. Many examples of positive and negative impacts have been documented, however until now there is no solution which allows the development of a relationship between food production and the environment. This solution must be adopting a new paradigm based in ecological concepts of extreme resource efficiency and the closing of nutrient and waste cycles, resource-use optimization [6]. As can be seen, this is not an easy task since it requires the creation of multidisciplinary teams which can see the problem holistically and try to give a solution that benefits all parties involved in the process. The efforts and the perception of the environment are different between countries, but ultimately the problem to be addressed holistically. However, studies on fish typically focus on species that currently have commercial value, causing species that lack such market value to be ignored. This is the case of several freshwater native species, which can be founded in central and South America. Some attempts to cultivate native species have occurred mainly in areas or rural communities, where in addition to enhance the conservation of species protein contributes to the diet of the community [7]. One of the most interesting case studies in Mexico is growing "white fish" (*Chirostoma estor*) with the aim repopulate some areas where the introduction of their populations has been declined [8]. Most documented is that of the native Central American cichlid (*Cichlasoma urophthalmus*), of which there have been many studies to support its culture [9]. In southern region some attempts to grow some silversides in Argentina (*Odontestes Basilichthys*), some Characidae family members have been grown in Brazil and more recently three species of carnivorous cichlids aquarium purposes. In Peru it has favored the cultivation of called piracucu (*Arapaima gigas*), one of the largest fish of fresh water. Possibly the Cichlidae family members are those that show the greatest potential for cultivation [10, 11].

The main purpose of this chapter is to show the experience of three studies with native species; one refers to a small native species located in the state of Querétaro and with a great ecological importance, *Girardinichthys multiradiatus* [12]. The study of this fish focused on the description of its habitat throughout a hydrologic cycle in which ecophysiological responses were determined in order to stablish guidelines for its management and to preserve its population. In this work population structure and dynamics were getting and trophic and ecophysiological responses to fluctuations in environmental factors were also identified in order to have the possibility of laboratory reproduction and growth. On the other hand, native mojarra *Herinchthys cyanogutattus* founded on the basin Pánuco river. In this case the purpose was to evaluate its useful in fishery and later in the aquaculture. The work consisted of two stages: First, the characterization of their environment in order to locate stable populations of the mojarra and to characterize ecologically its habitat. Second, the mojarra was moved to the laboratory to try different forms of acclimatization for its future use giving them tried food. Once acclimated, the stock was use to carried out density studies of individuals for culture (capacity of load), as well as of ideal thermal for its production. Finally, *Procambarus digueti* which faces severe ecological problems (over fishing, no control of heights neither of sexes, there is not articulated extraction methods neither fishing seasons and restrains), since they are captured as food and as curative remedies from pre-Hispanic eras. To this situation the strong environmental pressure is added by the disturbance of their habitat, what has carried a decrease in numerical abundance. The objective of this work was determined the optimum cultivation conditions with respect to the load capacity and diet in the growth of *P. digueti* in intensive production systems. The controlled production of this specie will reduce the fishing pressure and it will be able to serve to repopulate the sites where may have been decline the natural population.

To successfully achieve the cultivation of a native species, compared to the technological advantages offered by more exotic trading requires knowledge of the biology, ecology and aquaculture potential (ability to live at high densities, accept food encapsulation, and withstand high environmental variations) of each species. If aquaculture potential studies are performed with ecophysiological and bioenergetics approach may be developed predictive models of how to develop a population under different environmental factors, and even develop experimentally testable hypotheses [13].

2. THE EXPERIENCE WITH GIRARDINICHTHYS MULTIRADIATUS

Studies on the biological aspects of fish typically focus on species that currently have commercial value, causing species that lack such market value to be ignored. This is the case of several freshwater fish, specifically of several members of the Goodeidae family. This is a diversified and small family of cyprinodontoid fish, confined to the central plateau of Mexico where its dispersion center lies in the well-isolated Lerma basin. Four species of the Godeidae family have been reported in Querétaro: three species are distributed widely in the Lerma basin (*Goodea atripinis, Xenotoca variata, Goodea gracilis*), and one species (*Girardinichthys multiradiatus*) can only be found in one body of water in the municipality of Amealco [14]. Scientific knowledge about it focuses on sexual dimorphism, peculiar courtship rituals, and viviparity [15], taxonomic aspects [16, 17], ethology [18], biology [19], and trophic ecology [20].

2.1. Environmental Conditions

G. multiradiatus was founded on San Martín Dam, located at 60 km south of Queretaro city, near to Amealco municipality (100° 09’ 43’’ W; 20° 15’ 02’’ N), at 2600 meters above sea level. The climate is subhumid with summer rains (Cw1) with an average temperature of 15.1°C, the months of May and June are those with the highest temperatures. The average annual rainfall is 659.5 mm, occurring mainly during the summer [21]. The main contribution to the dam water is from rain. Sampling was made over a full hydrological cycle (one year, beginning in February) in which the *G. multiradiatus*population was monitored once every two months, at the same time physical factors were measure (T°, pH, dissolved O_2, turbidity, depth). The physicochemical parameters of water showed stable behavior during the studied hydrological cycle (pH = 7 to 9; dissolved oxygen = 6.5 to 7 ppm). On the other hand, the temperature showed significant variation, with the highest temperatures (20-25°C) recorded between April and August, with the lowest (10-18°C) recorded between October to February.

of *Dunaliella* populations in natural hypersaline environments (Oren 2005). In this part of the world *Dunaliella*has been poorly studied; only *Dunaliella viridis* has been described in some places of South America (Massyuk 1973; Ginzburg 1987). More recently new reports described new species in Chile (Azúa-Bustos et al. 2010; Demergasso et al. 2008). In the Atacama Desert a sub-aerial species of *Dunaliella* using the condensation of water on spider webs in the walls of a coastal cave could be illustrate one of the first strategies in land colonization by plants (Azúa-Bustos et al. 2010). Moreover, sequences related to *Dunaliella salina* into the Lake Tebenquiche in the Salar de Atacama (Demergasso et al. 2008) represent a clear evidence of *Dunaliella*distribution in the Altiplano. Additionally, Iltis et al. (1984) described in the Bolivian Altiplano a species inhabiting a homonymous hypersaline large lake (Laguna Colorada) located in Sud Lípez Potosí. Environmental conditions in this lake promote high algal biomass of *Dunaliella* sp. turning the water into a red–orange colored system. These large salars (Uyuni and Coipasa) have a close relation with chemical and physical conditions described in the Salar de Alconcha. Similar conditions were reported in Laguna Colorada (L) especially in total salt concentration (total salt concentration: 119.74 g l^{-1}) characteristic of terminal hypersaline systems with high nutrients concentration (Nitrogen and Phosphorous) and close to anaerobic condition (Oren 2005).

Several treatments have demonstrated that *Dunaliella* is capable of accumulating β-carotene and glycerol under conditions of high irradiance and other saline stress conditions (Arun and Singh 2012). Highly photosynthetic systems in the Altiplano represent exceptional conditions for these aspects such as salinity variations and high UV irradiance especially in aquatic systems over the salt crust (sunlight screen reflector). Intensive pigmentation and dynamic responses (e.g. aplanospore cysts formation) of *Dunaliella salina* to the seasonal changes is a clear example of well adapted species to these conditions. Moreover, some *Dunaliella* species not only reproduce by division of vegetative cells but show sexual reproduction as well (Oren2005). One of the most interesting adaptations in *Dunaliella* species is the formation of vegetative cysts (Hamburger 1905; Margulis et al. 1980; Borowitzka and Huisman1993; Oren 2005) during winter conditions of short and cold days (Borowitzka and Huisman 1993). These formations (aplanospore cysts) generate great resistance to long

Also, due to seasonal differences in water usage, the water level of the dam was low from April to August and high from October to March (Fig. 1).

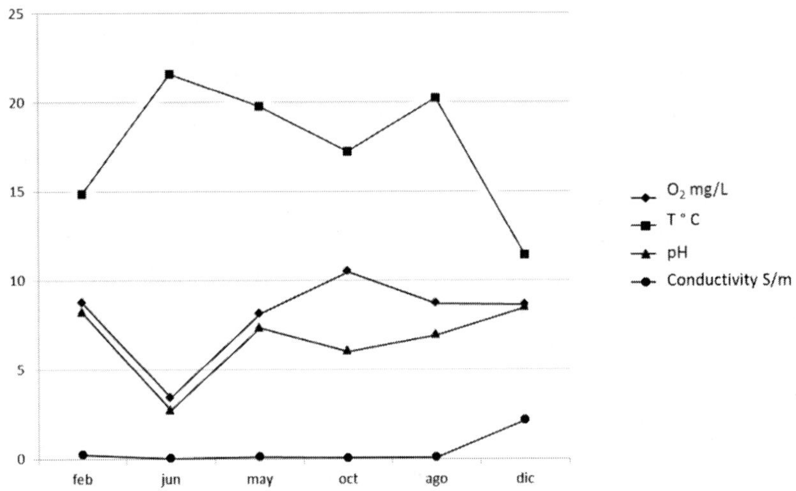

Figure 1. Environmental parameters during a hydrological cycle in San Martín Dam.

2.2. Population Ecology

According to the Cassie method, the population of *G. multiradiatus* consists of 12 classes, ranging from 8 to 48 mm standard length. Figure 2 shows the general structure of the mexcalpique population of San Martín. The numbers in parentheses indicate the percentage of each size class of the total population obtained through a year. Two of these size classes were found only in laboratory studies due to their small sizes; these sizes were smaller than what the nets in situ could catch.

Figure 3 shows the bimonthly structure of the population of *G. multiradiatus* in San Martín and the seasonal pattern of population growth, demonstrated in Von Bertalanffy equation with $L\infty = 47$; $K= 0.8870$ and, $t0 = -0.2103$. The variations throughout the year are present in both, in the structure of the population and its growth rate, with the shorter pattern cohorts having a higher growth rate and no increase in length during the period of February to June.

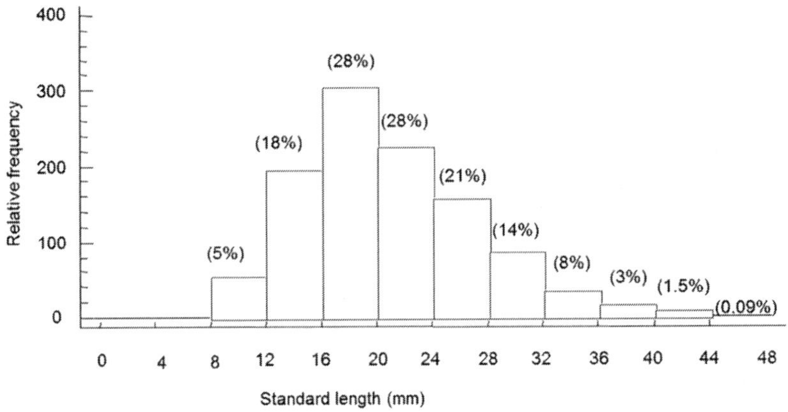

Figure 2. General structure of the population of San Martín mexcalpique. The numbers in parentheses indicate the percentage of each size class of the total population obtained through a year.

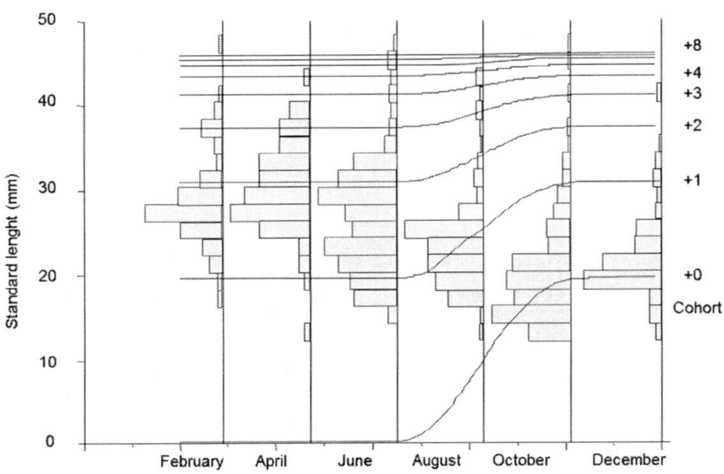

Figure 3. Bimonthly structure of the population of G. multiradiatus in San Martin and the curves of growth with the von Bertalanffy model for highly seasonal cycles. Upon reaching the asymptotic curve determines the final class of each age cohort.

2.3. Feeding Habits

To assess the daily feeding activity of the *G. multiraditus* in San Martín Dam samples were collected with spoon nets every four hours during a period of 24 hours (10:00, 14:00, 18:00, 22:00, 02:00 and, 06:00 hours). These catches allowed determining the feeding ecology of the species (feeding time, type of diet at different times of day, food components). The relative density in activity was measured using the catch per unit effort method (CPUE) based on the number of individuals caught by dragging. The fish were fixed in 70% alcohol and then transported to the laboratory (Nielsen et al., 1983). From 1022 stomachs analyzed, (1022), 18 food components were identified. The most abundant component found were insects (47%), especially the Diptera order, followed by detritus (24.0%) and Cladocera (17.5%), with the remaining components accounting for 10.7%. Unusual food components (less than 10% of the total), were only found at a specific times of the hydrological cycle, Table 1. The benthic review showed 20 components and trophic index was calculated indicating that *G. multiradiatus* is a polytrophic specie ($H \neq 0$). The maximum numbers of categories found in their stomachs were eight while minimum was two. Circadian sampling showed that this specie is polytrophic with two daily feeding periods (5:00 to 8:00 and 14:00 to 18:00 hours).

2.4. Bioenergetic

To quantify the aerobic metabolism and nitrogen excretion, animals were placed in a semi closed system with (0.5L) respirometric chambers, using a method which assumes that the reduction of oxygen and the increase of nitrogen in the chamber depend on the weight of the animal, the volume of water, the period of time in which no water circulated in the chamber and the ambient temperature [22]. After each cycle, sample was oven-dried in an electric oven between 70-80°C until the samples had constant weight. From each composite sample 2 g were measured and taken as analytical sample; the samples were digested with concentrated nitric acid. The determination of the percentage proximate composition was chemically analyzed according to the method of analysis described by the Association of Official Analytical Chemist [23]. While other 2 g sample tissue were combusted in a Parr bomb calorimeter to obtain oxycalorific measurements. The flow of energy that was used to determine the energy efficiency and assimilation is show in the next equation:

$$C = P + G + ER + EU + F$$

Table 1. Occurrence of food components (N = number sampled, n = number of organisms that have the category, % occurrence rate) by sex.

Category	Youth N = 654		Males N = 195		Females N = 173	
	n	%	n	%	n	%
Detritus	230	16.08	141	37.03	127	38.49
Diptera	517	55.31	146	34.32	142	34.41
Cladocera	231	21.5	51	13.73	38	13.73
Copepoda	53	3.78	25	2.26	31	2.26
Animal remains	41	1.27	29	2.56	47	2.56
Amphipoda	20	0.67	27	3.1	34	3.01
Hemiptera	16	0.52	27	2.96	38	2.96
Odonata	13	0.38	21	3.21	28	3.21
Himenoptera	1	0.06	4	0.11	12	0.66
Coleoptera	0	0	2	0.04	6	0.6
Crustacea	1	0.004	4	0.41	4	0.14
Ephemeroptera	3	0.06	0	0	5	0.38
Gasteropoda	3	0.05	1	0.02	6	0.17
Plecoptera	3	0.03	2	0.11	0	0
Thrichoptera	0	0	1	0.1	0	0
Homoptera	0	0	1	0.05	0	0

Where; C is the total energy content of food consumed, P and G are the energy equivalents of somatic and gonadal growth respectively, ER is the energy utilized in respiration, EU is the energy lost as nitrogenous and other waste compounds excreted in the urine, and F is the unabsorbed energy voided with the faeces (Bolduc et al., 2002; Bradshaw, 2003). All variables expressed in calories per gram of dry weight (cal/g).The ratio used to transform measured aerobic metabolism into calories was the standard oxycaloric coefficient for fish which mainly excrete N-NH_4 (Qox = 3.20 cal mg-1 O_2). Nitrogen excretion was estimated using literary references, taking into consideration the type of fish, size, feeding habits, and physiological status [24].

Physiological experiments showed higher energy expenditures in August with values of 2500 cal/g and minimal values for December with 200 cal/g. The increased energy expenditure was found in the early hours of the day (daylight hours) and then declined, reaching minimum values at night, with the exception of October, which displayed an inverse pattern. Calorimetric analysis did not show statistically significant difference between the energy provided by sex ($p > 0.05$). Main food energy intake was 4.8 ± 0.3 Kcal g-1 of dry weight, with the total weight of the mexcalpique consisting of, on average, 85.49 ± 2.49% organic matter, and 14.50 ± 2.49% mineral matter. By replacing the caloric values in the energy balance equation, was determined that *G. multiradiatus* uses approximately 81% of the energy consumed in the production of tissue and gametes (P and G), respiration process spent 5.7% (ER) and the rest 13.3% is invested in maintenance (EU and F). Multivariate analysis of environmental factors on the metabolism, showed no significant differences, however the temperature showed the lowest value of significance ($p = 0.08$).

3. HERINCHTHYS CYANOGUTATTUS

Texas Cichlids were formerly given the scientific name of *Cichlasoma cyanoguttatum*, but are now known by the name *Herichthys cyanoguttatus*. The genus *Herichthys* has been through several changes, and currently consists of nine species, native to lakes and rivers in south Texas and northern Mexico, making them the most northern naturally occurring species of cichlid in the world. It's the only native cichlid in the US and amongst the first cichlids imported to Europe, having first been imported in 1912. This species has also been introduced into areas they are not indigenous to, sometimes on purpose, but often by aquarium owners desperate to divest themselves of a fish the can no longer take care of. The areas of non-indigenous populations range from northern Texas to Florida, where it has become a popular game fish. This is due to having a tasty flavor similar to that of their distant relative, Tilapia.

In Mexico this species is called "Mojarra del norte" and It could be distinguished by a coupled of dark spots and a tiny blue circles on its sides. Adults show a olivaceus iridescent spots when viewed in the sun, there are also lines of the same color on the head, body and fins. During reproduction it is possible to see white region in the front part and a dark in the back especially in females, while males develop a prominent hump.

3.1. Biology

This fish could live in a wide range of temperature, between 5° and 30°C., [25, 26]. Trophic spectrum show variations between each population according to the region but in general is consider an omnivorous fish [27]. Many studies have described aspects of reproduction; the most relevant aspect is the monogamous behavior, when a male select a coupled it becomes aggressive and territorial [28].

3.2. Environmental Conditions

The first step in this research was to looking-for a population in order to do the ecological description of its habitat. H. *cyanoguttatus* was found in several places around Queretaro State, but only one was selected due to accessibility. This place is called Taxhido river and is located at 70km E, from the Queretaro capital 20° 35′ 18″O and 99° 40′ 47″N, the climate is subhumid with summer rains (Cw1) with an average temperature of 15.1°C, the months of May and June are those with the highest temperatures. The average annual rainfall is 659.5 mm, occurring mainly during the summer and the main contribution to the dam water is from a spring. Sampling was made over a full hydrological cycle (one year, beginning in February) in which the *H. cyanoguttatus* population was monitored once every two months, at the same time physical factors were measure (T°, pH, dissolved O_2, turbidity, depth). The physicochemical parameters of water showed stable behavior during the studied hydrological cycle (pH = 7.4 to 8.2; dissolved oxygen = 4.5 to 8.3 ppm). On the other hand, the temperature was constant between 29 and 31°C., Fig. 4.

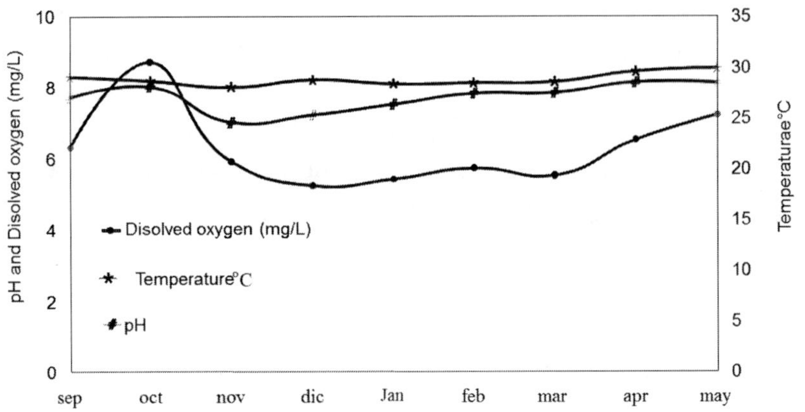

Figure 4. Environmental parameters tendency during an hydrologic cycle in Taxhido river.

3.3. Laboratory Studies

Organisms were collected in Taxhido and then transport to laboratory and maintained for acclimation during a month. After this time a group was used to optimal temperature determination and other group for an optimal density experiment.

3.3.1. Optimal Temperature

The fish were placed in 15 rectangular glass tanks distributed using a Latin square scheme in order to avoid spatial effects. The tanks´ dimensions were of 0.4.5 m depth, 0.8 m wide and 0.3 m long, with a water storage capacity of 100 L. Five triplicated treatments with 15 organisms were applied using environmental temperature as a medium value, so the treatments were; 24, 26, 28, 30, and 32°C. The handling of tanks involves, the feces removal and partial water change (30%) weekly. The fish were feeding with a commercial diet for Tilapia (Api-Tilapia 1, maltaCleyton® with 50% protein, 12% lipid, 13% ash, 3% fiber, 12 moisture) throughout the experiment. Feeding frequency was adjusted to three provisions offered three times daily starting at 8 AM, 1 PM and 6 PM. The results show that 28°C., is the best temperature for *H. cyanoguttatus*, fig. 5.

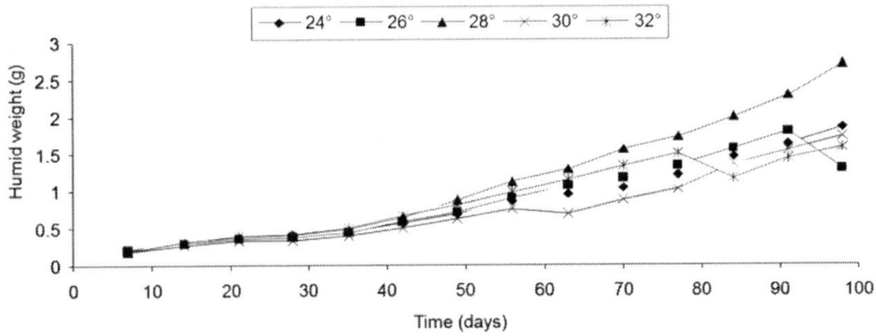

Figure 5. Humid weight behavior for the different temperature treatments

3.3.2. Optimal Density

Once the temperature was determined a similar experiment was carry out, but in this case the variable was the density. Five densities were probed, T1= 5, T2=10, T3=15, T4=20 ind/per aquarium. It is important to consider that control temperature was implemented in each aquarium in order to avoid an effect for spatial distribution. The results show that 15 individuals is the best density for *H. cyanoguttatus*, table 2.

3.3.3. Bioenergetics

Aerobic metabolism were determine in natural conditions with a semi closed system with (0.5L) respirometric chambers, using a method which assumes that the reduction of oxygen and the increase of nitrogen in the chamber depend on the weight of the animal, the volume of water, the period of time in which no water circulated in the chamber and the ambient temperature [22]. After each cycle, sample was oven-dried in an electric oven between 70-80°C until the samples had constant weight. From each composite sample 2 g were measured and taken as analytical sample; the samples were digested with concentrated nitric acid. The determination of the percentage proximate composition was chemically analyzed according to the method of analysis described by the Association of Official Analytical Chemist [23].

Table 2.

Performance parameters	T1= 5 ind	T2=10 ind	T3=15 ind	T4=20 ind
Initial number (n)	5	10	14	19
Final average number (n)	1	6	9	7
Survival rate (%)	26.66	56.66	61.7	38.88
Initial Total weight (g)	2.16	2.84	4.28	16.92
Initial individual average weight (g)	0.43	0.28	0.29	0.91
Final Total weight (g)	2.36	9.26	14.05	23.81
Final individual average weight (g)	0.59	1.09	1.49	2.05
Weight gain (%)	95.16	293.44	400.59	94.86

Physiological experiments showed higher oxygen consume at 15:00 hrs, while a minimum consumption was founded during the morning, Fig. 6.

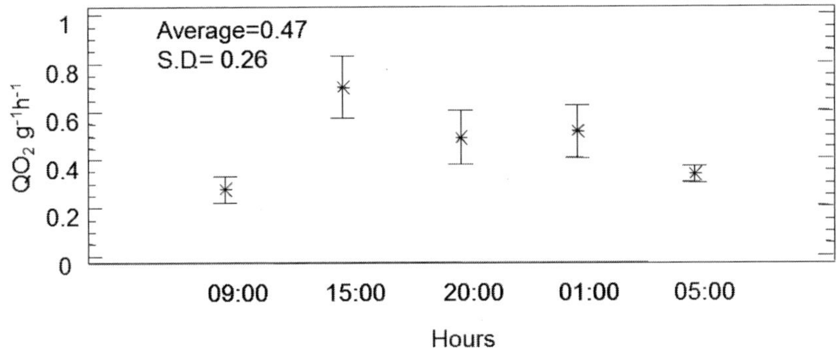

Figure 6. Oxygen consumption in natural condition for *H. cyanogutattus*

After three weeks in laboratory condition (acclimatization) oxygen consumption was measured to know if a metabolism was changed. The results show that higher consumption was found at 15:00 and minimal during the morning so the fish did not show modification in oxygen consumption.

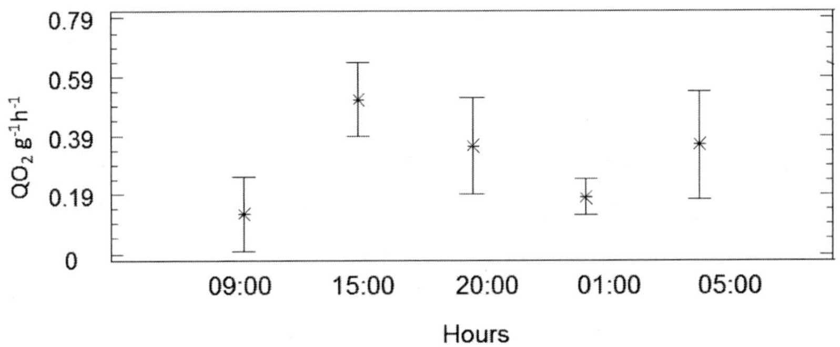

Figure 7. Oxygen consumption under laboratory condition for *H. cyanogutattus*

Finally, oxygen consumption was measured for each of the temperature treatments in order to know in which an alteration occurs. The results show that maximum values for oxygen were founded at T3= 30° minimal at T2 =26° treatments.

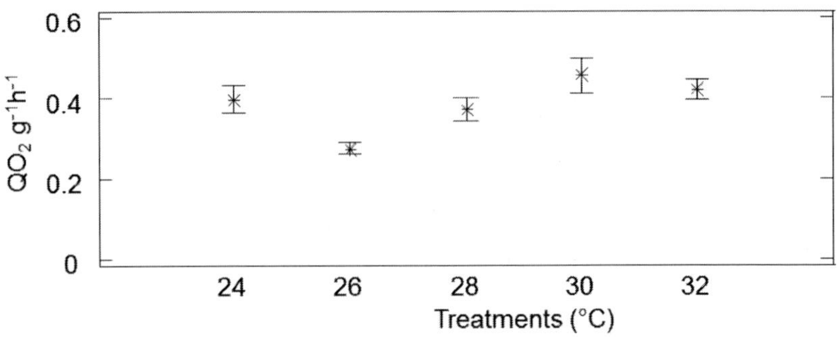

Figure 8. Oxygen consumption for each temperature treatment at the end of the experiment.

4. PROCAMBARUS DIGUETTI

The cambarids, known in Mexico as acociles (in Náhuatl), makaxil (in Mayan), chapos (in Purépecha), or freshwater crayfish, according to the region, are well known, and were regularly consumed by the Aztecs and other ethnic groups based around the Valle de Mexico already during prehispanic times [29]. *Procambarus digueti* is endemic to the Mexican Central-Occidental Plateau and is found only in certain parts of the Duero and Camécuaro rivers. The Duero River is 10 km long, whist the river Camécuaro is just 2 km long. The two rivers meet at Las Adjuntas. Furthermore, this species is known to occur in Lake Chapala, though samples were collected 40 to 50 years, so is believed to have naturally disappeared from some locations, though these local extinctions may in fact be unnatural This species is threatened by over-exploitation, habitat destruction and degradation, and the introduction of exotic species.These processes have already resulted in reduction in abundance of this species, and the extinction of some populations. Much of the natural habitat of this species has been altered by chemical pollution or by human activities such as canalization, clearing, dredging and embanking of rivers, construction of reservoirs, and the regulation of water levels and stream flows. Furthermore, this species is also threatened by the introduction of exotic crayfish such as *Procambarus clarkii* and*Cherax* species from Australia, which competes for food resources and refuges, and also alters the total production of the native ecosystems (Gutiérrez-Yurrita and Latournerié-Cervera 1999). In addition to over explosion of population, a cultural pressure exists, purepecha people attributes curative properties at *Procambarus* [30].

4.1. Study Area

The first part of the experiment was carry out in the national park of Camecuaro lake, which is found in Michoacan State over the municipality called Tangancícuaro, 19°54´10"N;102°12´20"O, at 1,700 meters above sea level. The national park has a spring called Camecuaro lake which is the main contribution for Camecuaro river, the depth is between 1.5 a 1.8 m; temperature 17.7 y 21°C; dissolved oxygen 7.3 y 7.5 mg/ L; visibility of 100%; hardness 138.8 y 145.5 mg/L $CaCO_3$. A handled extraction was practice in order to have a desirable sample size for the laboratory work. The organisms were transported in plastic bags with a supplemented oxygen and ice to avoid the over heat.

4.2. Laboratory Studies

Acclimatization of organisms was carrying out during a month in glass aquariums considering environmental conditions. The principal problems for the maintenance of the organism were the feed and the density, so in this case these two experiments were carry out.

4.2.1. Establishing The Diet

The organism were placed in 15 rectangular glass tanks distributed using a Latin square scheme in order to avoid spatial effects. The tanks´ dimensions were of 0.45 m depth, 0.8 m wide and 0.3 m long, with a water storage capacity of 100 L. Three commercial diets were probed T1= Trucha initial; T2= Camaronina, and T3=Tilapia initical, table 3. The handling of tanks involves, the feces removal and partial water change (30%) weekly. Feeding frequency was adjusted to two provisions offered three times daily starting at 8 AM and 6 PM.

Multiple condition factor (K) was calculate for each treatment, with this factor is possible to know the relative health for organisms [31].

$K=(10^2*W)/LbK=(102*W)/Lb$

Where;

K= Multiple condition factor

W= Weight

L= Length

b= exponent from $W = KL^b$

The type of growth was determine for the Ricker equation [32];

$W=aLbW=aLb$

Where;

W=weight

a = intercept

b= slope

As can be seen in

Table 3. Proximal Chemical Composition of the diets tested for growth

Diet Compounds	Trucha Iniciador (50:15)	Camaronina (35:8)	Tilapia Iniciador (32:4)
Protein (%)	50	35	32
Humidity (%)	12	12	12
Grass (%)	15	8	4
Crude fiber (%)	4	5	10
Ash (%)	12	10	10
Calcium (%)	2	1.4	
Phosphorous (%)	1.2	0.9	
E. L. N.	7	30	

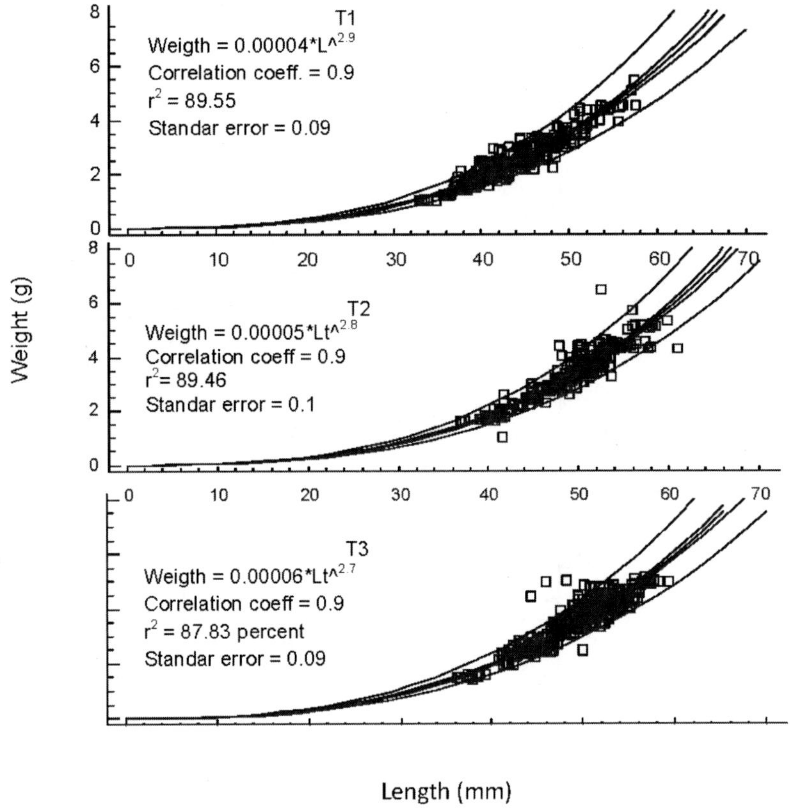

Figure 9. Regression for each one of treatments, values for the equation $W=KL^b$

Aerobic metabolism were determine in natural conditions with a semi closed system with (0.5L) respirometric chambers, using a method which assumes that the reduction of oxygen and the increase of nitrogen in the chamber depend on the weight of the animal, the volume of water, the period of time in which no water circulated in the chamber and the ambient temperature [22]. After each cycle, sample was oven-dried in an electric oven between 70-80°C until the samples had constant weight. After three weeks in laboratory under the diets treatment, oxygen consumption was measured to know if a metabolism was changed. The results show that higher consumption was found at 15:00 and minimal during the morning so the fish did not show modification in oxygen consumption, Fig. 10.

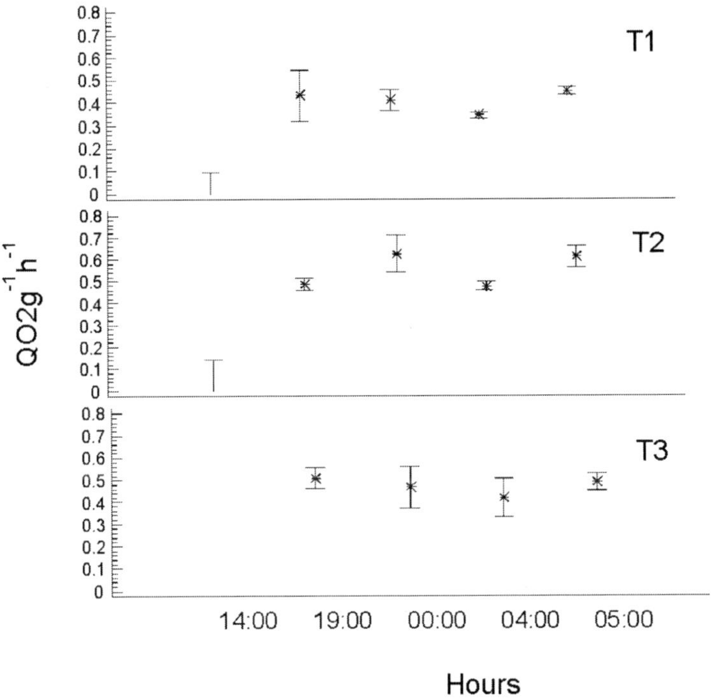

Figure 10. Oxygen consumption for each diet treatment at the end of the experiment

At the end production of total biomass was estimated, the results show that T1 is the best feed for*Procambarus digueti*, and T3 is the diet with a minor biomass production.

4.2.2. Establish an Optimal Density

The organism were placed in 15 rectangular glass tanks distributed using a Latin square scheme in order to avoid spatial effects. The tanks´ dimensions were of 0.4.5 m depth, 0.8 m wide and 0.3 m long, with a water storage capacity of 100 L. Five densities were probed, T1= 5, T2=10, T3=15, T4=20 ind/per aquarium. The handling of tanks involves, the feces removal and partial water change (30%) weekly. The fish were feeding with a commercial diet for trucha (50% protein, 10% humidity, 12% lipid, 12% ash, 4% fiber, 12% moisture) throughout the experiment. Feeding frequency was adjusted to three provisions offered two times daily at 8 AM, and 6 PM. The results show that a density of 15 organisms is optimal for *Procambarus* growth, table 4.

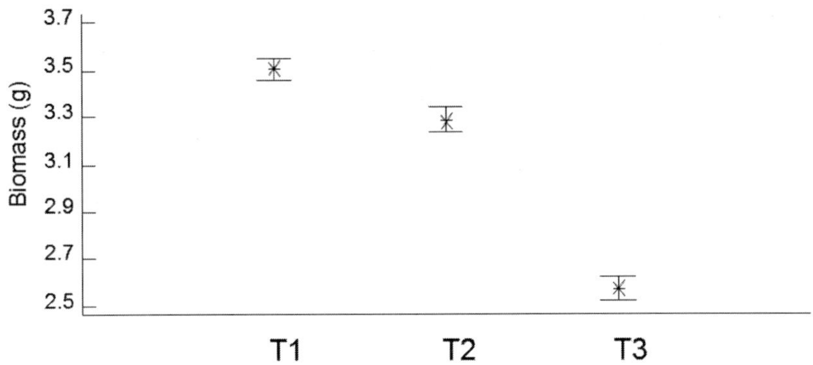

Figure 11. Biomass production at the end of experiment for each of the treatments

4.2.3. Bioenergetics

The flow of energy that was used to determine the energy efficiency and assimilation is show in the next equation:

$$C = P + G + ER + EU + F$$

Where; C is the total energy content of food consumed, P and G are the energy equivalents of somatic and gonadal growth respectively, ER is the energy utilized in respiration, EU is the energy lost as nitrogenous and other waste compounds excreted (50% protein) is the best food for *Procambarus digueti.* The energy balance is show in the table 5.

Table 4. Performance growth for the densities treatment in *Procambarus digueti*

Treatment	5	10	15	20
Initial number (n)	5	10	15	20
Final average number (n)	5	8	9	6
Survival rate (%)	100	80	60	30
Initial individual average weight (g)	4.8	5.23	5.53	5.18
Final individual average weight (g)	7.12	5.67	6.36	2.82
SGR (%/día)	1.94	1.71	1.83	1.018
Weight gain (%)	612.9	467.2	536	182
Production	0.302	0.158	0.091	0.063

Table 5. Energetic balance for *Procambarus digueti*

Diet	C	P	R	U	F
	Consume cal/mg	Biomass cal/mg	Oxygen cal/mg	Nitrogen cal/mg	Feces g
Tilapia (32:4)	5423.44	3.664	1.309	0.450	0.014
Camaronina (34:8)	5545.03	3.845	1.175	0.525	0.009
Trucha (50:15)	5706.273	3.873	1.264	0.569	0.006

5. CONCLUSION

Aquaculture has been supporting human demands for fish products for centuries and is an important industry worldwide. Global production from aquaculture has been increasing steadily, having more than doubled in the last decade; aquaculture now supplies one third of seafood consumed worldwide. With the massive increase in world aquaculture production in 1990s, the current aquaculture industry is one of the fastest growing sectors in world food production. However, the expansion of aquaculture has been accompanied by degradation of the natural environment, especially on marine aquaculture. Directly impacts of fisheries and aquaculture are habitat modification, collection of wild seedstock, changes of food webs, introduction of nonnative fish species and diseases that harm wild fish populations, and nutrient pollution. According to the FAO, major issues that need to be addressed are problems with access to proper technology and financial resources, together with environmental impacts and diseases. Another argues that further increases in aquaculture production will come mainly from further investment in biotechnology. The development of new strategies or technologies does not imply that the old one disappears; to the contrary the main idea is to use the experience and improve existing technology.

These three experiences and review of similar cases that have been developed in Mexico allow us to establish a general methodology in order to know the aquaculture potential for native species.

1 Knowledge.- The first step consist in to know and get the information above the specie. In the case of *G. multiradiatus* the knowledge practically doesn't exists so the research was oriented to the basic biology and ecology aspects. For example, with *G. multiradiatus* the main objective was to obtain reproduction but to reach these aspects it is required the maintenance under laboratory conditions and for this the knowing of food habits are essential. In the case of *H. cyanogutattus* the main problem was its taxonomic status, so the principal problem to obtain the basic knowledge is that some aspect were publish under the scientific name *Cichlasoma cyanoguttatum* and others aspects with the actual name *Herinchthys cyanogutattus.*

2 Environmental prospections.- It is necessary to know the basic physicochemical parameters (T°, pH, DO_2) in order to establish the strategy for transportation and laboratory maintenance. Field observations and ecological description is necessary in order to know the feeding habits, interspecific competence and the disposition of resources. With these parameters is possible to start the research.

3 Feed.- Under laboratory conditions feeding is the main problem in order to continue so the proofs needed is the acceptance of commercial food. The three species show an acceptance for commercial food but the problem here is the metabolism aspects such as assimilation and performance growth. This kind of problems could be solved with a bioenergetics approach.

4 Bioenergetics.- Ecophysiological basis of species should take into consideration the physiological characteristics and ecological role of the organism in question. Physiological analysis will reflect the conditions which affect population characteristics, such as population growth, intraspecific competition, and functional and numerical responses. These studies can do more than being simply descriptive, since they enable the development of scenarios that can be tested either through strictly controlled laboratory experiments or field experiments.

The new aquaculture research must be consider to add native species, this work is an effort to get the basic information in order to development of biotechnology and a link between the basic and applied science.

ACKNOWLEDGEMENTS

This work was partially finance by CONACYT.

REFERENCES

1. FAO, editor. Review of the State of the World Marine Fishery Resources2011; Rome: FAO.
2. Pillay TV. Aquaculture and the environment. second ed. Iowa: Blacwell publishing; 2004.
3. Bailey C. Blue revolution: the impact of technological innovation on Third World fisheries. The Rural Sociologist. 1985;5(4): 259-66.
4. Costa-Pierce B. Ecological Aquaculture: the evolution of the Blue Revolution second ed. Iowa: Blackwell Science; 2002. p. 383.
5. Saborio Coze A, Flores Nava A. Review of environmental impact assessment and monitoring of aquaculture in Latin America Environmental impact assessment and monitoring of aquaculture. Rome: FAO; 2009. p. 395–454.
6. García-Trejo F, Soto- Zarazúa GM. La nueva ecología, una perspectiva integradora hacia un paradigma de sustentabilidad. CIENCIA@UAQ. 2012;5(1).
7. Ross LG, Beveridge MC. Is a better strategy necessary for development of native species for aquaculture? A mexican case of study. Aquaculture research. 1995[29]:539-47.
8. Martínez-Palacios CA, Barriga-Tovar E, Taylor JF, Ríos-Durán G, Ross LG. Effect of temperature on growth and survival of Chirostoma estor estor, Jordan 1879, monitored using a simple video technique for

remote measurement of length and mass of larval and juvenile fishes. Aquaculture. 2001(1-4):369-77.

9. Ross LG, Martinez-Palacios CA, Morales E. Developing native fish species for aquaculture: the interacting demands of biodiversity, sustainable aquaculture and livelihoods. Aquaculture Research. 2008[39]:675–83.

10. Bardach JE, Ryther JH, Mclarney WO. Acuacultura, crianza y cultivo de organismos marinos y de agua dulce. México: AGT 1986.

11. Martinez-Palacios CA, Ross LG. The reproductive biology and growth of the Central American cichlid Cichlasoma urophthalmus (Günther) Journal of Applied Ichthyology. 1992;8:99–109.

12. Garcia-Trejo F, Hurtado Gonzalez S, Soto Zarazua G, Alatorre-Jácome O, Rico-García E, Gutiérrez-Yurrita PJ. Ecophysiological responses to the effect of annual management on an endemic viviparous fish in central plateau of México Neotropical ichthyology 2013;11(1):117-23.

13. Gutiérrez-Yurrita PJ, Montes C. Environmental factors controlling the crayfish Procambarus clarkii activity in Doñana National Park temporary freshwater marsh, SW. Spain Comparative Biochemistry and Physiology, Part A 1998;120:713-21.

14. De la Vega-Salazar MY. Estado de conservación de los peces de la familia Goodeidae (Cyprinodontiformes) en la mesa central de México. Revista Biología Tropical. 2006;54(1):163-77.

15. Macias GC. Social behavior and operational sex ratios in the viviparous fish Girardinichthys multiradiatus. Copeia. 1994;4:919-25.

16. Álvarez del Villar J. Peces Mexicanos (Claves). In: Pesca. CNCd, editor. Instituto Nacional de Investigaciones en Biología Pesquera Mexico 1970. p. 166.

17. Miller RR, Fitzsimmons JM. Ameca splendens, a new genus and species of goodeid fish from Western Mexico, with remarks on the classification of the Goodeidae. Copeia. 1971;1:1–13.

18. Uyeno T, Miller R, Fitzimons J. Karyology of cyprinodontoid fishes of the Mexican family Goodeidae Copeia. 1983;2:497–510.

19. Macias-Garcia C, Burt de Perea T. Ultraviole-based female preferences in a viviparous fish Behavioral Ecology and Sociobiology 2002;52(1):1-6.

20. Trujillo-Jiménez P, Viveros EE. La ecología alimentaria del pez endémico Girardinichthys multiradiatus (Cyprinidontiformes: Goodeidae), en el Parque Nacional Lagunas de Zempoala, México. Revista de Biología Tropical 2006;54 1247-55.

21. García E. Modificaciones al Sistema de Clasificación Climática de Koppen. México: Enriqueta García; 1988.

22. Steffensen JF. Some errors in respirometry of aquatic breathers: how to avoid and correct for them Fish Physiology and Biochemistry. 1989;6:49-59.

23. AOAC. Official Methods of Analysis of AOAC International,. sixteen ed. Gaithersburg, MD, USA.: AOAC; 1998.

24. Klekowsky RZ, Duncan A. Physiological approach to ecological energetics. In: Grodzinzky H, Groves TDD, editors. Methods for ecological bioenergetics. Oxford: Blacwell Scientific Publication; 1975. p. 227-61.

25. Hubbs CT, Lucier GP, Garrett RJ, Edwards SM, Dean E, Marsh E, et al. Survival and abundance of introduced fishes near San Antonio, Texas. Texas Journal of Science. 1978;4[30]:369-76.

26. Shafland PL, Pestrak JM. Lower lethal temperatures for fourteen non-native fishes in Florida. Environmental Biology of Fishes 1982(7):149–56.

27. Lee DS, Gilbert CR, Hocutt CH, Jen RE. Atlas of North American freshwater fishes. Raleigh, NC: North Carolina State Museum of Natural History; 1980.

28. Itzkowitz M, Nyby J. Field Observations of Parental Behavior of the Texas Cichlid Cichlasoma cyanoguttatum. American Midland Naturalist. 1982;2[108]:364-8.

29. Gutierrez-Yurrita PJ, Morales-ortiz A, Oviedo A, Ramirez-Perez C. Distribution, spread, habitat characterisation and conservation of the crayfish species (Cambaridae) in Querétaro (central México). Freshwater Crayfish. 2001;13:349-58.

30. Gutiérrez-Yurrita PJ, R LCJ, editors. Preliminary studies on exploitation of two native crayfish species of the genus Procambarus (CRUSTACEA: DECAPODA) by purépecha people in Michoacán, México. Proceedings of the III International Congress of Ethnobiology; 1992.

31. Kuri-Nivon E. Manual para la producción de trucha y salmón.. Idaho: Universidad de Idaho; 1976. Manual para la producción de trucha y salmón. Universidad de Idaho.

32. Richter JB. Anfangsgründen der Stöchiometrie oder Messkunst chemischer Elemente (Foundations of stoichiometry or surveying chemical elements). Breslau1792.

Chapter 4

Extremophiles: photosynthetic systems in a high-altitude saline basin (Altiplano, Chile)

Alejandro Angel[1] , Irma Vila[2], Venecia Herrera[1]

[1]Centro de Investigacio´n de Medio Ambiente (CENIMA), Universidad Arturo Prat, Casilla 121, Iquique, Chile

[2]Departamento de Ciencias Ecolo´gicas, Facultad de Ciencias, Universidad de Chile, Casilla 653, Santiago, Chile

ABSTRACT

Studies of different hypersaline systems have revealed various types of limitation. We evaluated the phototrophic microbial assemblages over a whole seasonal cycle (wet vs dry) in the Salar de Alconcha, a high-altitude (4250 m altitude) saline basin. Phototrophic microbial communities were obtained from contrasting ecotypes, and examined for the effects of proximity and salinity variations. We also analyzed pigment profiles pointing to photosynthetic activity. While taxonomic diversity was limited to three algal groups (chlorophytes, diatoms, and cyanobacteria) ecological preferences were highly variable. Physical limitations when the photosynthetic system turn drier (maintaining viability and stability) appear to be the most successful adaptation (constrained assemblages) to the extreme condition in the Altiplano. This suggests that phototrophic microorganisms rarely achieve optimal growth, and could only do so when rain events reduce salinity (e.g. austral summer). However, environmental condition over the salt crust (total salt concentration: 119.74 g l^{-1}) was an important driver to algal biomass. Overall, dominated by *Dunaliella salina* (≈18,000 cell ml^{-1}) turning the water into a red–orange colored system (β-carotene); aplanospore cysts were observed only during the driest season (austral winter). Our results suggest specific restrictive environments (e.g. environmental dissimilarity in the physical landscape) for phototrophic microbial colonization in high-altitude saline systems quite dependent upon water availability (system on the edge). The present study is a contribution for a better understanding of both the ecology of extreme environments and polyextremophiles communities (phototrophs) that inhabit them. Active salars are considered to be useful analogs of ancient photosynthetic systems, currently very pressured by groundwater extraction. Thus, altering the volume of water through the basin would have negative consequences on the structure and dynamics of local communities, and also in the stability of ecosystem functions.

KEYWORDS

Active salars, Aplanospore cysts, β-Carotene, Chlorophytes, Cyanobacteria, Diatoms, Hypersaline systems

1. INTRODUCTION

It has been known that microbial communities tend to be the dominant biota in hypersaline environments. Indeed, strongly saline environments are inhabited by specific halophilic species (Oren 2005). Microbial communities from strongly saline environments (salinity range 50–150 g l^{-1}) have potentially interesting physiology (Oren 2002), and some authors have suggested these systems as relevant models for a better understanding of both the early stages of life on Earth and possible extraterrestrial life (Kunte et al. 2002; Mancinelli et al. 2004). Highly saline wetlands in the Altiplano are inhabited by only few higher organisms such as brine shrimps, crustacean and insects, some lakes support zooplankton communities (e.g. amphipod, copepods, chironomid, ostracod, cladoceran, leech, rotifer, nematode and trichopteran) and surrounding macrophytes (Zúñiga et al. 1991, Scott et al.2015). The dominant phototrophs are halophilic and halotolerant algae and cyanobacteria as well as anoxygenic phototrophic bacteria (Demergasso et al.2008; Dorador et al. 2008; Thiel et al. 2010).

Active salars in the Altiplano are considered to be remnants of the extensive paleolakes that once occupied the plains; these aquatic habitats developed since the end of the Pleistocene vary from fresh water systems to lakes and perennial salt lakes with high levels of sodium, sulfates, and chlorides (Vila and Muhlhauser 1987; Risacher et al. 2003). In general, perennial salt lakes and permanent meromictic saline systems tend to have relatively constant salinities over extensive periods of time (e.g. dry season, Oren 2005). On the contrary, temporal and permanent terrestrial wetlands borderline the salt crust can frequently experience fluctuating salinities, water availability, and also can be connected with hypersaline systems during intensive rain events (El Niño-Southern Oscillation ENSO, Risacher et al.2003). Active salars are truly extreme environments in that resident organisms must deal with not only high salt concentrations, but also dynamic changes in salinity, temperature, and water availability, making these organisms polyextremophiles.

Recent evidence indicates the importance of hypersaline microbial mats to interpreting isotopic biosignal in geological records (Finke et al. 2013). In this aspect, interactions between high-altitude saline systems quite dependent upon water availability and microbial mats formation revealed the presence of unique clusters of *Archaea* (not previously reported) in the Salar del Huasco (Dorador et al.2010). Moreover, examinations in the Salar del Huasco suggest that cyanobacterial communities are unique, related to others previously described from the Antarctic, along with others from diverse, but less extreme environments (Dorador et al.2008). Furthermore, an important novel lineage of *Gammaproteobacteria* within the community structure indicates the importance of anoxygenic phototrophic communities as primary producers from different systems in the Salar de Atacama (Thiel et al. 2010). These lakes described a high microbial diversity and spatial variability (Demergasso et al. 2004) both in time and space related to the salinity variation (Demergasso et al. 2008). Although aquatic microhabitats occur extensively through the Altiplano, particularly in salars, dynamics of algal communities have not been specifically studied and almost nothing is known about comparison between contrasting conditions (e.g. bofedales vs. salt crust systems).

The Altiplano is considered one of the most extreme environments on the planet (Alpers and Brimhall 1988). Active salars (water available) are characterized by high UV radiation, high heavy metals content, wide daily temperature variation (temperature range −20 to 25 °C, approximately), and aquatic systems with variable salt concentration (Dorador et al. 2008; Demergasso et al. 2008; Risacher et al.2003). Visually, environmental dissimilarities in the physical landscape into the Salar de Alconcha (4250 m altitude) exhibit the presence of contrasting ecotypes. One area receives inputs of water running along the salt crust between vegetation boundary (Bofedal). On the contrary, over the salt crust a terminal lagoon (Laguna Colorada) very much dependent upon water availability can coexist nearly. Our objectives were to characterize the phototrophic microbial assemblages in the Salar de Alconcha (salt flat area 3.8 km^2) by analyzing both community dynamics and composition as related to the environmental conditions. The influence factors evaluated between both systems that may explain their behavior are discussed.

2. MATERIALS AND METHODS

2.1 Study area

Salar de Aconcha is a (total basin area 120 km^2) high-altitude (4250 m of altitude) saline basin located in the southern Altiplano (Fig. 1). The region comprises a network of high-altitude salars associated with wetland vegetation (e.g. Bofedales) that may have underlying peat layers (Maldonado 2014), inhabiting in margins of shallow saline systems. Salar de

Alconcha is an extreme environment (temperature mean 3.5 °C) and exhibits visual dissimilarity between two areas (north and south) with two different systems (total water area 0.75 km^2) quite dependent upon water availability (no more than 10 cm of water column); potential evaporation is greater (annual average 1620 mm) than precipitation (annual average 200 mm), falling mostly during the austral summer and some snow in winter (Risacher et al. 2003). Both systems studied, Bofedal (S 21°03′14″, W 68°29′32″) and Laguna Colorada (S 21°03′68″, W 68°29′26″), are located approximately 150 m apart (Fig. 1) and were selected based on proximity and notable environmental dissimilarities (Fig. 2). Laguna Colorada (L) is a terminal lagoon situated over the salt crust (Fig. 2a), lacking any superficial channels approximately 100 m from the nearest vegetation. On the contrary, the Bofedal (B) is situated in a terrestrial system that borderlines the salt crust and less than 10 m from the nearest vegetation (Fig. 2b). However, the connection between both systems during intensive rain events is highly possible.

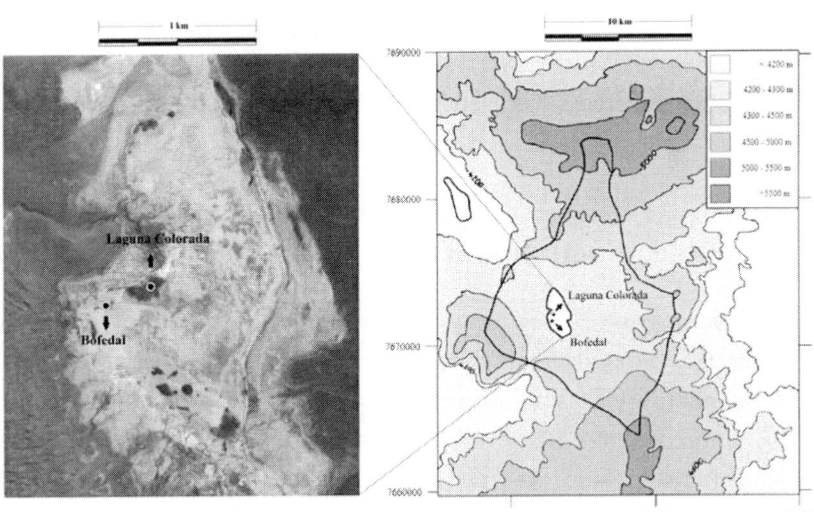

Figure 1. Geographic location of the two permanent wetlands studied in the Salar de Alconcha, Northern Chile. *Solid circles* represent sites studied

Figure 2. Habitat description. **a** Laguna Colorada, salt crust system with intensive *red–orange water* can be observed;**b** Bofedal, inflow waters running along the salt crust and borderline a vegetation boundary (Bofedales); associated biota (Flamingos) can be observed

2.2 Sample collection and biological analysis

Samples were collected over a whole seasonal cycle (wet vs dry) in 2010, approximately 3-month intervals (January and March: wet season; August and November: dry season). Microbial samples were collected mixing two bottles until complete 10 l in the same spot (samples were taken in the morning after surface ice was melted) and filtered in situ with 20 µm mesh, then fixed with lugol-acetic solution (Parsons et al. 1984). Likewise, other two unfiltered bottles were taken for smaller cells evaluation; these samples were filtered (sample volume 100 ml) with an equipment to concentrate phytoplankton on polycarbonate membrane (pore size 5 µm, Fournier 1978). The filters were deposited in sedimentation chambers for 24 h. Abundance values and taxonomic identification were analyzed after sedimentation following the Utermöhl method (1958). The phototrophic microbial species were identified and their abundance estimated with an inverted microscope (Olympus CK-2) at 1000× and 400× magnification. Each cell was counted as an individual. For colony-forming species, each colony was also counted as an individual. Microbial concentration (cells ml^{-1}) was expressed in Table 2 as the total percentage in relation with the total abundance for each station. Predominance and frequency were calculated using the procedure described by Bodenheimer (1955) in Avendaño and Saíz (1977). Ecological indices (dominance, Shannon diversity and equitability) were calculated using procedures described by Harper (1999). Taxonomic identification was based on species and keys described in Teodoresco (1905), Simonsen (1987), Lange and Mary Ann (2002), Cadima et al. (2005) and Diaz and Maidana (2005).

2.3 Chemical analysis and HPLC chromatography

All the samples (1 l bottle per each sample) were kept cold until return to the laboratory. Analyses of B, NO_2^-, NO_3^- and PO_4^{3-} were performed using absorption spectrophotometer (APHA 2005). The elements Cu, Fe, Mn and Mg were quantified by atomic absorption spectrophotometer FAAS. Measurements were made in duplicate for each analysis. Pigment profiles were estimated using homogeneous suspensions of 4 ml taken from each sample after thorough mixing. Cells were centrifuged at 5000 rpm for 5 min and the washed cell pellet was mixed with 4 ml of acetone/water (80:20, v/v). The mixture was left for 10 min at room temperature to ensure complete extraction. The extract was centrifuged for 5 min at 5000 rpm and the colourless biomass was discarded (Jin et al. 2003). Pigment profiles were obtained by separation with HPLC chromatography (Shimadzu LC-10AV) provided with a Shimadzu SPD 10-AV UV–visible detector, using a LichroCart RP-18 (5 μm) column of 250 × 4 mm. These profiles were lutein, β-carotene, chlorophyll-a, chlorophyll-b, lycopene, astaxanthin, and zeaxanthin. The detection of β-carotene was adjusted to 454 nm (standards of DHI Water and Environment).

2.4 Statistical analysis

Datasets were analyzed with CANOCO software (ter Braak 1988). A preliminary Detrended Correspondence Analysis DCA (following recommendations by ter Braak and Šmilauer 1998) indicated a unimodal response of species variance with log transformation data. DCAs were used to evaluate affinities and differences between species and stations to avoid the arc-shaped distribution when there is a single strong gradient affecting the samples (Gauch 1982). On the contrary, Nonmetric multidimensional scaling (MDS) was applied to ordinate samples in three dimensions according to their distances. The number of three dimensions was chosen to keep the stress value for dimensional downscaling below the recommended threshold of 0.1 for an ideal preservation of the original distances between samples (Clarke 1993).

3. RESULTS

3.1 Limnological features

Despite the fact that the Salar de Alconcha is a high-altitude extreme environment still support a rich number of well-adapted species. According to the nature of the habitat the Bofedal (B) and Laguna Colorada (L) clearly differ from each other (Table 1). Environmental dissimilarity in the physical landscape defines specific limnological conditions (e.g. salinity range, physical condition of water availability and phototrophic microbial composition). Although both habitats in the Salar de Alconcha endured these fluctuating environmental conditions, microbial assemblages (phototrophs) in Laguna Colorada (L) were more stable to the effects of seasonal changes than the Bofedal (B). Limnological conditions clearly differ among both systems studied especially in conductivity, ionic composition and dissolved oxygen (Table 1). Laguna Colorada (L) reported an extreme hypersaline condition (total salt concentration: 119.74 g l^{-1}) whereas salinity condition in the Bofedal (B) was clearly lower (total salt concentration: 0.51 g l^{-1}). These environmental conditions result in less restrictive environments for phototrophic microbial colonization in the

Bofedal (although resident time was highly variable and decline abruptly in dry seasons) than Laguna Colorada (Table 2).

Table 1. Limnological variables studied in the Salar de Alconcha

Salar de Alconcha	pH	DO	Conductivity	$N\text{-}NO_2^-$	$N\text{-}NO_3^-$	$P\text{-}PO_4^{3-}$	B	Cd	Cu	Mg	Mn	Fe
Laguna Colorada	8.21 ± 0.45	1.96 ± 0.6	187.1 ± 2.2	0.007 ± 0.01	33.57 ± 2.4	0.38 ± 0.01	318.2 ± 15.7	185.3 ± 5.8	0.01 ± 0.001	40.1 ± 2.7	33.0 ± 3.7	1.8 ± 0.02
Bofedal	7.75 ± 0.35	8.01 ± 0.9	1.03 ± 0.16	0.014 ± 0.03	1.21 ± 0.57	0.14 ± 0.03	2.4 ± 0.42	1.1 ± 0.14	0.25 ± 0.32	80.65 ± 4.1	637.5 ± 9.4	16.15 ± 9.3

Dissolved oxygen: DO (mg l^{-1}); Conductivity (mS cm^{-1}); Ionic composition (mg l^{-1})

Table 2. Distribution and composition of phototrophic microorganisms identified in the Salar de Alconcha. (L) Laguna Colorada; (B) Bofedal; GC: general classification

Number code	Species/season	January		March		August		November		%Pred	% Freq	GC
		L1	B1	L2	B2	L3	B3	L4	B4			Dom/Con
Cyanobacteria												
1	*Aphanocapsa elachista* West & West	N	3.15	N	6.11	N	0.68	N	N	1.02	37.5	Acc/O
2	*Oscillatoria anguina* (Bory) Gomont	N	3.62	N	0.52	N	N	N	1.71	0.50	37.5	Acc/O
3	*Oscillatoria bornetii* Zukal	N	N	N	0.47	N	0.34	N	1.24	0.11	37.5	Acc/O
4	*Oscillatoria curviceps* Agardh	0.05	N	0.65	0.55	0.73	0.34	N	N	0.29	62.5	Acc/C
5	*Oscillatoria splendida* Greville	N	N	N	1.27	N	0.34	N	N	0.15	25	Acc/O
6	*Oscillatoria sp1*	0.39	N	1.06	0.28	1.22	N	0.82	N	0.62	62.5	Acc/C
Chlorophytes												
7	*Dunaliella salina* Teodoresco	95.90	N	95.15	N	91.62	N	97.39	N	67.66	50	D/O
8	*Monoraphidium sp1*	N	0.09	N	N	N	N	N	1.09	0.05	25	Acc/O
9	*Mougeotia tumidula* Transeau	N	1.22	N	N	N	N	N	N	0.13	12.5	Acc/R
10	*Pediastrum sp1*	N	0.24	N	N	N	N	N	N	0.03	12.5	Acc/R
11	*Planktosphaeria gelatinosa* G.M. Smith	N	0.52	N	0.61	N	N	N	N	0.12	25	Acc/O
Diatoms												
12	*Achnanthes cf. rossii* Husted	N	0.33	N	0.98	N	N	N	N	0.14	25	Acc/O
13	*Amphora atacamae* Frenguelli	N	N	N	N	N	N	N	0.78	0.03	12.5	Acc/R
14	*Amphora boliviana* Patrick	N	N	N	N	N	N	N	0.62	0.02	12.5	Acc/R
15	*Amphora coffeaeformis* (Agardh) Kützing	N	2.35	N	3.03	N	9.46	N	12.74	1.41	50	Acc/O
16	*Amphora subrobusta* Hustedt	N	N	N	N	N	N	N	2.18	0.07	12.5	Acc/R
17	*Amphora ovalis* (Kütz.) Kützing	0.37	N	0.35	N	0.41	N	N	N	0.19	37.5	Acc/O
18	*Amphora sp1*	0.05	N	0.48	0.21	0.49	N	0.10	N	0.20	62.5	Acc/C
19	*Amphora sp2*	N	5.18	N	6.47	N	N	N	3.73	1.36	37.5	Acc/O
20	*Brachysira sp1*	0.17	1.08	0.13	1.46	0.45	1.35	0.05	0.16	0.46	100	Acc/C
21	*Caloneis limosa* (Kütz.) R. M. Patrick	N	0.19	N	N	N	N	N	N	0.02	12.5	Acc/O
22	*Craticula sp1*	0.20	N	0.18	N	0.65	N	N	N	0.15	37.5	Acc/O
23	*Cymatopleura solea* Bourrelly	N	0.14	N	0.08	N	N	N	N	0.02	25	Acc/O
24	*Cyclotella sp1*	0.44	0.05	0.68	0.56	0.08	0.41	0.03	0.25	0.32	100	Acc/C
25	*Cymbella helvetica* Kützing	N	N	N	1.13	N	N	N	1.09	0.16	25	Acc/O
26	*Cymbella minuta* Hilse	N	0.30	N	5.45	N	N	N	0.31	0.62	37.5	Acc/O
27	*Cymbella tumida* (Brébison) Van Heureck	N	N	N	1.03	N	N	N	11.81	0.49	25	Acc/O

28	*Denticula lauta* Kützing	0.07	0.52	0.07	0.41	0.20	1.46	0.04	2.49	0.31	100	Acc/C
29	*Denticula elegans* Kützing	N	0.42	N	0.21	N	0.23	N	0.16	0.08	50	Acc/O
30	*Denticula thermalis* Kützing	N	0.14	N	0.21	N	0.99	N	0.31	0.09	50	Acc/O
31	*Denticula sp2*	N	0.09	N	0.23	N	1.13	N	0.47	0.10	50	Acc/O
32	*Fallacia sp1*	N	0.52	N	2.35	N	0.23	N	2.49	0.40	50	Acc/O
33	*Fragilaria sp1*	N	0.09	N	N	N	5.41	N	4.04	0.38	37.5	Acc/O
34	*Fragilaria sp2*	N	N	N	6.16	N	3.04	N	5.91	0.98	37.5	Acc/O
35	*Frustulia rhomboides var rhomboides* (Ehr.) De toni	N	1.08	N	0.75	N	N	N	0.78	0.22	37.5	Acc/O
36	*Frustulia rhomboides* Ehrenberg	N	N	N	N	N	5.41	N	N	0.24	12.5	Acc/R
37	*Hantzschia amphioxys* Grunow	0.37	0.33	0.17	0.33	0.73	0.11	0.05	0.25	0.29	100	Acc/C
38	*Mastogloia similis cf.* Hustedt	N	3.06	N	6.11	N	2.30	N	2.74	1.17	50	Acc/O
39	*Mastogloia sp1*	N	N	N	0.11	N	N	N	4.35	0.15	25	Acc/O
40	*Mastogloia sp2*	N	6.59	N	0.31	N	N	N	N	0.74	25	Acc/O
41	*Microcostatus andinus* Lange-Bertalot & Rumrich	N	0.52	N	2.07	N	N	N	N	0.28	25	Acc/O
42	*Navicula pupula var. rectangularis* (Gr.) Grunow	0.68	N	N	N	2.04	N	0.28	N	0.45	37.5	Acc/O
43	*Navicula sp1*	0.39	N	0.41	N	0.41	N	0.70	N	0.34	50	Acc/O
44	*Navicula sp2*	0.32	12.70	0.13	16.92	0.69	9.58	0.14	6.68	4.01	100	O/C
45	*Navicula sp3*	0.20	5.83	0.10	8.93	0.18	0.34	0.21	1.40	1.76	100	Acc/C
46	*Navicula sp4*	0.24	N	0.11	N	N	N	0.05	N	0.08	37.5	Acc/O
47	*Navicula sp5*	N	2.68	N	5.17	N	N	N	1.55	0.89	37.5	Acc/O
48	*Navicula sp6*	N	6.12	N	N	N	4.62	N	N	0.86	25	Acc/O
49	*Navicula sp7*	N	21.17	N	N	N	7.44	N	4.66	2.74	37.5	O/O
50	*Nitzschia baccata* Hustedt	N	0.83	N	0.41	N	0.41	N	0.16	0.16	50	Acc/O
51	*Nitzschia halloyii* Maidana & Herbst	N	0.61	N	0.99	N	N	N	N	0.17	25	Acc/O
52	*Nitzschia hybrida* Grunow	0.17	N	0.34	N	0.10	N	0.13	N	0.13	50	Acc/O
53	*Nitzschia lacuarum* Hustedt	N	6.26	N	3.65	N	1.13	N	0.87	1.14	50	Acc/O
54	*Nitzschia sp1*	N	N	N	N	N	N	N	1.24	0.04	12.5	Acc/R
55	*Pinnularia maior* (Kützing) Raben	N	0.56	N	0.73	N	1.13	N	1.12	0.23	50	Acc/O
56	*Pinnularia episcopalis* Cleve	N	0.33	N	N	N	0.23	N	N	0.05	25	Acc/O
57	*Pinnularia viridis* (Nitzsch) Ehrenberg	N	5.18	N	1.13	N	2.30	N	2.98	0.87	50	Acc/O

58	*Pinnularia spl*	N	1.74	N	3.01	N	1.24	N	4.20	0.70	50	Acc/O
59	*Rhopalodia brebissonii* Krammer	N	0.66	N	0.85	N	1.35	N	0.93	0.25	50	Acc/O
60	*Rhopalodia wetzelii* Hustedt	N	0.42	N	0.92	N	28.39	N	3.57	1.52	50	Acc/O
61	*Surirella striatula* Turpin	N	0.80	N	0.52	N	N	N	N	0.14	25	Acc/O
62	*Surirella chilensis var. tumida* Hustedt	N	0.09	N	0.11	N	4.06	N	0.87	0.23	50	Acc/O
63	*Surirella robusta* Ehrenberg	N	0.08	N	0.05	N	3.27	N	0.62	0.18	50	Acc/O
64	*Surirella sella* Hustedt	N	0.19	N	N	N	N	N	N	0.02	12.5	Acc/R
65	*Surirella wetzeli* Hustedt	N	N	N	0.26	N	0.90	N	0.16	0.07	37.5	Acc/O
66	*Stauroneis aff atacamae* Hustedt	N	0.66	N	0.08	N	0.11	N	2.33	0.16	50	Acc/O
67	*Stauroneis anceps f. gracilis* (Ehrenberg) Hustedt	N	0.14	N	0.08	N	0.32	N	2.02	0.10	50	Acc/O
68	*Synedra spl*	N	1.13	N	6.77	N	N	N	2.95	0.94	37.5	Acc/O
	Total cells ml^{-1}	20,450	10,100	18,450	10,650	12,300	4450	19,400	3200			

(*D* dominant, *O* occasional, *Acc* accidental, *C* constant, *R* rare, or *N* not present). In predominance (Pred %), the species are classified as: dominant (D: >5 %), occasional (O: 2.5–5.0 %), or accidental (Acc: <2.5 %). In frequency (Freq %), the species are classified as: constant (C: >50 %), occasional (O: 50–25 %), or rare (R: <25 %)

3.2 Community structures, biological factors and dynamics

While diversity was limited to three taxonomic groups (chlorophytes, diatoms, and cyanobacteria), the ecological preferences were quite variable. Three divisions were identified with 30 genera and 68 species; *Cyanophyceae* (6 species), *Bacilliariophyceae* (57 species), and *Chlorophyceae* (5 species). Marked shift were report among these groups both in time and space as consequences of fluctuating environmental condition (Fig. 3). As could be expected, the great number of genera and species in the Bofedal (H′: 2.98 ± 0.24) compared with Laguna Colorada (H′: 0.31 ± 0.13), which indicates less restrictive environment for colonization here (phototrophs), is an example of natural reservoir for diversity (phototrophs) within the basin. In general, the highest peak in abundance and diversity were reported in wet season (diatoms: 10,300 ± 180 cells ml^{-1}; cyanobacteria: 1250 ± 110 cells ml^{-1}; green algae: 18,750 ± 1600 cells ml^{-1}), and decline when the environmental conditions turn drier (diatoms: 8550 ± 1250 cells ml^{-1}; cyanobacteria: 625 ± 5 cells ml^{-1}; green algae: 15,100 ± 5450 cells ml^{-1}). Seasonal changes were also reflected in assemblages composition mainly dominated by permanent residents although ecological preferences were quite variable. Diatoms were positively correlated with poor-water salinity concentration whereas chlorophytes were negatively correlated to these conditions (Fig. 5). The most dominant groups in the Bofedal (B) were diatoms and cyanobacteria especially in wet season (Fig. 3a). Laguna Colorada (L) was clearly dominated by green algae *Dunaliella salina* (Fig. 3b). The most diverse group in both systems were diatoms (pennate and lesser centric) followed by large filamentous cyanobacteria (500–2000 μm), and unicellular green algae (Table 2).

In the Bofedal, a total of 57 species and 25 genera were identified; the most abundant genera were *Navicula*, *Amphora*, *Rhopalodia*, *Nitzschia*, *Pinnularia*, and*Mastogloia* (Fig. 4a). In this aspect, *Navicula* was the most dominant genus (33.3 % total abundance) with five species very sensitive to seasonal changes (Table 2). Microbial biomass (phototrophs) was highest when the system receive precipitations (10,250 ± 450 cells ml^{-1}), and abruptly decline when the conditions turn drier (3800 ± 850 cells ml^{-1}).

However, Laguna Colorada (L) was more stable and significantly more restrictive for species composition than the Bofedal (B) with a total of 16 species and 10 genera (Table 2). The most abundant genera in Laguna Colorada were *Dunaliella*, *Navicula*, and *Oscillatoria* (Fig. 4b). Although, abundance values were not statistically different ($p = 0.11$) maximum rates were influenced by seasonal rains (19,500 ± 1450 cells ml^{-1}). These values were considerably influenced by algal biomass of *Dunaliella salina* (95.4 % total abundance) only associated with few accidental species (Table 2). This condition resulted in less equitability (0.11 ± 0.047) and high dominance (0.90 ± 0.046). On the contrary, the Bofedal reported high equitability (0.78 ± 0.050) and lower dominance (0.082 ± 0.027).

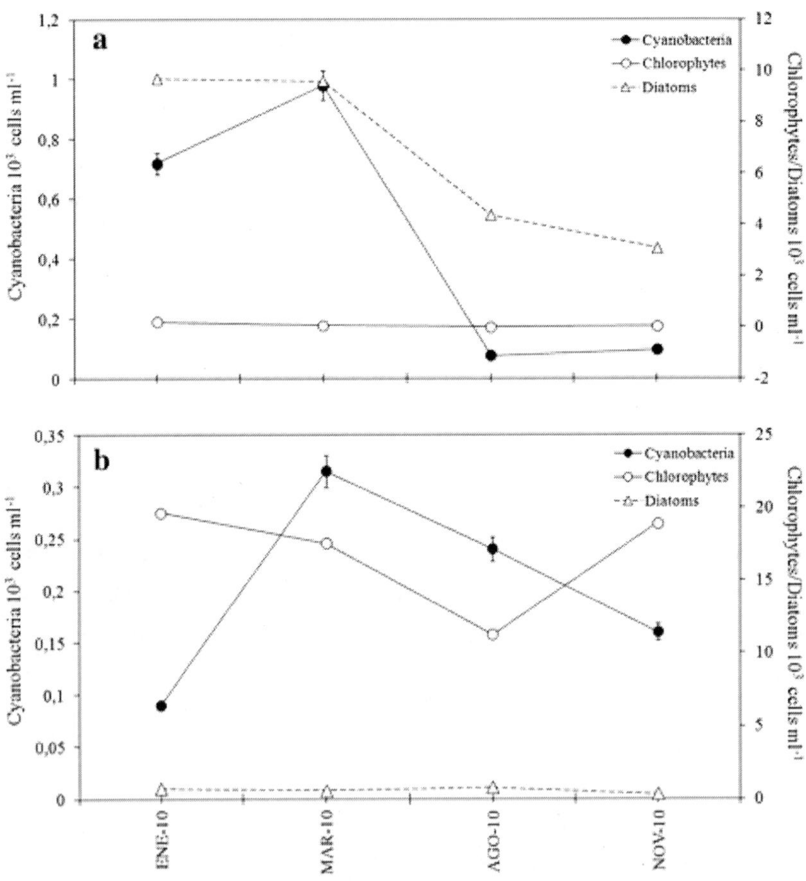

Figure 3. Time course of division groups (chlorophytes, cyanobacteria, and diatoms) in the Salar de Alconcha. **a** Bofedales; **b** Laguna Colorada

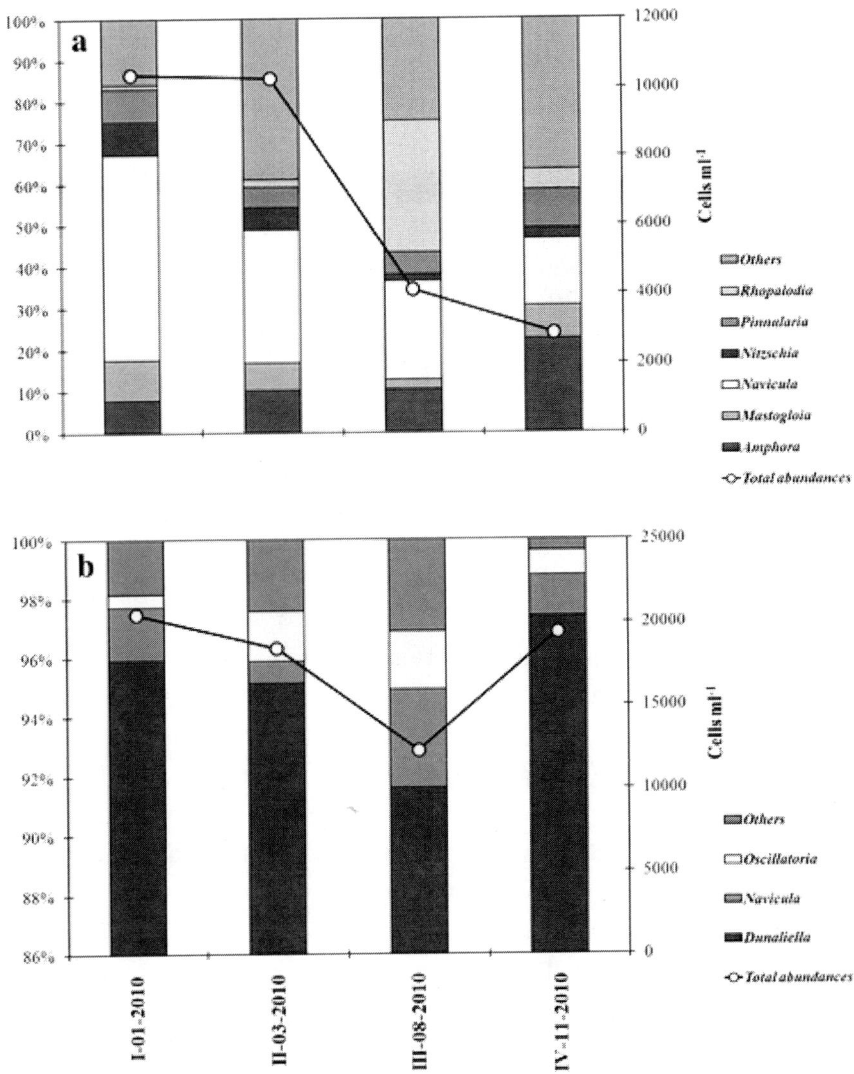

Figure 4. Relative (percentage) and absolute abundances of the main genera in both systems studied. **a** Bofedal; **b**Laguna Colorada

Results from DCA analysis for coverage values and stations are represented in Fig. 5. All the samples were distributed along the plane defined by the two first axes. Axes I and II account for 72.8 % of total variance from the data set (57.6 and 15.2 % for axes I and II, respectively). Comparative analysis revealed specific ecological preferences. On the first axis, samples dominated by *Dunaliella salina* were situated nearly to the ordination axis and were opposed to samples dominated by pennate diatoms and large filamentous cyanobacteria, with the exception of two species*Oscillatoria curviceps* and *Oscillatoria sp1*. Laguna Colorada (L) was associated separately from the Bofedal (B), which also formed two different groups (wet and dry seasons). With respect to distribution DCA analysis formed two different groups with*Dunaliella salina* and others 15 accidental species limited to Laguna Colorada (L). Additional supplementary variables represented by three divisions were summarized by *Chlorophyceae* associated to Laguna Colorada (L) due to higher algal biomass of *Dunaliella salina*. On the contrary, diatoms and filamentous cyanobacteria resulted in preferences of *Bacilliariophyceae* and *Cyanophyceae* (included as supplementary variable in the DCA analysis) to the Bofedal (B). Additionally, distribution of species and stations resulting from the analysis of non-metric multidimensional scaling (MDS) with constrained ellipse 95 % species level using Bray Curtis Index (Fig. 6) was running to contrast the information with DCA analysis. The non-metric MDS distribution for stations was similar to DCA analysis (Fig. 6a), and indicates that phototrophic microbial communities from both systems clearly differ from each other. Marked shift among *Dunaliella salina* and the entire community structure (including both systems studied) was clearly expressed in the MDS analysis resulting far away from the species into the ellipse (Fig. 6b). Some light microscopy photographs of the most important species found in both systems are shown in Fig. 7.

3.3 Life stages of Dunaliella salina and HPLC chromatography

Seasonal stress and salinity variations lead to a series of changes in pigmentation, morphology, and abundance of *Dunaliella salina* in Laguna Colorada (Fig. 8). While most of the cells were red (Fig. 8a) with elongated ovoid shapes (characteristic of red cells) in wet season (austral summer), round non-motile cells (aplanospore cysts) were observed only during the driest season in austral winter (Fig. 8b). Algal biomass increased considerably in November (austral spring) with most of the cells olive and green-olive (Fig. 8c). Pigment profiles by HPLC chromatography only reported β-carotene overcome the limit detection without sample concentration (Fig. 9); negative results for other pigments in concentrated samples over the limit detection corroborated the dominant concentration of β-carotene within the system.

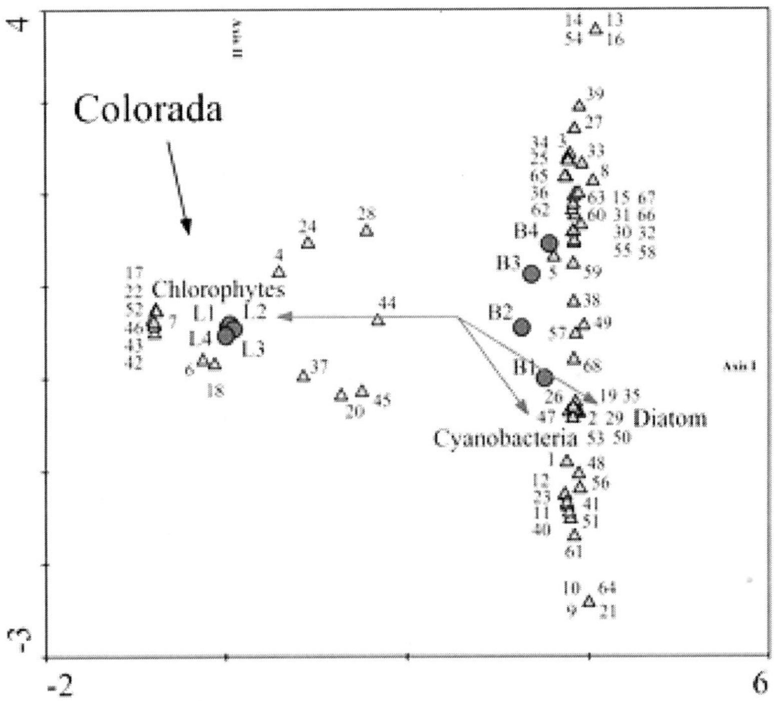

Figure 5. Detrended correspondence analysis (DCA) based in samples and species data (*L*: Laguna Colorada; *B*: Bofedal). *Symbols* in *upper panel* (*triangle*: species; *circle*: samples). *Solid arrows* indicate displacement of division groups (chlorophytes, cyanobacteria, and diatom). *Numbers* correspond to species code described in Table 2

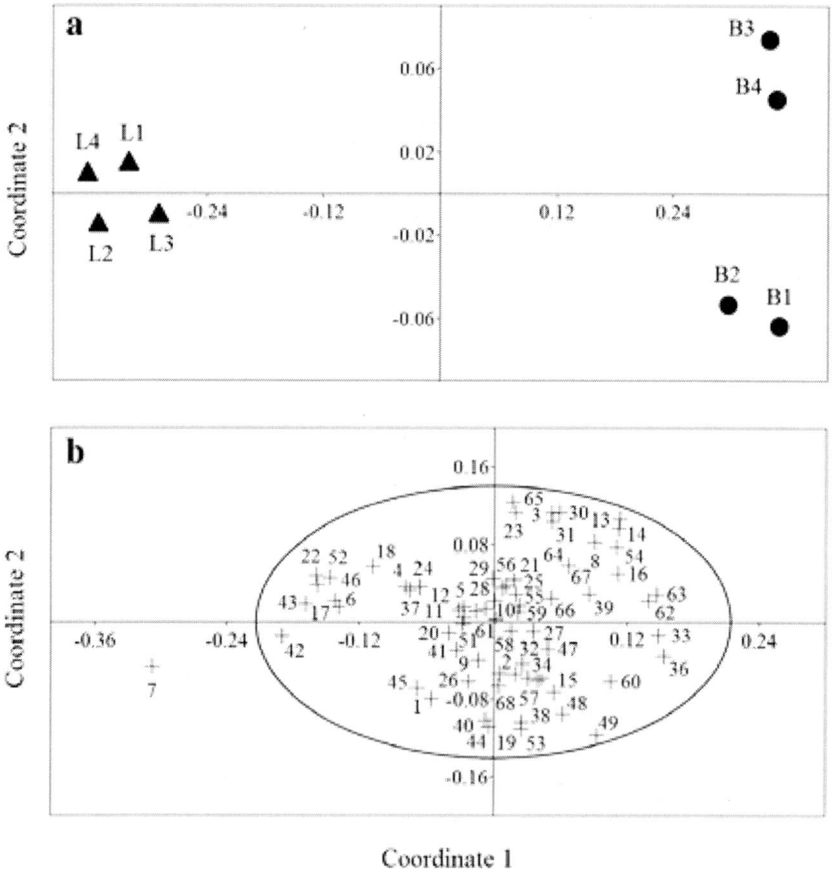

Figure 6. Non-metric (MDS) plot based on seasonal abundance data from Laguna Colorada (*triangle*) and the Bofedal (*circle*) in the Salar de Alconcha. Ordination was based on the Bray–Curtis coefficient (**a**) samples (**b**) species. The stress value for dimensional downscaling was 0.06 (**a**) and 0.18 (**b**), respectively. In the first plots (**a**), data points from Laguna Colorada (L) and the Bofedal (B) are separated by both dimension axes. This indicates that phototrophic microbial communities of the two sampling locations clearly differed from each other as well as *D. salina* was situated far away from the total species (*ellipse group*) identified in both systems. *Numbers* in *figure* **b** correspond to species code described in Table 2

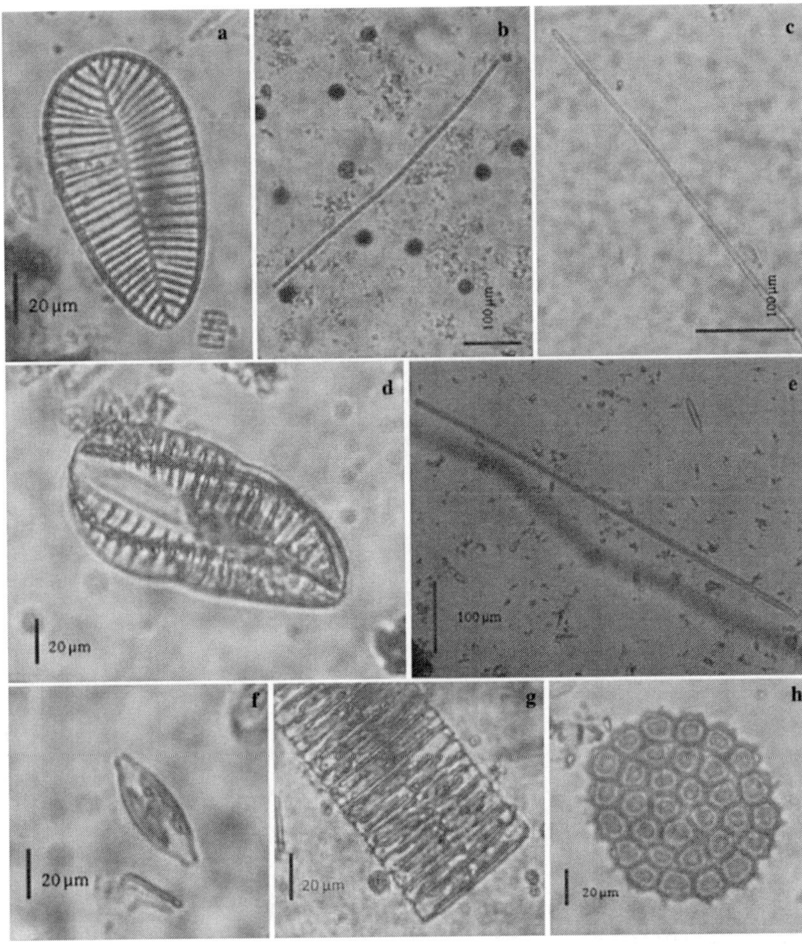

Figure 7. Photographs of different species observed by light microscopy. *Panel* correspond to **a** *Surirella wetzeli*(Bofedal); **b** *Oscillatoria* sp., and Aplanospore cells (*Dunaliella salina*, Laguna Colorada); **c** *Oscillatoria splendida* (Bofedal); **d** *Rhopalodia wetzelii* (Bofedal); **e** *Oscillatoria curviceps* (Laguna Colorada); **f***Stauroneis anceps f. gracilis* (Bofedal); **g** *Fragilaria* sp. (Bofedal); **h** *Pediastrum* sp. (Bofedal)

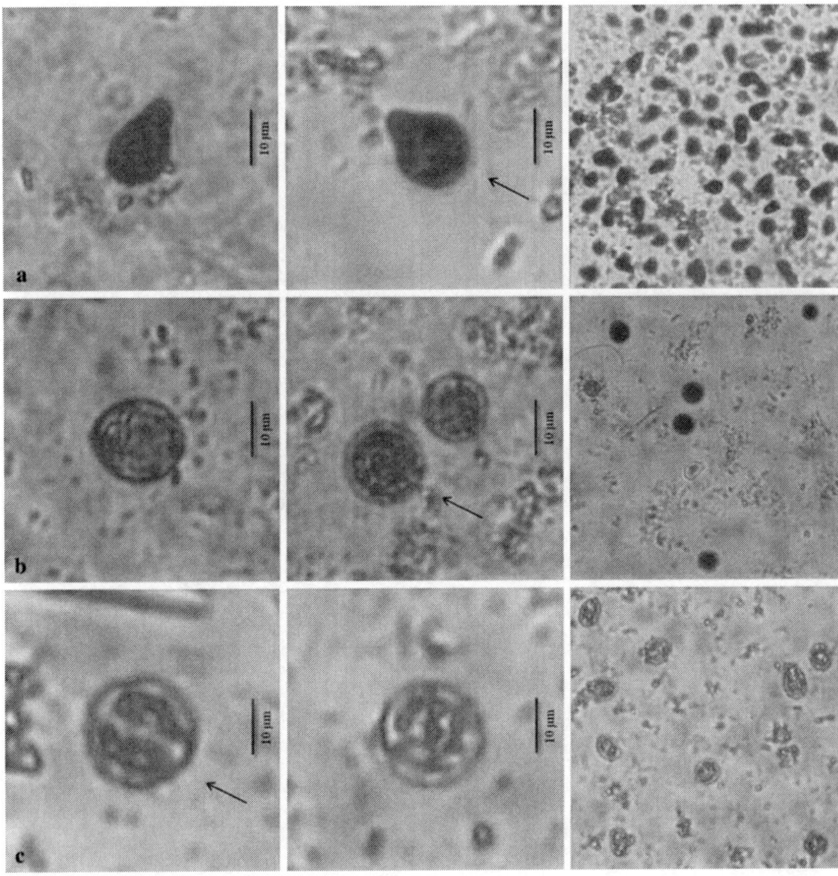

Figure 8. Light microscopy examination of *Dunaliella salina* (Laguna Colorada) in different seasons. **a** January, indicating intense *red pigmentation*; **b** August, indicating aplanospores with *thick cells wall*; **c** November, indicating cell division and *olive-green pigmentation*

4. DISCUSSION

4.1 Extreme photosynthetic communities

Phototrophic microbial communities do not have a significant association with chemical and physical variables in high-altitude wetlands, probably because species ensembles have renovation cycles longer than one year

(Márquez-García et al.2009). However, specific analysis in each ecosystem is necessary to understand this aspect, mainly due to high spatial and temporal heterogeneity as well as influences of complex environmental factors. Photosynthetic microbial communities described in the Salar de Alconcha differ from other large high-altitude lakes with lower salinity such as the Lago Chungará (Dorador et al. 2003). Environmental dissimilarity in the physical landscape defines specific environments for phototrophic microbial assemblages. Several species identified in the Salar de Alconcha has been described as benthonic and typical of hypersaline, mesohaline, or brackish systems, such as *Dunaliella salina* (Fig. 10a), *Rhopalodia wetzelii*, and *Hantzschia amphyoxis* (Fig. 10b), respectively (Cadima et al. 2005).

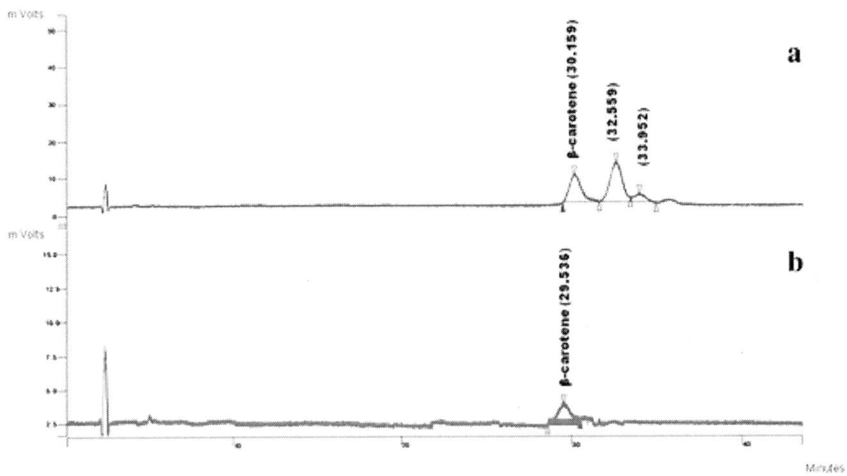

Figure 9. Chromatographic analysis by HPLC in microbial extracts. **a** chromatographic profile of Laguna Colorada; **b**chromatographic profile of β-carotene standard

An interesting association between Bofedales (B) and some phototrophs groups (e.g. diatoms or large filamentous cyanobacteria) seems to evidence an important linkage into the ecosystem function. The persistence and survival of diatoms species (e.g. species inhabiting nitrogen-poor systems) and large filamentous cyanobacteria (sizes from 700 to 2000 μm) under natural conditions seem to depend on a variety of physical factors provided by the Bofedal (e.g. temperature balance (reduce daily amplitude in surface water), soil quality or water retention). Some of these species [e.g. *Oscillatoria anguina*, *Oscillatoria bornetii*, and*Oscillatoria splendida* Fig. 7c)] were strongly restricted to these conditions whereas others two large filamentous cyanobacteria [*Oscillatoria curviceps* (Fig. 7e) and*Oscillatoria sp1* (Fig. 7b)] were also distributed over the salt crust (Laguna Colorada). Distribution of cyanobacteria in the Altiplano describes the presence of*Oscillatoriales*, *Pleurocapsales*, *Chroococcales*, and *Nostocales* in the Salar del Huasco (Dorador et al. 2008). Furthermore, cyanobacteria also describe the presence of unique stratified photosynthetic microbial mats (e.g. *Oscillatoriales*,*Synechococcus*, *Cyanothece* and *Gloeocapsa* groups and genus *Gloeobacter*) in the salar de Llamara, a salt flat basin (Salar de Llamara) located in the Atacama desert (Demergasso et al. 2003). Additionally, photosynthetic pigments were detected being more diverse on the upper photosynthetic layer associated to evaporites domes of gypsum in the Salar de Llamara, where cyanobacteria and diatoms species were also detected (Rasuk et al. 2014). However, there is little information about cyanobacterial communities under hypersaline condition (Dorador et al. 2008). Our results describe the presence of two filamentous cyanobacteria surviving under extreme hypersaline condition (total salt concentration: 119.74 g l^{-1}; conductivity: 187.1 ± 2.2 mS cm^{-1}). Several species found in the Salar de Alconcha represent unique examples to observe in situ the seasonal adaptation to a wide-range of environmental condition. For example, *Hantzschia amphyoxis* are common inhabitant of shallow ecosystems, slightly brackish habitats including Antarctic saline lakes, or even saline soils (Germain 1981; Margalef 1983; Sabbe et al. 2002). Furthermore, *Rhopalodia wetzelii* (Fig. 7d) was a very sensitive species to the seasonal changes (Fig. 10c); probably because this species has an important role in saline, mesohaline, or brackish systems (e.g. Bofedales). This species has been described in several systems in the Altiplano (Eastern cordillera and alluvial plains in Bolivia), classified as euryhalobous, mainly benthic or epiphyte (Cadima et al.2005). An interesting strategy of this species is the association with a cyanobacterial endosymbiont, which allows it to grow into nitrogen poor ecosystems; such as the condition described for species associated to the Bofedal (B) in the Salar de Alconcha.

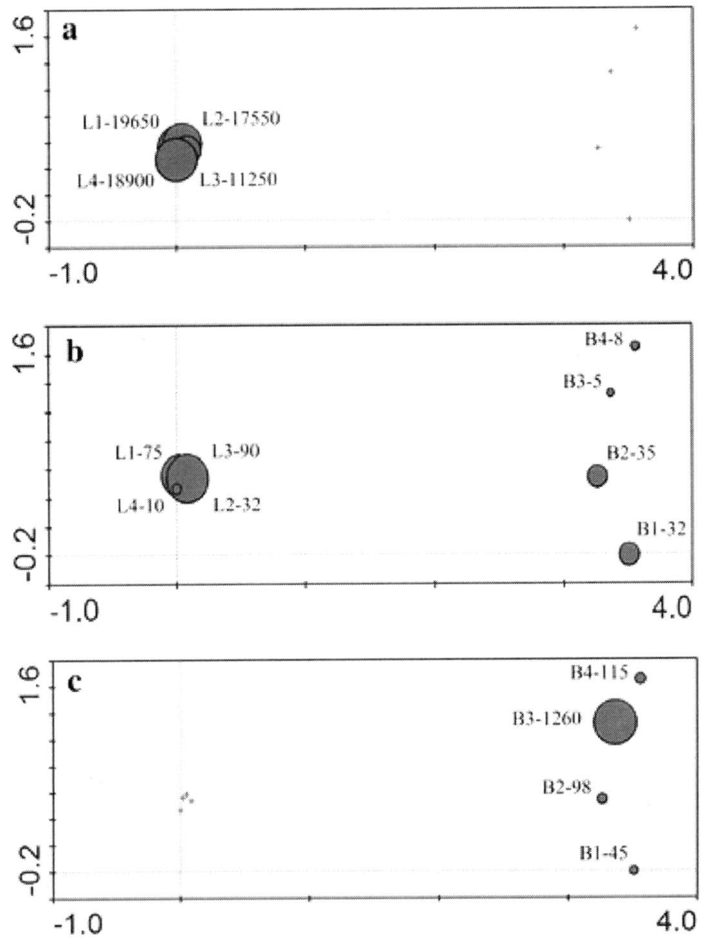

Figure 10. Plots of the main species based on DCA analysis in the Salar de Alconcha. (L) Laguna Colorada; (B) Bofedal. Numbers indicate absolute abundance in cells ml^{-1}. **a** *Dunaliella salina*; **b** *Hantzschia amphioxys*; **c***Rhopalodia wetzelii*

4.2 Physical limitations and Dunaliella species

Dunaliella has been described as the main or the only primary producer in diverse hypersaline ecosystems worldwide, although it is quite surprising that only few quantitative studies have been published on the dynamics

periods of dryness, with two extremely resistant thick coats which form a rough cell wall (Leonardi and Cáceres 1997); germinations process occurs in shallow sediments that had become exposed to diluted waters (Oren and Ben-Yosef 1997). This behavior has not been described yet in the Altiplano. Interestingly, our results reported aplanospore formations only during the driest season in winter condition (Fig. 8b), and germination process in the austral spring when the systems could receive intermittent rains events (Fig. 8c). Furthermore, our result in Laguna Colorada detected cadmium concentrations (Table 1) over the tolerance limit described for toxic levels (plants tolerance) in nature (Dong et al. 2007).

5. CONCLUSION AND PERSPECTIVES

Life in extreme ecosystems represent both eukaryotic and prokaryotic organisms, comparative analysis of all taxa in complex biocoenosis of each ecosystem will provide answers to the fundamental questions of the origin, distribution, and evolution of life (Pikuta et al. 2007). In contrast to the idea that extreme environments have relatively low biological diversity, our results support the hypothesis that contrasting conditions may actually promote phototrophic microbial diversity even in extreme environments. Our results were similar with other studies on phototrophic microbial communities from natural hypersaline environments, where the composition is limited to three taxonomic groups (i.e. chlorophytes, diatoms and cyanobacteria, Oren 2005). However, intermittent or intensive rains events may actually promote the colonization of several temporal species (Acc: accidental species, Table 2). On the contrary, the limited composition detected during the driest seasons appears to be the most successful adaptation to extreme hyper-arid conditions. The scatter in this relationship may in part be explained as an example of tolerance process from slightly brackish habitats to mesohaline or hypersaline ecotypes. Our results show that the structure and composition of phototrophic microbial communities from contrasting systems in the Salar de Alconcha were determined by different factors such as total salt concentration, conductivity, dissolved oxygen and ionic composition.

Salar de Alconcha is an highly variable active salar depending on the groundwater inputs in different zones, and thus, inflows water with very different salinities can coexist nearly. These conditions result in variable habitats with different forms of life (chlorophytes, cyanobacteria, and diatoms) existing at the edge of biological limit under extreme physiological conditions. Additionally, their natural conditions represent an exceptional opportunity to study microbial communities experiencing wide-range of environmental conditions. Our results suggest specific restrictive environments (e.g. environmental dissimilarity in the physical landscape) for phototrophic microbial colonization in active salars with variable saline systems (highly dependent upon water availability). Overall, these ecosystems represent fragile natural reservoirs under high pressure for groundwater associated to copper and others mineral extraction (intensive mining industry). Thus, technical biological support and habitat descriptions are essential for the efforts to preserve these unique ecosystems.

ACKNOWLEDGMENTS

This work was supported by CENIMA (Centro de Investigación de Medio Ambiente), UNAP institutional research project (IQUF03PRO-01020201007). The authors would like to express their sincere thanks to Walter Sielfeld Kowald for his assistance in collecting samples and his critical comments which helped to improve this manuscript

.

REFERENCES

1. Alpers CN, Brimhall GH (1988) Middle Miocene climatic change in the Atacama Desert, northern Chile: evidence from supergene mineralization at La Escondida. Geol Soc Am Bull 100:1640–1656

2. APHA-AWWA-WEF (2005) Standard methods for the examination of water and wastewater, 21st edn. American Public Health Association, Washington DC

3. Arun N, Singh DP (2012) Microalgae: the future fuel (review). J Algal Biomass Utln 3(1):46–54

4. Avendaño V, Saíz F (1977) Taxocenosis de rotíferos limnéticos de la laguna El Plateado. Ans Mus Hist Nat Valparaíso 10:121–133

5. Azúa-Bustos A, Gonzales-Silva S, Salas L, Palma RE, Vicuña R (2010) A novel subaerial *Dunaliella* species growing on cave spiderwebs in the Atacama Desert. Extremophiles 14:443–452

6. Borowitzka MA, Huisman JM (1993) The ecology of *Dunaliella salina* (Chlorophyceae, Volvocales): effect of environmental conditions on aplanospore formation. Bot Mar 36:233–243

7. Cadima M, Fernandez E, Lopez L (2005) Algas de Bolivia con énfasis en el Fitoplancton: Importancia, Ecología, Aplicaciones y Distribución de Géneros. Centro de Ecología Difusión Simón I. Patiño, Santa Cruz

8. Clarke KR (1993) Non-parametric multivariate analysis of changes in community structure. Aust J Ecol 18:117–143

9. Demergasso C, Chong G, Galleguillos P, Escudero L, Martínez-Alonso M, Esteve I (2003) Microbial mats from the Llamará salt flat, northern Chile. Rev Chil Hist Nat 76:485–499

10. Demergasso C, Casamayor E, Chong G, Galleguillos P, Escudero L, Pedrós-Alió C (2004) Distribution of prokaryotic genetic diversity in athalassohaline lakes of the Atacama Desert, Northern Chile. FEMS Microbiol Ecol 48:57–69

11. Demergasso C, Escudero L, Casamayor EO, Chong G, Balagué V, Pedrós-Alió C (2008) Novelty and spatio-temporal heterogeneity in the bacterial diversity of hypersaline Lake Tebenquiche (Salar de Atacama). Extremophiles 4:491–504

12. Diaz C, Maidana N (2005) Diatomeas de los salares Atacama y Punta Negra II Region—Chile. Centro de Ecología Aplicada, Universidad de Chile

13. Dong J, Mao WH, Zhang GP, Wu FB, Cai Y (2007) Root excretion and plant tolerance to cadmium toxicity—a review. Plant Soil Environ 53(5):193–200

14. Dorador C, Pardo R, Vila I (2003) Variaciones temporales de parámetros físicos, químicos y biológicos de un lago de altura: el caso del lago Chungara. Rev Chil Hist Nat 76:15–22

15. Dorador C, Vila I, Imhoff JF, Witzel KP (2008) Cyanobacterial diversity in Salar de Huasco, a high altitude saline wetland in northern Chile: an example of geographical dispersion? FEMS Microbiol Ecol 64:419–432

16. Dorador C, Vila I, Remonsellez F, Imhoff JF, Witzel KP (2010) Unique clusters for Archaea in Salar de Huasco, an athalassohaline evaporitic basin of the Chilean Altiplano. FEMS Microbiol Ecol 73:291–302

17. Finke N, Hoehler TM, Polerecky L, Buehring B, Thamdrup B (2013) Competition for inorganic carbon between oxygen and anoxygen phototrophs in hypersaline microbial mats, Guerrero Negro, Mexico. Environ Microbiol 15(5):1532–1550

18. Fournier RO (1978) Membrane filtering. In: Sournia A (ed) Phytoplankton manual. Monographs on oceanographic methodology, vol 6. Unesco, Paris, pp 108–112

19. Gauch HG (1982) Multivariate analysis in community structure. Cambridge University Press, Cambridge

20. Germain H (1981) Flore des diatomées d'éau douces et saumatres. Sociéte´ Nouvelle des Éditions Boubée, Paris

21. Ginzburg M (1987) Dunaliella: a green alga adapted to salt. Adv Bot Res 14:93–183

22. Hamburger C (1905) Zur Kenntnis der Dunaliella salina und einer Amöbe aus Salinenwasser von Cagliari. Arch Protistenkd 6:111–131

23. Harper D.A.T. (ed. 1999) Numerical Palaeobiology. John Wiley and Sons, New York

24. Iltis A, Risacher F, Servant S (1984) Contribution a l'etude hidrobiologique des lacs sales du sud de l' Altiplano bolivien. Rev Hydrobiol Trop 17:256–273

25. Jin ES, Feth B, Melis A (2003) A mutant of the green algae *Dunaliella salina* constitutively accumulates zeaxanthin under all growth conditions. Biotechnol Bioeng 81:115–124

26. Kunte H, Trüper H, Stan-Lotter H (2002) Halophilic microorganisms. In: Horneck G, Baumstark-Khan C (eds) Astrobiology, the quest for the conditions of life. Springer, Koln, pp 185–200

27. Lange B, Mary Ann T (2002) The diatom flora of the Salton Sea, California. Hydrobiologia 473:179–201

28. Leonardi PI, Cáceres EJ (1997) Light and electron microscope observations of the life cycle of *Dunaliella salina*(Polyblepharidaceae, Chlorophyceae). Nova Hedwigia 64:621–633

29. Maldonado F (2014) An introduction to the bofedales of the Peruvian high Andes. Mires Peat 15(5):1–13

30. Mancinelli R, Fahlen T, Landheim R, Klovstad M (2004) Brines and evaporates: analogs for Martian life. Adv Space Res 33:1244–1246

31. Margalef R (1983) Limnología. Omega, Barcelona

32. Margulis L, Barghoorn ES, Ashendorf D, Banerjee S, Chase D, Francis S, Giovanonni S, Stolz J (1980) The microbial community in layered sediments at Laguna Figueroa, Baja California, Mexico: does it have Precambrian analogues? Precambrian Res 11:93–123

33. Márquez-García M, Vila I, Hinojosa L, Mendez M, Carvajal J, Sabando M (2009) Distribution and seasonal fluctuations in the aquatic biodiversity of the southern Altiplano. Limnologica 39:314–318

34. Massyuk NP (1973) Morfologija, Sistematika, Ekologija, Geograficheskoe Rasprostranemie Roda *Dunaliella*Teodoresco. "Naukova Dumka", Kiev, pp 1–244

35. Oren A (2002) Molecular ecology of extremely halophilic Archaea and Bacteria. FEMS Microbiol Ecol 39:1–7

36. Oren A (2005) A hundred years of *Dunaliella* research: 1905–2005. Saline Syst 1:2

37. Oren A, Ben-Yosef N (1997) Development and spatial distribution of an algal bloom in the Dead Sea: a remote sensing study. Aquat Microb Ecol 13:219–223

38. Parsons TR, Maita Y, Lalli CM (1984) a manual of chemical and biological methods for sea water analysis. Pergamon, Elmsford

39. Pikuta EV, Hoover RB, Tang J (2007) Microbial extremophiles at the limits of life. Crit Rev Microbiol 33:183–209

40. Rasuk MC, Kurth D, Flores MR, Contreras M, Novoa F, Poire D, Farias ME (2014) Microbial characterization of microbial ecosystems associated to evaporites domes of gypsum in Salar de Llamara in Atacama Desert. Microb Ecol 68(3):483–494

41. Risacher F, Alonso H, Salazar C (2003) The origin of brines and salts in Chilean salars: a hydrochemical review. Earth-Sci Rev 63:249–293

42. Sabbe K, Verleyen E, Hodgson DA, Vanhoutte K, Vyverman W (2002) Benthic diatom flora of freshwater and saline lakes in the Larsemann Hills and Pauer islands, East Antarctica. Antarct Sci 15:227–248

43. Scott S, Dorador C, Oyanedel JP, Tobar I, Hengst M, Maya G, Harrod C, Vila I (2015) Microbial diversity and trophic components of two high altitude wetlands of the Chilean Altiplano. Gayana 79(1):45–56

44. Simonsen R (1987) Atlas and catalogue of the diatom Types of frederich Hustedt, J. Cramer. Gerbrüder Borntraeger, Berlin, Stuttgart

45. Teodoresco EC (1905) Organisation et développement du *Dunaliella*, nouveau genre de Volvocacée-Polyblepharidée. Beihz Bot Centralbl, Bd. XVIII, p 215–232

46. ter Braak CJF (1988) CANOCO—a FORTRAN program for canonical community ordination by [partial] [detrended] [canonical] correspondence analysis, principal 18 components analysis and redundancy analysis (version 2.1). Report LWA-88-02, Agricultural Mathematics Group, Wageningen

47. ter Braak CJF, Šmilauer P (1998) CANOCO reference manual and user's guide to Canoco for Windows: software for canonical community ordination (version 4) microcomputer power. Ithaca, New York, pp 352–353

48. Thiel V, Tank M, Neulinger SC, Gehrmann L, Dorador C, Imhoff JF (2010) Unique communities of anoxygenic phototrophic bacteria in saline lakes of Salar de Atacama (Chile): evidence for a new phylogenetic lineage of phototrophic Gammaproteobacteria from pufLM gene analyses. FEMS Microbiol Ecol 74:510–522

49. Utermöhl H (1958) Neue wege in der quantitativen erfassung des plankton (mit besonderer Berücksichtigung des Ultraplanktons). Ver Int Ver Theor Angew. Limnol 5:567–596

50. Vila I, Muhlhauser H (1987) Dinámica de lagos de altura. Perspectivas de investigación. Arch Biol Med Exp 20:95–103

51. Zúñiga LR, Campos V, Pinochet H, Prado B (1991) A limnological reconnaissance of lake Tebenquiche, Salar de Atacama, Chile. Hydrobiologia 210:19–24

Chapter 5

Impacts of Aquaculture on Habitats and Best Management Practices (BMPs)

Gulnihal Ozbay[1], Grant Blank[1] and Taworn Thunjai[2]

[1] Department of Agriculture and Natural Resources, Delaware State University, Dover, USA

[2] Department of Fisheries, Kasetsart University, Ladyao, Jatujak, Bangkok, Thailand

1. INTRODUCTION

The demand for food must be met as the human population reaches an estimated nine billion people by the year 2050. This means we must increase overall food production by 70% and this increase must be sustainable and food price affordable (United Nations FAO 2009). Most of the population growth is expected to continue in underdeveloped countries with limited technologies and venues (United Nations FAO 2009). As a popular high protein food source, seafood contains omega-3 fatty acids that are required for healthy human development (UMD Medical Center

2013). Seafood is low in calories, total fat, and saturated fat, while high in vitamins and minerals including vitamins A and D, phosphorus, magnesium, selenium, and iodine (FAO FOCUS 2013). Fish have been shown to have numerous health benefits (Table 1). Seafood is a healthy, low-fat alternative to beef, poultry, and pork and significant omega-3 fatty acids much higher than vegetable-based diets (FAO FOCUS 2013). Specifically, omega-3 fatty acids contained within fish oil are critically important for infants and babies to develop a normal brain (FAO FOCUS 2013).

Population growth and economic development trends are the most important drivers for the demand for high quality and nutritional seafood products (Ewart 2013). With wild capture fisheries exceeding the maximum sustainable harvest capacity, aquaculture has become a bridge in closing the gap between rising demand and traditional seafood sources (Figure 1). Today, farmed seafood accounts for about 50% of overall production in the global marketplace (Bush et al. 2013). The United States aquaculture industry, valued at over $1.1 billion, produces a variety of fish and shellfish species for food, recreation, and industrial needs (Ewart 2013). However the United States is in a seafood deficit, importing more seafood to meet the demands for seafood consumption than it can produce (NOAA Office of Aquaculture 2013).

Commercial aquaculture is a young and rapidly expanding industry in the United States and the need for information on sustainable growth and development has increased dramatically during the past few decades (Wilson et al. 2002; FAO FOCUS 2013). Aquaculture in the simplest terms is the farming of aquatic plants and animals. Furthermore, the National Oceanic and Atmospheric Administration (NOAA) Office of Aquaculture (2013) describes aquaculture on a broader scale as the breeding, rearing, and harvesting of plants and animals in all types of water environments, including ponds, rivers, lakes, and the ocean. Similar to agriculture, aquaculture can take place in the natural environment or in a manmade environment where controlled cultivation and husbandry of aquatic plants and animals are achieved. Using aquaculture techniques and technologies, researchers, aquaculturists and the aquaculture industry are "growing," "producing," "culturing," and "farming" all types of freshwater and marine species (NOAA Office of Aquaculture 2013). According to Ewart (2013), aquaculture has a long history dating back a few thousands of years in China and Egypt. Aquaculture within the United States dates back to the late 1800s, when hatchery technologies were utilized to cultivate fish for restoration of depleted inland freshwater fishes (Ewart 2013). Ewart (2013) stated with a short commercial history (about 50 years), the United States aquaculture industry has a current annual farm gate value of $1.9 billion. Included in the domestic aquaculture production are variety of fish and shellfish species for food, recreation (stock enhancement, restoration, ornamental fish, aquatic plants, live bait), and industrial applications (food additives).

Aquaculture can benefit more than human economies and diets. Oyster shellfish aquaculture provides many of the same ecosystem services as natural oyster reefs (Dealteris et al. 2004; Erbland and Ozbay 2008). Unlike some finfish farming practices, rearing shellfish in high densities in shallow water with abundant phytoplankton concentrations can have positive effects on the environment and may promote biodiversity (Shumway et al. 2003; Dealteris et al. 2004; O'Beirn et al. 2004; Tallman and Forrester 2007; D'Amours et al. 2008; Erbland and Ozbay 2008; Taylor and Bushek 2008).

As stated by Emerson (1999), the process of aquaculture has been under increasing scrutiny as the world tries to supply food for a population which is currently over seven billion. This criticism is happening regardless of how aquaculture is perceived as an economic windfall for developing countries or potential food industries. Aquaculture is the fastest growing food production sector in the world but its sustainability is not fully satisfied (FAO 2013). This chapter will reassure the ultimate question we ask ourselves: is sustainable aquaculture our solution?

Emerson (1999) discussed how pollution, destruction of sensitive coastal habitats, threats to aquatic biodiversity and significant socio-economic costs must be balanced against the substantial benefits and how aquaculture has great potential for food production and the alleviation of poverty for people living in coastal areas where most of the poorest in the world live. He also stated a delicate balance between food security and the environmental costs of production must be achieved. This leads us to our second question: how do we make the world's fastest growing food sector environmentally and socially responsible?

Table 1. Nutrition facts on various seafood species (The United States Food and Drug Administration 2008).

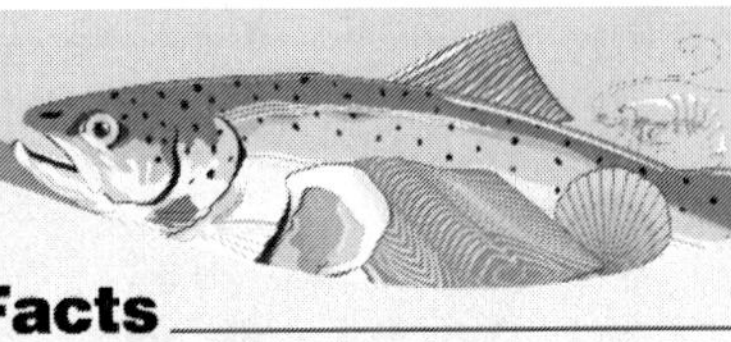

Seafood
Nutrition Facts

Cooked (by moist or dry heat with no added ingredients), edible weight portion. Percent Daily Values (%DV) are based on a 2,000 calorie diet.

Seafood Serving Size (84 g/3 oz)	Calories	Calories from Fat	Total Fat g	Total Fat %DV	Saturated Fat g	Saturated Fat %DV	Cholesterol mg	Cholesterol %DV	Sodium mg	Sodium %DV	Potassium mg	Potassium %DV	Total Carbohydrate g	Total Carbohydrate %DV	Protein g	Vitamin A %DV	Vitamin C %DV	Calcium %DV	Iron %DV
Blue Crab	100	10	1	2	0	0	95	32	330	14	300	9	0	0	20g	0%	4%	10%	4%
Catfish	130	60	6	9	2	10	50	17	40	2	230	7	0	0	17g	0%	0%	0%	0%
Clams, about 12 small	110	15	1.5	2	0	0	80	27	95	4	470	13	6	2	17g	10%	0%	8%	30%
Cod	90	5	1	2	0	0	50	17	65	3	460	13	0	0	20g	0%	2%	2%	2%
Flounder/Sole	100	15	1.5	2	0	0	55	18	100	4	390	11	0	0	19g	0%	0%	2%	0%
Haddock	100	10	1	2	0	0	70	23	85	4	340	10	0	0	21g	2%	0%	2%	6%
Halibut	120	15	2	3	0	0	40	13	60	3	500	14	0	0	23g	4%	0%	2%	6%
Lobster	80	0	0.5	1	0	0	60	20	320	13	300	9	1	0	17g	2%	0%	6%	2%
Ocean Perch	110	20	2	3	0.5	3	45	15	95	4	290	8	0	0	21g	0%	2%	10%	4%
Orange Roughy	80	5	1	2	0	0	20	7	70	3	340	10	0	0	16g	2%	0%	4%	2%
Oysters, about 12 medium	100	35	4	6	1	5	80	27	300	13	220	6	6	2	10g	0%	6%	6%	45%
Pollock	90	10	1	2	0	0	80	27	110	5	370	11	0	0	20g	2%	0%	0%	2%
Rainbow Trout	140	50	6	9	2	10	55	18	35	1	370	11	0	0	20g	4%	4%	8%	2%
Rockfish	110	15	2	3	0	0	40	13	70	3	440	13	0	0	21g	4%	0%	2%	2%
Salmon, Atlantic/Coho/Sockeye/Chinook	200	90	10	15	2	10	70	23	55	2	430	12	0	0	24g	4%	4%	2%	2%
Salmon, Chum/Pink	130	40	4	6	1	5	70	23	65	3	420	12	0	0	22g	2%	0%	2%	4%
Scallops, about 6 large or 14 small	140	10	1	2	0	0	65	22	310	13	430	12	5	2	27g	2%	0%	4%	14%
Shrimp	100	10	1.5	2	0	0	170	57	240	10	220	6	0	0	21g	4%	4%	6%	10%
Swordfish	120	50	6	9	1.5	8	40	13	100	4	310	9	0	0	16g	2%	2%	0%	6%
Tilapia	110	20	2.5	4	1	5	75	25	30	1	360	10	0	0	22g	0%	2%	0%	2%
Tuna	130	15	1.5	2	0	0	50	17	40	2	480	14	0	0	26g	2%	2%	2%	4%

Seafood provides negligible amounts of *trans* fat, dietary fiber, and sugars.

U.S. Food and Drug Administration
(January 1, 2008)

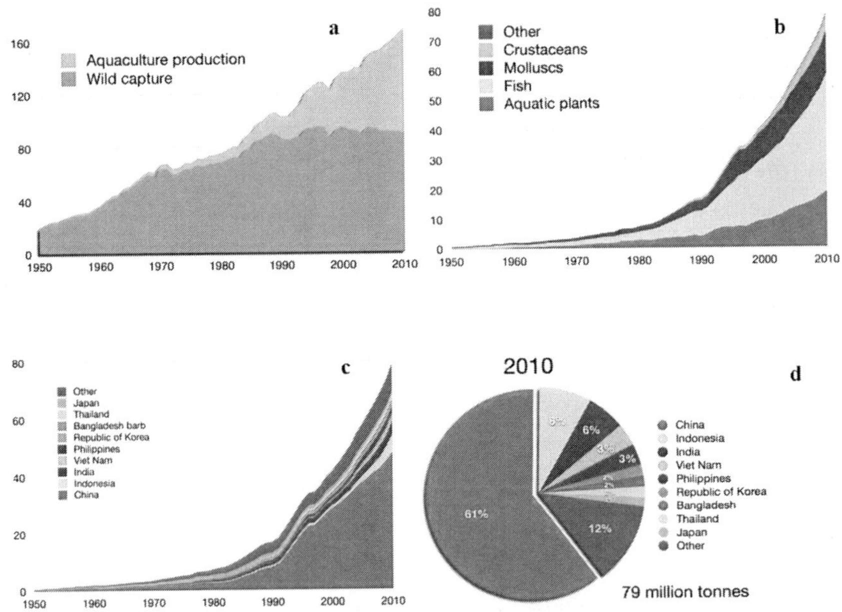

Figure 1. Global harvest of aquatic organisms in millions of tons between 1950-2010 (FAO 2011); b. Global aquaculture production in millions of tons between 1950-2010 (FAO 2011); c. Main aquaculture countries between 1950-2010 (FAO 2011); and d. Main aquaculture countries in 2010 (FAO 2011).

As we search for answers to these questions, the World Wildlife Federation (WWF 2013) gives additional reasons for why aquaculture must become more responsible. According to the WWF (2013), over 53% of the fisheries worldwide are exploited when over 32% are either depleted, overexploited or recovering including are top ten marine fisheries and as much as 30% of all capture fisheries production are either fully exploited or overexploited. Over 90% of large fish were overfished including are several important commercial fishes (i.e. tuna, skipjack, cod, sturgeon) to the point their survival is threated. Whether it is fully exploited or overexploited, by 2048 fish species harvested for food will be collapsed unless urgent management practices are taken to improve the current conditions (WWF 2013). Unwanted fish (by-catch), like many other animals, die due to inefficient, illegal, and destructive fishing practices every year. This destructive fishing practice along with overfishing largely results in poor fisheries management, pirate fishes, subsidies, and unfair fisheries partnerships (WWF 2013).

Over the past 50 years in the Unites States, the demand for seafood has increased as the population reached over 300 million people (NOAA 2011). Seafood import is over 86% of total seafood demand in the United States (NOAA 2010). Unfortunately, many economically and ecologically important fish species are disappearing from our oceans through over-harvest, loss of habitat, and pollution as we stated earlier. As our most important fisheries are collapsing, fishermen and seafood processors are losing their businesses. A solution to this issue lies in aquaculture, particularly marine aquaculture. Although technological advances enable safe, profitable, and environmentally sustainable culturing of aquatic organisms, sustainability seems to be the key to long lasting aquaculture practices that are profitable and environmentally sound.

Considering the majority of fish we consume are farm-raised fish with over 100 species cultured globally, various culturing technique have been used including traditional earthen ponds to high-tech tank systems, each culture method/technique yields its own environmental foot print (Monterey Bay Aquarium Foundation 2013). Although most aquaculture facilities manage with the best intent for stress reduction, beneficial health and fast growth, many larger intensive aquaculture systems are managed where the stock is raised under stressful environmental conditions where there is little ecological balance. In compensating for poor environmental and health condition, managers have often relied heavily upon chemical, antibiotic and water treatments to get their fish to harvest before the system becomes too stressful for optimum growth. Managers risk rising production costs, stock mortalities, and the degradation of habitats that receive liquid waste discharges (Briggs and Funge-Smith 1994). Dalsgaard et al. (1995) explained the concept of ecological sustainability as maximization of internal feedback within a culture system. They refer this maximization as the integrated resources management practice similar to the agro-ecological engineering approach to integrated agriculture-aquaculture farming used in China. Such a system would minimize inputs and wasted outflows of resources and maximize profits.

As the aquaculture industry grows, the use of treatments unapproved by the Food and Drug Administration (FDA) and/or the misuse of chemicals and treatment strategies administered to culture seafood also grows (FDA 2008). To protect consumers, it is important to ensure that both imported and domestic aquaculture seafood products are free from potentially harmful drug, microbial, and heavy metal residues. These residues in food can cause acute, chronic or microbial effects on people. An acute response from hypersensitivity or allergenicity may occur (FDA 2008). Chronic effects can be long term and they are difficult to detect because these events are typically underreported. Cancer is a potential chronic long term effect (Virtanen et al. 2008). Microbial effects caused by drug residues have an effect on human intestinal flora which limits the activity of intestinal bacteria (FDA 2008). Moreover, antibiotic drug residues can affect the development of resistant bacterial populations. FDA (2008) provides one such example "the unapproved use of fluoroquinolones, such as ciprofloxacin, poses the risk of increasing antibiotic resistant bacteria with the potential for serious human health consequences from untreatable infections. In addition, chronic dietary exposure to high concentrations of fluoroquinolone residues, particularly during early growth, may result in a number of toxicities including joint and testicular lesions." The use of unapproved compounds or misuse of FDA approved new animal drugs, will impact the safety of aquaculture products for the consumers in the United States.

In order to identify potential ways to decrease unnecessary outflows from aquaculture systems in the United States, the National Pollutant Discharge Elimination System (NPDES) permitting is utilized by the United States Environmental Protection Agency (EPA) on a case-by-case basis, typically in larger aquaculture operations. It is difficult to make correlations between aquaculture effluents and environmental impacts without accurate records from each facility. It is crucial to examine the aquaculture practices not only in the United States, but also practices world-wide that can minimize our impacts on aquatic ecosystems and while simultaneously increasing food production.

Some other areas of concern regarding aquaculture include, but are not limited to: eutrophication, benthic enrichment, habitat alteration, erosion, disease, water quality, and effective implementation of best management practices (Coastal Habitat Protection Plan 2005). With the addition of nutrients and phytoplankton, bacteria and viruses, become even more important in regards to aquaculture water quality concerns. There is a direct correlation between bacterial diversity and nutrient content. Naturally occurring bacteria from the environment and the guts of cultured fish stocks thrive in nutrient-rich waters and the surface layers of sediments (Garland Science 2011). There are various pathways humans can be infected by the zoonotic pathogens including food and contact with contaminated environments (Friend 2006). Viruses are a special concern in non-native stocks, where introduced species and hybrids may bring new viral strains into an area. Even if the potential for introduced viruses is reduced, periodic outbreaks of viruses are not uncommon (Yanong and Erlacher-Reid 2012)Y. *Vibrio*bacteria are major fish pathogens that are particularly problematic in aquaculture settings (Chatterjee and Haldar 2012). Uncontrolled proliferation within farm operations appears to have made a direct contribution in the dispersion of *Vibrio* pathogens in receiving water bodies (Yanong and Erlacher-Reid 2012).

Integrated aquaculture may provide solutions to many of aquaculture's problems. Since no organism lives naturally in a vacuum, stocking production facilities with complementary species is a logical way to integrate multiple species while simultaneously increasing production for a given area. For instance, to control algae and plant growth, grass carp or other herbivores may be raised along with primary stock. Suspension feeding bivalves are useful organisms in filtering phytoplankton. Mori (1979) found that phytoplankton concentrations decreased by 94% after water was passed through eleven oyster rafts. Not only are the secondary stocks beneficial in controlling water quality, they often are valuable food products as well. Integrated multi-trophic aquaculture systems yield not only greater profit and lower cost but also enhance economic stability and provide more acceptable management practices (Bastin 2013).

Various fish farming techniques have been used depending on the species and their growth stage. Some of these include but are not limited to: ponds, open net pens or cages, hatchery, bag and rack, raceways, recirculating systems, shellfish culture, submersible net pens, suspended culture, tuna ranching, and aquaponics. Although our discussion is limited primarily to inland aquaculture practices with particular emphasis on pond aquaculture in this chapter, recirculating aquaculture and aquaponics systems are also discussed as popular aquaculture practices that are frequently employed to eliminate potential nutrient loads to the surrounding environment. More specifically, our discussion on recirculating aquaculture and aquaponics systems is due to use of recirculating aquaculture systems for commercial aquaculture species of high market value or application of aquaponics for their roles in minimizing nutrient loads from aquaculture water discharge and increase farm profits by growing alternate crops.

In order to address Best Management Practices (BMPs) in this chapter, we will explore studies from various countries including Thailand, South Africa, United States, Canada, and Australia. These studies address issues regarding how to best manage an aquaculture operation, while minimizing environmental effects and maintain profitable output. BMPs reflect the most technically practical and economically feasible methods which reduce environmental impacts and limit operation costs at aquaculture facilities. One primary goal is to discuss effluent treatment systems that reduce loads of organic matter, suspended solids, and nutrients to prevent polluting receiving waters. The best method to prevent soil and water quality problems includes selecting a site with appropriate soils and an adequate water supply, and maintaining moderate organism densities and feeding rates (Boyd 1989). Secondary management techniques to prevent soil and water quality imbalances include liming, fertilization, and aeration (Boyd 1989). Agricultural irrigation, created wetlands, settling basins, and biological filters also are practical methods for improving the quality of effluents from ponds that will be discussed within this chapter.

In the challenging area of integrating aquaculture Best Management Practices (BMPs), it is imperative that older, proven methods be incorporated with new and innovative ideas. Nearly 40 years ago at Woods Hole, MA and Fort Pierce, FL, Ryther et al. (1975) developed working integrated waste recycling systems utilizing commercially valuable mariculture stocks. Their systems proved so efficient that the final effluent of their system was incapable of supporting further growth or contributing to eutrophication. They suggested that similar systems can be developed for other aquaculture operations to desired needs and purposes. At the Eilat Laboratory in Israel, Neori et al. (1998) established a land-based integrated system that attempted to eliminate external food sources and water exchanges.Avnimelech (2012) provided detailed information on biofloc technology and how this technology can be used to increase farm profit and reduce the nutrient loads of the system. This manual discussed super-intensive biofloc shrimp farming and effects of biofloc technology on the sexual development of shrimp broodstock and other practices. The University of Virgin Islands Aquaculture Program (2013) established a biofloc system that produces tilapia every six months by using biologically active and suspended solids serving as the primary waste treatment in the farm. Additional management practices include good aeration, settleable solid removal, pH adjustments and anaerobic denitrification in this system. Even if this type of system may not be as profitable as growers would like, it is easy to see how the basic principles may be applied to a wide range of aquaculture systems. Unfortunately, there is little impetus to develop such systems unless discharge regulations are increased or the systems are shown to be profitable. Coupled with recirculating systems, aquaponics is described as a synergistic growing technique by the Aquaponics Association (2013) by growing fish and plants together in the same systems. The logic behind this system is that nitrate-nitrogen in fish waste serve as a fertilizer to grow the plants. Once the plants such as lettuce, basil, parsley remove nitrate-nitrogen, this water returns to the fish environment so no water has to be discharged to the environment (The Aquaponics Association 2013).

Integrated aquaculture is nothing new and has been used for thousands of years although their uses commercially are most recent (Bennett et al. 2012). With the demand for high protein food diet, limited resources and environmental concerns, integration provides a solution to maximize profits and reduce potential impacts on the surrounding habitat. By culturing multiple species, farmers can offset the negative impacts in the environment. In China, farmers have been using integrated farming practices for years, although not at the commercial scale at the present time, and have maximized the resource uses to feed their growing population (Bennett et al. 2012).

In this chapter, we will further discuss water quality, eutrophication and disease causing organisms concerns along with effective treatment methods for aquaculture effluents and best farm management strategies in the interest of giving aquaculture professionals, educational professionals, students, and decision-makers a better perspective on how to move forward in a rapidly-changing global market.

Aquaculture will play very important role in feeding about nine billion people by the year 2050 (Nutreco 2011). Meeting this demand can only be possible if seafood is farmed in a sustainable way, both environmentally and economically. As we work together we will find better ways to improve quantity, quality, and sustainability of food supply within the aquaculture sector.

2. ISSUES OF SPECIAL CONCERN IN AQUACULTURE

2.1. Water Quality and Eutrophication

The highly variable nature of any aquatic environment is often held in a delicate balance by several mechanisms which are common in undisturbed habitats. When anthropogenic stressors (e.g. discharge from aquaculture, farming practices) are introduced into the environment this delicate balance can be disturbed. As a result of an increased aquaculture activity and related farming practices, the effects of seepage and discharge off farms can disturb the healthy conditions of aquatic ecosystems within entire watersheds As described by SAMS (2013), high concentrations of nutrients may lead to deleterious effects, especially in receiving water bodies with the limited water exchange such as lochs. The harmful effects that occur come about as a result of changes in microbial growth and community composition. These changes often result in toxic conditions arising from harmful algae blooms, de-oxygenation of water and sediments from excessive microbial growth, and the transfer/concentration of toxic compounds through the food web. Dissolved and particulate materials in estuaries and coastal environment increase from both natural and anthropogenic sources such as rivers, sewage outfalls, agriculture, and fish farms. These dissolved and particulate materials provide nutrients for phytoplankton and bacteria because they are sources rich in carbon, nitrogen, and phosphorus. Particulate and dissolved materials can also be carriers of heavy metals and drug residues harmful to aquatic life (SAMS 2013). Science Daily (2013) describes eutrophication as the enrichment of an ecosystem with nitrogen or phosphorus, or a mixture of both chemicals. Regarding eutrophication and healthy aquatic system, the water quality variables with the highest concentrations in pond effluents relative to the normal criteria allowed by National Pollution Discharge Elimination System (NPDES) permits are our major concern and discussed in detail throughout the chapter. This includes total dissolved solids, total phosphorus, and biochemical oxygen demand. Eutrophication is the leading problem associated with nutrient runoff of phosphorus (Boyd 2001). Resulting phytoplankton blooms often create an increase in organic matter by two to four times the original amount of metabolic wastes, multiplying the negative effects (Boyd and Queiroz 1997).

As we previously stated, total dissolved solids, total phosphorus, and biochemical oxygen demand are the water quality variables that have the highest concentrations in pond effluents relative to NPDES permits for standard water quality for effluents (Shireman and Cichra 1991; Schwartz and Boyd 1994a). These variables have especially high concentrations in the final 25% of effluent when ponds are completely drained (Boyd 1978; Schwartz and Boyd 1994b; Seok et al. 1995). According to Boyd et al. (2000), total suspended solids and total phosphorus are water quality variables consistently higher in concentration in aquaculture effluents than the typical concentration in effluents of other industries in the southern United States. In comparable studies of the effects of aquaculture effluents on water quality from catfish facilities between Alabama and Mississippi, Hariyadi et al. (1994) found greater concentrations of suspended clay, turbidity, dissolved inorganic phosphorus, total ammonia, and nitrite concentrations. Although effluents from aquaculture facilities with less commonly cultured species have not been studied as thoroughly as channel catfish pond effluents, it is reasonable to assume discharge off aquaculture ponds with other benthic species will have similar nutrient concentrations because of feeding and intensive culture methods. However, the methods of management will vary depending on the species cultured and life stage being cultured. Methods need to be developed for reducing effluent volume and improving the quality of aquaculture effluents in general. Developing specific procedures for removing or reducing suspended solids, total phosphorus and biochemical oxygen demand from pond effluent are especially important. The goal is to develop methods to treat aquaculture discharge so that the materials meet NPDES water quality criteria for effluents (Boyd et al. 1998).

Many techniques have been developed that can be effective in reducing the volume and enhancing the quality of aquaculture pond effluent. These methods include but are not limited to the use of proper site evaluation and design procedures, good construction practices, use of high quality feeds and good feed management, attention to erosion control, moderate stocking densities, reduction in water exchange, seine harvest, and the use of settling basins (Boyd and Tucker 1998). Suitable methods for removing aquaculture waste within effluents include sedimentation, filtration, and mechanical separation using screens, chemical and biological amendments, and using high quality fish-meal (Wheaton 1977, Boyd et al. 1998, Coloso et al. 2001). Boyd and Tucker (1998) summarized methods for using and improving effluents from ponds. These methods have advanced over the years and include hydroponics, irrigation, the development of culture medium for other aquatic organisms, constructed wetlands, settling basins, biological filters, nutrient removal by water hyacinths or other floating macrophytes, and fluidized-bed filters. Queiroz et al. (1998) tested the effectiveness of various bioorganic catalysts[1] - on water quality, soil organic carbon, and channel catfish production and recorded higher concentrations of dissolved oxygen and a slight increase in phytoplankton productivity.

According Boyd and Tucker (1998), the most efficient procedures for treating effluents appear to be irrigation, settling basins, and wetlands. Filter-feeding fish, mollusks and certain plants have been successfully cultured in aquaculture pond effluents to reduce nutrient and organic matter loadings.Tucker et al. (1996) reported that harvesting fish without draining ponds between fish crops maintained water storage potential and reduced average annual nutrient and organic matter discharge by over 60% relative to annually drained ponds.

Unlike nitrogen or carbon, phosphorus can only enter the watershed via land-use runoff and coastal areas (Thompson and Polz 2006). Release of phosphorus into the aquatic environment is dependent on soil type, landscape slope, rainfall intensity, and the particle trapping capabilities of the watershed in question because phosphorus is considered a particle bound nutrient. Soluble Reactive Phosphorus (SRP) is a biologically available inorganic form of phosphorus often measured in estuarine systems to better assess the available phosphorus used by the aquatic organisms (Mitsch and Gosselink 2007). Through their bioactivity, oysters transport more phosphorus to sediments than they re-mineralize through metabolism (Dame et al. 1989). Mitsch and Gosselink (2007) stated that phosphorus removal within a system occurs through algal cell absorption and co-precipitation of phosphates in high pH waters. Therefore oysters and algae, both of which have been raised in an aquaculture setting, may provide an economical solution to improving the condition of certain effluents.

One of the most efficient methods for removing excess nutrients in water is seaweed culture. Seaweeds absorb the dissolved nutrients, nitrate and phosphate through their whole plant body. The nutrient absorption is very efficient as seaweeds are immersed and waste no energy for uptake and transport of either water or nutrients (SES 2013).

The most severe consequence of eutrophication on estuarine ecosystems is the depletion of dissolved oxygen (Becker et al. 2008) (Figure 2). Oxygen is consumed during the decomposition of organic matter, resulting in hypoxic and/or anoxic conditions unless dissolved oxygen is replaced. Excess organic matter increases microbial populations which utilize the available dissolved oxygen in order to break down the organic matter. Along with the increase in microbial populations, increases in nutrients from organic matter result in phytoplankton blooms. Phytoplankton cannot produce oxygen at night but instead uses up dissolved oxygen in the system that might otherwise be needed by various other resident organisms. If these conditions are sustained over time this can lead to low levels of dissolved oxygen referred to as a hypoxic condition (NOAA 1998). Low dissolved oxygen levels, specifically less than 5mg/L, can result in large fish kills in estuarine waters with limited tidal or water exchange activities and can have a detrimental impact on various commercially important species (Becker et al. 2008). Boesch et al. (2001) stated "in addition to the obvious requirements for fish and shellfish growth, lack of oxygen also limits nitrification and subsequent denitrification, compounding the effects of eutrophication. Rivers, lakes, estuaries, and coastal areas receiving the nutrient rich water with low dissolved oxygen become impaired and ecosystem health is compromised more often." More often dissolved oxygen is the limiting condition in waters of intensive pond aquaculture facilities and this condition is mostly as a result of poor management and bad planning (Boyd 1998).

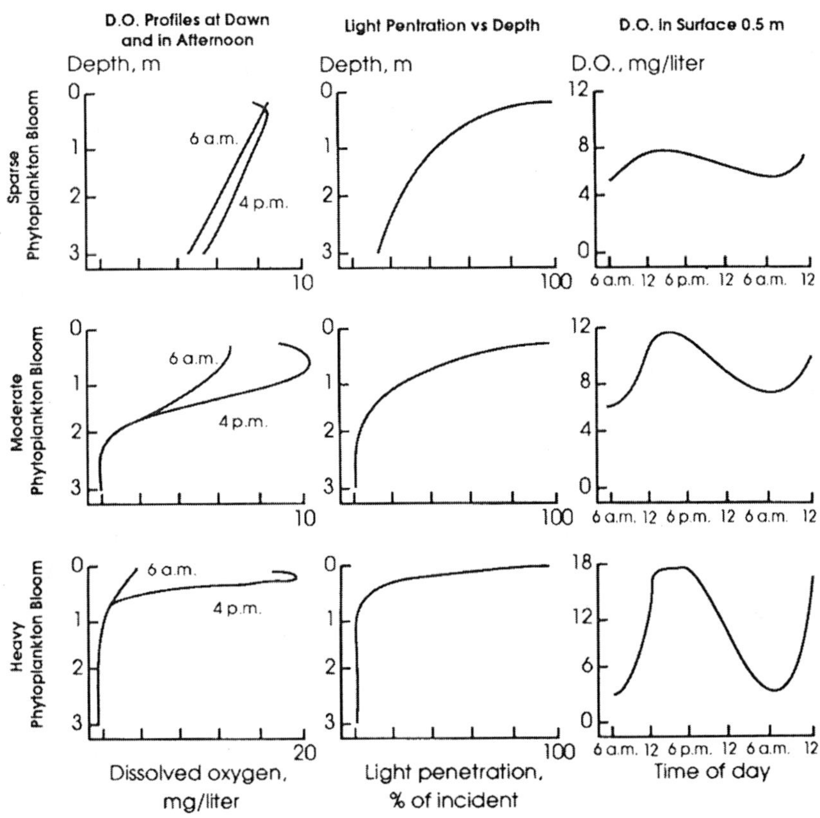

Figure 2. Relationships among phytoplankton density, dissolved oxygen, and light penetration in fish ponds (Boyd 1990).

Boyd and Musig (1992) summarized that the discharge of effluents below the permitted limits are very important. Effluent discharges by one farm may contaminate the water source of another farm downstream. Therefore, if intake water used for filling pond and for water exchanges are highly polluted, water quality problems can occur at even very low feeding rates. Poor water quality in incoming water may increase the risk of disease transfer and intensity of any potential diseases. Pollution load created by aquaculture should not exceed the assimilative capacity of the ponds and water supply of that area. Boyd and Queiroz (1997) stated that receiving stream waters assimilate pollutants through various physical, chemical, and biological processes. As long as the pollution load in the pond effluents does not exceed the assimilative capacity of a water body, adverse environmental changes should not occur.

Boyd (1995a) suggests that the best method to prevent soil and water quality problems is by selecting a site with good soils, an adequate supply of high quality water and to maintain moderate levels of fish densities and feeding rates. Secondary management techniques to prevent soil and water quality imbalances include liming, fertilization and aeration. Sedimentation basins may still needed to be considered to prevent ponds from discharging excess sediments.

Similar to nitrogen, phosphorus and dissolved oxygen, pH, alkalinity, hardness, salinity and ammonia are a few other water quality variables that require constant monitoring in modern aquaculture systems because these variables may become a threat to the habitat in receiving waters (Ozbay 2002). The Best Management Practices section at the end of this chapter describes in detail how the impacts of aquaculture farming are minimized and how striving for sustainability is the key for the long term profitable and environmental friendly farming practices.

2.2. Pathogens and Disease Risks

As we stated previously, aquaculture refers to culture of organisms (animals or plants) under controlled or semi-controlled conditions. In order to be commercially successful, aquaculture establishments generally have to operate at high density and under conditions which facilitate fast growth. Whatever the species or the type of aquaculture operation (i.e. pond, recirculation, aquaponics, and raceways) in question, maintaining good stock health is the key to successfully operating a profitable aquaculture facility (Bowser 2012). Even when present in low numbers, most disease-causing agents including bacteria, viruses, parasites, and fungi can cause problems and have significant impacts on the fish and associated habitat (Bowser 2012).

The presence of bacteria or viruses in the aquaculture system can be detrimental to the overall operation and surrounding environment. As Pietrak et al. (2010) stated, infection and disease can invade from multiple sources of water, wild fish or shellfish, newly-introduced farmed fish or shellfish, contaminated equipment, predators (i.e. birds, turtles), and human visitors. Newly introduced disease-causing pathogens can lead to production loss from mortality, lost marketability of products, and an inability to transport the product to other locations and farms (Pietrak et al. 2010).

Most diseases and related issues can be prevented by using proper management techniques. It is easier and more cost-effective to prevent disease-causing pathogens from entering the systems than it is treating the pathogens after they have already been introduced into the facility (Bowser 2012). AsBowser (2012) stated, maintaining optimum water quality conditions and keeping the facility clean and well organized are some of the key factors to reduce various stressors which fish are exposed to and will reduce the likelihood of a disease problem. Water quality problems listed as critical are: temperature, dissolved oxygen, pH, alkalinity, hardness, un-ionized ammonia-NH_3, nitrite, and potentially toxic substances including heavy metals, drug residues, pesticides, and CO_2 (Bowser 2012).

Limited to the few intensively studied commercial aquaculture species, there is currently a large gap in our knowledge concerning diseases associated with other species with potential commercial and ecological importance. Included in this group are Enterobacteriaceae and fecal *Streptococci*, which threaten swimming beaches as well as wild fauna and can easily spread and persist in natural environment (Figueras et al. 2000). Viruses are a special concern in non-native stocks, where introduced species and hybrids may bring new viral strains into an area. Pathogenic bacteria such as ones belonging to the genus *Vibrio* have caused devastating disease outbreaks in shellfish larviculture (Thompson et al. 2004). These outbreaks resulted in substantial financial losses for commercial hatcheries and culture facilities (Austin 2010).

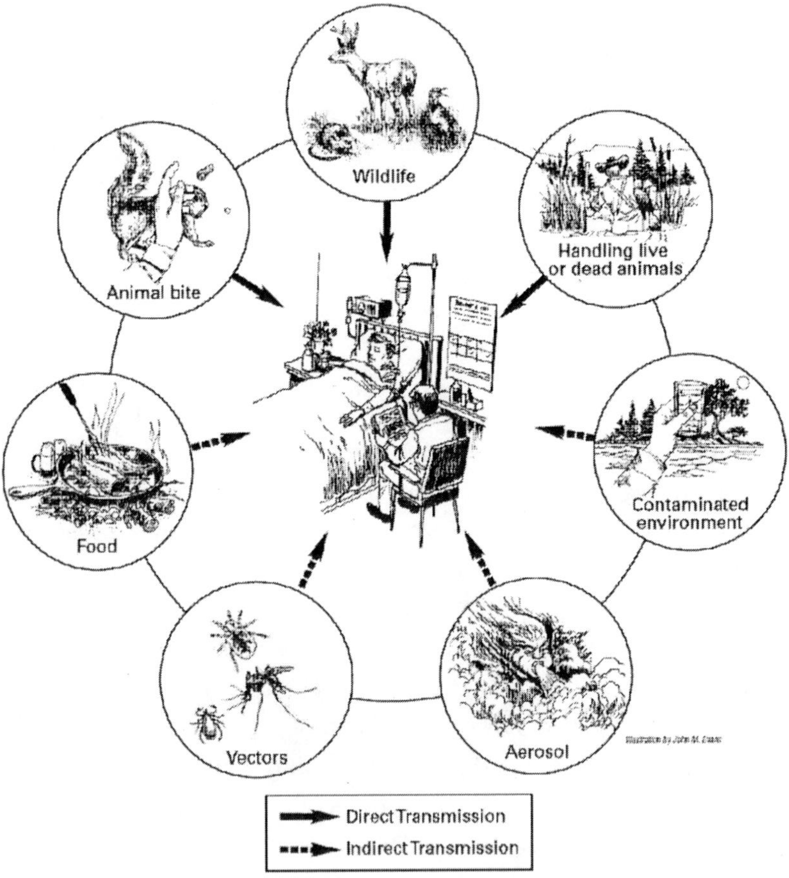

Figure 3. Common routes for potential transmission of infectious diseases and how they are transferred from animals to human and water to human and animal and others (Friend 2006).

Numerous *Vibrio* species present in the aquatic environment are also common human pathogens including *V. cholerae*, *V.vulnificus*, and *V. parahaemolyticus* and can cause wound infections and gastro-intestinal disease (Austin 2010). Urakawa and Rivera (2006) and Austin (2006) reported other species such as *V.anguillarum*, *V. logei*, and *V. tapetis* as finfish and bivalve pathogens known to cause vibriosis, disease, and in some cases mortality in aquaculture facilities and hatcheries.

Similar to fish and shellfish pathogens, *V. shilonii* and *V. coralliilyticus* are a few *Vibrio* species linked to coral reef bleaching events, having detrimental impacts on the health and biodiversity of highly productive ecosystems (Thompson et al. 2004). Thompson and Polz (2006) reported that *Vibrio* play an important role in nutrient cycling in the aquatic environment by excreting different chitin-degrading enzymes when attached to zooplankton. This could also be devastating to commercial species such as crabs and lobster which rely on chitin exoskeletons. Thompson and Polz (2006) reported *V. cholerae* to occur as a free living form in the water column and attached to zooplankton. The direct relationship between nutrient enrichment via eutrophication and the occurrence of *V. cholerae* in estuarine and coastal environments calls for further investigations (Grimes 1991). Threats of shellfish-borne disease from *V. parahaemolyticus* (Vp) and *V. vulnificus* (Vv) are of significant public health concern in the United States (Baker-Austin et al. 2010). Increased *Vibrio*-related disease incidence and changes in*Vibrio* populations are a likely consequence of changing environmental conditions (Lipp et al. 2002). As Friend (2006) show in Figure 3, humans can be contaminated by eating seafood grown in infected waters. Escaped *Vibrio* from aquaculture can disrupt natural systems and can be a potential threat to wildlife or livestock. Infected wildlife or livestock entering into systems can be a threat the aquatic health as well (Friend 2006)Although our discussion in this chapter is limited to bacterial pathogens, other pathogens causing diseases in fish includes viral infections, fungal infections, water mold infections, such as *Saprolegnia* sp., metazoan parasites, such as copepods, unicellular parasites, such as *Ichthyophthirius multifiliis* (Moyle and Cech 2004).

Preventive measures are the most cost effective and practical ways to minimize disease outbreaks in these types of establishments. The common problems can be avoided by strictly adhering to the following practices: avoiding the movement of animals and farm traffic, having a good background check of the stocks brought into the farm, or certified stocks "pathogen free", utilizing good quality pasteurized feeds and tools to monitor water quality, and keeping good farm records (Pietrak et al. 2010).

Water quality has a direct and vital impact on the transmission of pathogens. Good water quality reduces the risks of transmission and mortality rates. Regardless of outbreak history at a farm, each farm should develop a biosecurity plan and the plan must be adapted to the specific farm and operation, location and culture method, consider existing threats in the area and avoid environmental contaminant risks (Pietrak et al. 2010).

Yanong and Erlacher-Reid (2012) stated biosecurity in aquaculture as the best practice to minimize the risk of introducing an infectious disease into a facility. Likewise biosecurity minimizes the risks where a diseased fish or infectious agents leaves the facility and is able to spread to other facilities and infect other susceptible species. The biosecurity goals they discussed include: animal management, pathogen management, and people management. —According to Yanong and Erlacher-Reid (2012), the main management practice is to obtain healthy animals (eggs, fry, juveniles, brood stocks) and optimize their health and immunity through good husbandry practices. Pathogen management primarily includes prevention, reduction and elimination of pathogens. While preventative practices can be cost-effective and easy to follow through, pathogen reduction and elimination can be very expensive and may cause further environmental and economic damages if the methods fail.

People management practices include educating everyone involved including visitors and suppliers. Well planned and coordinated facility work schedules and periodic worker trainings are the keys to ensure that people follow tight biosecurity plans and keep it in their minds as they complete daily tasks (Yanong and Erlacher-Reid 2012). There are various factors which play important roles for facilitating pathogen entries into a facility, spreading from unit to another, from one species to another in the facility and finally infect the whole facility. These factors depend on the species of concern, their immune status, life stages and susceptibility to pathogens, husbandry practices, and water quality conditions. In addition, understanding further characteristics of a particular pathogen (i.e. biology and life cycle of pathogen, reservoir potential), its survivorship in the facility, on the tools and equipment, application of the approved treatment options, understanding regulatory status and compliance with biosecurity protocols are additional biosecurity measures to minimize disease outbreak risks in a facility.

One application particularly useful for treating and disinfecting pond bottoms is to dry out ponds for one or two weeks, or longer if necessary (Boyd et al. 2012). As Boyd et al. (2012) states, parasites and disease organisms and their vectors survive in areas where puddles and wet areas remain or when the area has constant rain. When those areas cannot be dried they can be treated with burnt lime, calcium oxide, hydrated lime or calcium hydroxide. The purpose of these various chemical applications is to raise the pH above 10 to kill potentially harmful organisms. Boyd et al. (2012) suggested until natural food organisms have re-established, stocking shrimp or fingerling fish in the ponds should be avoided due to the toxicity risks of lime residues. Coagulation with alum, limestone or polyelectrolytes is effective in reducing virus counts (Boyd and Tucker 1998).

A well thought out biosecurity plan is necessary to minimize the potential for catastrophic losses from infectious disease in the facility. Knowledge is the key to understanding the risks associated with disease outbreaks. Knowing your animal, where your fish comes from, water source of the facility, how pathogens may potentially enter, live and persist in the facility, good husbandry practices, diagnostic tools and legal treatment options (Yanong and Erlacher-Reid 2012). Further practices include: having experts aid in the development of a biosecurity plan (production specialist, animal health professional, engineer, scientists...etc.), planning the facility sanitation, disinfection and system management schemes, good water quality monitoring, separating populations by their life stages, good planning of disposal and facility re-arrangement if necessary. Keeping good records of every operation in the farm (i.e. hazardous waste disposal, chemical use, water quality, fish growth and survivorship, feeding, vaccine application) is also critical in maintaining a productive and healthy facility (Yanong and Erlacher-Reid 2012).

Depending on the type of aquaculture operation (ponds, raceways, cages, or recirculating systems) specific biosecurity measures and management practices should be used. Although fundamental practices are generally the same for many biosecurity plans, practices may vary depending on the species cultured, life stage of the animals, pathogen, type of operation and many others listed earlier. The most important aspect of this plan is to prevent disease outbreaks so that economic and environmental risks are reduced. A biosecurity plan, along with the facility sanitation and disinfection practices, is part of the best management practices used in various successful modern aquaculture settings.

3. METHODS OF MINIMIZING ENVIRONMENTAL IMPACTS

3.1. Wetlands

There are various definitions on what wetlands are and what best describes wetlands. Kalff (2002) described wetlands as the transition zones between terrestrial and aquatic systems where the soils are waterlogged for at least part of the year or covered by shallow water, and which are typically occupied by rooted aquatic vegetation (*macrophytes*); not all wetlands are physically connected to lakes or lotic systems. Occupying three times the surface area of lakes, wetlands cover about 8.6 million km^2, or 6.4%, of Earth's land area (Shine and de Klemm 1999). There are tremendous benefits associated with the presence of wetlands (USEPA 2006).Figure 4a shows a healthy wetland and Figure 4b demonstrate the schematic representation of nutrient cycling in the soil-water column of a wetland. Many biogeochemical transformations occur in wetlands and mostly anaerobic conditions exist at the soil water interface. The plants also create an aerobic zone near the roots and different oxidation reduction mechanisms occur in the soil leading to nutrient cycling (USEPA 2008). Within the aerobic zone surrounding plant roots, ammonia is oxidized to nitrate by a process called nitrification; nitrate is then readily diffused into adjacent anaerobic soil. Nitrate is reduced to molecular nitrogen through denitrification, or may be reduced to ammonium under certain conditions through the dissimilatory nitrate reduction process (Ruckauf et al. 2004; Reddy and Delaune 2008). The nitrogen cycle is shown in Figure 4c.

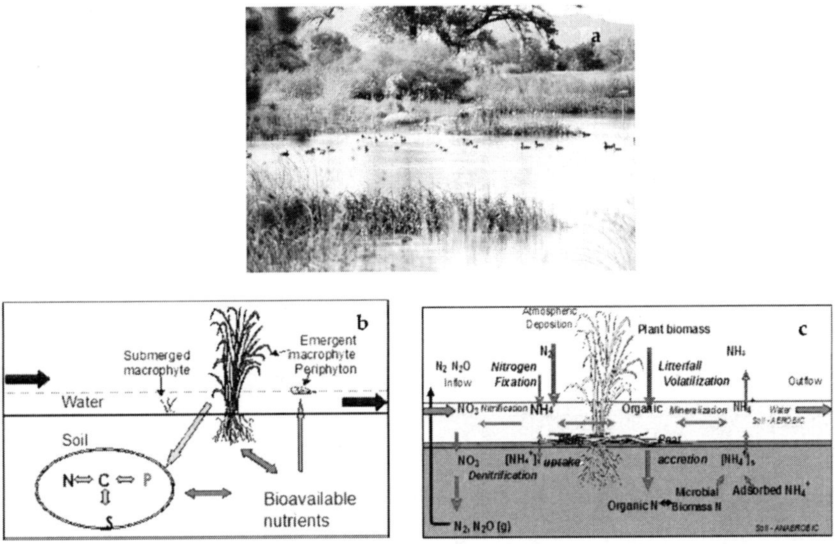

Figure 4. a. Healthy Wetland. Courtesy of http://www.newbedfordguide.com/wp-content/uploads/2011/02/wetlands1.jpg; b. Basics nutrient cycles in soil-water column of a wetland (USEPA 2008); and c. Nitrogen cycle in wetlands (USEPA 2008).

Phosphorous enters wetlands in different forms (PO_4^-, PO_3^-...etc.); both biotic and abiotic mechanisms regulate accumulation and transformation of phosphorous compounds within the water column and soil. Biotic processes include assimilation by vegetation, plankton, periphyton, and microorganisms and abiotic processes include sedimentation, adsorption by soils, precipitation, and exchange processes between soil and the overlying water column (Reddy and Delaune 2008). Transformations of nitrogen, phosphorous, sulfur, iron, manganese, and carbon occur in the anaerobic environment and are mostly microbial mediated. Transportation and translocation of transformed constituents occur in the oxidized layer, providing a barrier to translocate some reduced constituents (USEPA 2008).

The value of wetlands for flood control, water storage, and water purification are estimated to be $15,000 ha/year (Kalff 2002). Their value as for fish and wildlife habitat, recreation, or maintaining biodiversity must also be considered (Mitch and Goselink 2007).

Unfortunately, half of these wetlands are disappearing and being converted for agricultural uses such as rice monoculture and aquaculture (Kalff 2002; Figure 5). Many nations restrict development in the wetlands because of the ecological value placed on the wetlands (Boyd and Tucker 1998). Considering their significant roles in removing excess nutrients, breaking down harmful metals and toxic substances via microorganisms living in soil, preventing soil erosion controls, capturing solids in flowing waters, providing habitat for many wildlife species, many countries insist restoration of degraded wetlands or the mitigation the creation of the new ones (Mitsch and Gosselink 1993). Mitsch and Gosselink (1993) stated that natural wetlands and constructed wetlands are very effective in reducing nutrient and organic matter concentrations in wastewater. Wetlands act as biological filters by removing suspended minerals and organic matters from water. Natural and constructed wetlands can be used for treating agricultural, municipal, and industrial wastewaters (Moshiri 1995). Boyd and Tucker (1998) describe the removal processes of suspended minerals and organic matter from water by a wetland as: sedimentation of suspended particles, filtration of suspended particles by plant materials, uptake of nutrients by plants and bacteria, decomposition of organic matter, denitrification, nitrification, and adsorption of ions by the soil. Macrophytes in wetland systems play a key role as substrate for periphyton and actively transport oxygen to the rhizosphere, which serves to facilitate chemical transformations in the sediment (Schwartz and Boyd 1995).

Schwartz and Boyd (1995) passed pond effluent through constructed wetlands which drastically reduced concentrations of potential pollutants in channel catfish effluents. Concentrations of total settleable solids, total suspended solids, and total phosphorus were reduced 50% or more by the constructed wetlands except for total ammonia nitrogen due to low hydraulic residence time (HRT) of the wetlands in their study. The greatest removal of total phosphorus (TP), 84% and nitrate-nitrogen (NO_3-N), 75% were obtained in wetlands with a four-day HRT (Schwartz and Boyd 1995).

Passing water through wetlands was more effective in removing pollutants than simply holding water in the ponds in their study.

Figure 5. a. Degraded mangroves in Vietnam, courtesy © EJF/Thornton; and b. Shrimp Farm in South America courtesy WWF (http://www.worldwildlife. org/cci/aquaculture_photos.cfm)

There are two basic types of ponds used to raise channel catfish; levee ponds and watershed ponds (Boyd 1985). Levee ponds discharge little water following rains because of their limited watershed area. However, watershed ponds discharge larger water volume following heavy rains due to their larger watershed areas (Schwartz and Boyd 1994a). Most channel catfish farming is conducted in levee ponds where ponds consist of the inside slopes and tops of levees, resulting in high seepage rates especially from rain during the winter. Watershed ponds are usually located much farther apart than levee ponds, so it is typically not feasible to transfer water between ponds for reuse. Boyd and Tucker (1998) suggested that large wetlands could be used to treat effluent when ponds are drained. A smaller wetland could be used for treating the last 25% of highly concentrated effluent from watershed ponds. In a study by Shpigel et al. (2013), the authors demonstrated that nitrogen, phosphorous, and total suspended solids were efficiently removed using *Salicorna* as a biofilter within a constructed wetland. In another study by Lymbery et al. (2013), wetlands removed 60-90% of total nitrogen loads and at least 85% of total phosphorus, and orthophosphate loads from the aquaculture effluent.

Some advantages of utilizing wetlands in the wastewater treatment process include the elimination of chemical treatments, an inexpensive construction process, and wetlands contribute to wildlife habitat and plant communities and to local hydrologic processes. Therefore, using natural wetlands for aquaculture should be minimized to prevent them from disappearing (Schwartz and Boyd 1995; Kalff 2002). Because of the need for large areas, concern arises over the feasibility of using wetlands for treating aquaculture effluents (Schwartz and Boyd 1995; Boyd and Tucker 1998). Integration with pond effluent management procedures might reduce the area of wetland needed for treating fish farm effluents (Schwartz and Boyd 1995). One of best management options allows for the maximization of fish production while maintaining a good pond environment with minimal impacts on the adjacent coastal system including maintaining good stocking densities to improve food assimilation efficiency in a biogeochemical energy model (Serpa et al. 2013).

3.2. Settling Basins and Retention Ponds

Settling basins can be built to remove turbidity and suspended solids from pond water supplies. Sediment ponds should be fairly deep to minimize land requirements and to provide hydraulic residence time. In general, a hydraulic residence time of at least 6-8 h is necessary but 2-3 days of retention is preferred (Boyd 1995b). Preliminary sedimentation studies on catfish pond effluents suggested that settleable solids and total phosphorus could be removed as effectively in settling basins as in wetlands (Seok et al. 1995; Boyd et al. 1998). Sedimentation can reduce biochemical oxygen demand by 40 to 50% (Boyd et al. 1998). Schwartz and Boyd (1994a) obtained information on the quality of effluent released from channel catfish ponds during pond draining and fish harvest in watershed ponds. The concentrations of total nitrogen, ammonia nitrogen, soluble reactive phosphorus, total phosphorus, and biochemical oxygen demand started increasing as early as the seining phase (Boyd et al. 2000).Schwartz and Boyd (1994a) suggested that the best way to minimize the pollution potential of aquaculture pond effluents is to harvest ponds as quickly as possible and not to discharge water during seining or to discharge fairly contaminated water into a settling basin or retention pond (Figure 6).Cathcart et al. (1999) suggested that harvesting catfish during late summer/ early fall can significantly decrease effluent discharge from the production ponds due to low water level. This may apply to other species cultured in the ponds such as shrimp and tilapia; however this practice may not be the right fit for other culture systems. Boyd and Musig (1992) found that settleable solids were seldom present in measurable quantities in effluents discharged at shrimp harvest, as seines are not used.

Figure 6. a. Settling basin. Picture courtesy of Auburn University; b. Constructed wetlands in Mississippi (**Tucker 2009**) (BMP No. 6 2002).

The maximum instantaneous settleable solids rate allowed by the EPA (1979) is 1 ml/L for 30-day average and 2ml/L for daily maximum (USEPA NPDES 2010). Boyd and Tucker (1998) found that the effluents from catfish ponds might contain settleable matter higher than the allowable limit, and have a moderate oxygen demand at harvest. Boyd et al. (1998) demonstrated that a settling time of 8 hours sufficiently reduce total suspended solids and total phosphorus by 75%, and turbidity and 5-day Biochemical Oxygen Demand (BOD_5) by more than 40% in catfish pond effluents. Teichert-Coddington et al. (1999) studied the final effluent from draining shrimp ponds to settling ponds, and obtained near maximum sedimentation of most variables within 6 hours residence, with a removal of 88% of total suspended solids, 100% of settleable solids, 63% of BOD_5 and 55% of total phosphorus.

Using separate settling ponds to treat aquaculture effluent can be a problem on commercial fish farms because of the land requirements and construction costs. Production ponds can be utilized as settling ponds, but this would result in a loss of production capacity. Seok et al. (1995) suggested holding the last fraction of pond water for several days in production ponds before discharging it to the environment as a practical way to allay effluent impact.

The characteristics of solids in pond effluents have been studied to provide information for designing and operating pilot sedimentation basins to test their efficiency for treating pond effluents (Ozbay and Boyd 2004; Ozbay and Boyd 2003a; Ozbay and Boyd 2003b; Boyd 1999). Recommendation is made to lower water 25% of its full volume and settle pond effluents for minimum of 2 to 4 hours and more if necessary to remove over 90% of settleable matters, 75% of total suspended solids and over 50% phosphorus loads in the nearby pond used as a settling basin. Cathcart et al. (1999) studied the reduction of effluent discharge and groundwater use in catfish ponds in Mississippi. Deepening the settling ponds receiving overflows from adjacent production ponds reduced the effluent discharges of ponds by 40 – 90%. Hargreaves et al. (2003) summarized in a SRAC Report that over 50% of total suspended solids, total nitrogen, total phosphorus, and biochemical oxygen demand are related to particles less than 5 micrometers in diameter. Boyd (2000) suggested from the estimate of runoff from watershed studies that settling basins used to treat storm runoff from typical watershed type catfish ponds would need to have volumes of 30 to 40% of pond volume in order to provide a retention time of 8 hours. Thus, because of the large volume required, settling basins do not appear to be a feasible solution for treating storm runoff. Settling basins for treating intentional discharge for partial or complete draining would need to be around 10 to 20% of the volume of the largest pond on the farm (Boyd 2000).

Among the frequently applied practices for treating pond effluents such as coagulant application, water exchanges, and settling basins, many farmers advocate settling ponds for effluents. Even though settling has certain benefits in removing solids stirred into water during catfish harvesting, at other times, nutrients and organic matter in effluents are likely to be phytoplankton or dissolved substances, which do not settle easily (Boyd 2000). Schwartz and Boyd (1996) suggested that after seining, the last 25% of effluent water can be held in the pond for two to three days (depending on the farm operation and timing) to allow solids to settle before they are drained completely. This reduced the discharge of settleable solids by 96%, total nitrogen by 74%, and total phosphorus by 69% and organic matter by 59%. This level reduction is very effective but may not be feasible considering limited space availability in most aquaculture farms. Settling basins are not recommended to treat storm runoff of watershed type ponds because of the large volume of pond water required to reach desirable effluent qualification with a retention time of 8 hours (Hargreaves et al. 2003). Lutz et al. (2011) suggest additional buffer strips to allow plants to pick up excess nutrients and allow water to further slow down before it reaches any downstream creeks. Table 2 provides application discharge data on the wastewater treatment plant and recommended maximum daily loads on water quality parameters as main concerns to EPA (USEPA NPDES 2010). Depending on the type of operation and inflow or existing water conditions, outflow water quality parameters are recommended not to exceed the concentrations provided in the table 2 for NPDES Permit.

According to Tucker and Hargreaves (2008), uneaten feed and fecal wastes are the primary producer of solids that potentially degrade environmental conditions at a farm. Solid accumulation can deteriorate the conditions in a facility and create a threat to the aquatic species cultured. Solids can damage fish gills or block shellfish from filtering and increase dissolved oxygen demand due to increased microbial activity in the accumulated organic materials (Tucker and Hargreaves 2008). Excess phosphorus and nitrogen in the sediment with high solids accumulation has a drastic effect in receiving water bodies causing eutrophic conditions (Tucker and Hargreaves 2009). As the most frequently applied tool for solids removal of pond effluents, settling basins or retention ponds are used to mitigate aquaculture effluent or overflows (depending on size and availability) where the water from ponds or other types of aquaculture facilities are treated by the natural processes to minimize or eliminate pollutants (Setty 2013). Particles settle if given enough time by gravity and microbial community which break down excess nutrients and other pollutants into a less harmful or harmless form (Setty 2013). Coupled with other practices suggested for the best management practices, settling basins and retention ponds remove a significant portion of contaminants and excess nutrients as we discussed in this section and have been recommended for pond aquaculture facilities (Boyd and Tucker 1995).

Table 2. Application discharge data (EPA NPDES 2010).

Parameter	Units	Discharge Data[(1),(2)]	
		Maximum Daily Discharge	Average Daily Discharge
Flow	MGD	--	0.06
pH	Standard Units	7.7-8.2 (min-max)	
Biochemical Oxygen Demand, 5-day (BOD_5)	mg/L	4.9	0.6
Total Suspended Solids (TSS)	mg/L	25	7
Ammonia (as N)	mg/L	6.76	4.26
Total Residual Chlorine	mg/L	0.03	0.01
Nitrate and Nitrite N	mg/L as N	0.68	0.35
Total Dissolved Solids (TDS)	mg/L	553.00	496.00

3.3. Physical Amendments

Sedimentation and filtration are two of the most commonly used particle removal techniques in aquaculture. The applications of this technology have become priceless because untreated effluents or discharges may pose a threat to the environment by carrying various materials in excess quantities including soils, nutrients, and minerals (Ozbay 2002). According to Ebeling and Vinci (2013), total suspended solids (TSS), settleable solids, 5-day biochemical oxygen demand (BOD_5), and total phosphorus (TP) are the four major pollutants found in aquaculture effluents/discharge. These pollutants are regulated to ensure that their concentrations can be minimized through the removal of solids containing feces and uneaten feed.

It is important that pollutant concentrations associated with specific particle size ranges in the effluents are considered during the physical removal stage and knowledge on the characteristics of these particles make the removal process more successful (Ozbay and Boyd 2003a). Water quality requirements are frequently discussed but physical characteristics and particle size distribution of the pollutants in the water are not known well. Analytical technology including size fractionation using sieves, laser diffraction, size fractionation using membranes, and characterization using the Coulter registered method have all improved and have been applied in different industries depending on the effluent and discharge characteristics of the particles in question (Cripps 1993). Boyd (2000) reported that about 40% of total suspended solids (TSS), total phosphorus (TP), total nitrogen (TN), and biochemical oxygen demand were associated with particles 51 µm or larger in sizes in catfish ponds.Table 3 shows differences in concentrations of water quality parameters of the pond effluents before and after filtration (Ozbay and Boyd 2003a).Cripps (1995) found that aquaculture wastewaters typically have low TSS concentrations, compared to various industrial and municipal wastewaters, and numerous small particles which clog the 45 µm membranes used to filter the solids out of suspension in aquaculture waters. Suitable treatment techniques should separate particles from the primary effluent flow. Cripps (1995) indicated that by using filters with pore sizes ranging between 200 – 5 µm, increased treatment effects were achieved through sequential decrease in pore size, which removed more nutrients. It is important for a treatment process to remove relatively larger particles, resulting in a reduction in nutrient loading of the effluent.

Cripps (1995) found lower concentration of TN than TP after filtration, and an increase in the filtration effort reduced both nutrients. Most of the plankton/ solids of eutrophic pond waters, with over 50% of the TSS, are found in particles smaller than 10 μm. Ozbay and Boyd (2003a) found that removing particles down to a very small size provided the required benefits targeted for the pond effluents. They recorded percentage removal of total phosphorus (TP), total nitrogen (TN), 5-day biochemical oxygen demand (BOD_5), and particulate organic matters (POM) associated with total suspended particles (TSS) removal using filters of different pore sizes (41, 30, 20, 10, 8 and 5μm). Most water quality parameters except total nitrogen were substantially reduced except nitrogen after the effluent water was passed through the filter with 41μm pore size. Further removal was achieved with the consecutive filtration using filters with smaller pore sizes. They only monitored a noticeable reduction of total nitrogen with successively finer filter sizes.

Time required to remove different size particles was also studied by Ozbay and Boyd (2003a) and they suggested a settling time of 24 hours to remove about 30% of TSS and TP and 35% of POM, 25% BOD_5 and 20% of TN. Considering the length of time, they do not recommend using settling basins for treating storm overflow and pond draining effluent. In their later study, Ozbay and Boyd (2003b) used turbidity and found it was strongly correlated with total suspended solids and inorganic suspended solids. The relationship was stronger between total suspended solids and inorganic suspended solids as compared to total suspended solids and particulate suspended solids due to the fact that fluctuations in phytoplankton concentration over time changes particulate organic matter concentration.

Ackefors and Enell (1994) found the majority of the phosphorus from fish farms is bound to the particles while nitrogen is not bound to the particles but more in a dissolved form in water. Similar to the phosphorus most of the biodegradable organic matter producing biochemical oxygen demand was in the particles in their study. Cripps (1992) studied the distribution of total phosphorus and total nitrogen in six serially filtered aquaculture effluent samples. Only the fraction containing particles smaller than 5 µm pore size added a disproportionately high nutrient load to the effluent in his study. Reduction rates of 60% for total phosphorus and 34% for total nitrogen were achieved using 5 µm pore size filter. Total nitrogen concentrations were greater in smaller particles than the large particles in his study. However,Cripps (1992) found 69% of the total phosphorus was associated with particles larger than 45 µm diameter. In the effluent the majority of suspended particles produced by the farms were within the size range of 4 – 120 µm diameters. Depending on the particle characteristics and farm effluent, excess nutrient removal can be achieved using the correct diameter in filtration. Cripps (1995) summarized the distribution of the particles for each successive size group and found phosphorus levels depend on the particle size distribution in the pond effluent. Removal of the relatively larger particles separated by the filters had little effect on the size distribution; hence changes in mean diameter were small but the effects were consequential. The phosphorus content in both suspended solids and particles increased significantly with decreased particle sizes (Cripps 1995). Bergheim et al. (1991) used screens with 200 µm or less pore sizes to remove particles. He found further reduction of phosphorus by using the screen with the filter pore size smaller than 5 µm produced negligible results, and in practice he found it difficult to implement. However, the phosphorus content of smaller particles (based on the total phosphorus concentration of water before the effluent was filtered) was significantly greater than larger particles. This difference may appear small but actually represents a large difference in filtration effort (5-200 µm pore size).

Particle size analysis, if used in conjunction with other techniques such as fractionation and nutrient analyses, can be used for the characterization of aquaculture wastes and for monitoring the improvement in wastewater treatment efficiency (Cripps 1994). Membranes can be used for fractionation; however these techniques on their own are limited in practical application. But when combined with other forms of analyses, such as nutrient concentration studies in a known volume of water, the determination of proper treatment techniques is simplified (Cripps 1996).

Table 3. Percentage of total suspended particles removed by filters of different pore sizes (41, 30, 20, 10, 8 and 5μm). Percentages of total phosphorus, total nitrogen, 5-day biochemical oxygen demand, and particulate organic matters are removed with the removal of total suspended solids (**Ozbay and Boyd 2003**a).

Filter Pore Size	Average Percentage Removal (cumulative)				
(μm)	TSS	POM	TP	TN	BOD_5
41	22.5	28.8	21.5	12.9	22.9
30	28.0	30.7	27.2	14.0	24.3
20	32.0	34.9	28.7	17.8	25.4
10	34.5	35.4	31.0	18.5	28.1
8	38.7	40.1	36.0	22.4	33.8
5	47.9	51.6	37.9	23.5	34.0

Although we provided a detailed overview on the feasibility of using filters with various pore sizes, specifically the effectiveness of filters with 5 μm or smaller pore sizes, for the purpose of removing phosphorus and nitrogen, sedimentation is probably the most practical application to remove the large particles in the effluent before further filtration is applied to remove the particles bound to smaller particles which cannot be effectively removed via sedimentation. Sedimentation is discussed in detail in the settling basin section of this chapter. Commonly, screens are placed in front of pond discharge areas to prevent fish, leaves, twigs, or other large debris from escaping in the pond effluent.

3.4. Chemical Amendments

The smaller particles of colloidal clay settle slowly and they may impart unwanted turbidity to pond water (Boyd 1998). Some of the chemicals applied to aquaculture ponds to remove this undesired turbidity in pond waters include coagulants like alum (aluminum sulfate), ferric chloride, gypsum, lime and polymers. Although using organic matter to reduce turbidity has advantages from an environmental standpoint, this is difficult to obtain and apply to ponds (Boyd 1990). Coagulants are most often added to alter the physical state of dissolved and suspended solids, thereby facilitating their removal by filtration and sedimentation (Boyd and Tucker 1998; Pepper et al. 1996) (Figure 7). Coagulants destabilize colloids, thereby permitting suspended particles to form aggregates that can settle out of solution. Coagulation with alum, limestone or polyelectrolytes is very effective in removing suspended matter and phosphorus from water (Boyd and Tucker 1998).

Figure 7. Application of the chemical amendment, lime, to an aquaculture pond in Auburn, Alabama (Photo courtesy of Ozbay).

Boyd (1995) demonstrated that phosphorus precipitates from pond water as insoluble iron, aluminum, or calcium phosphates. Alum and ferric chloride are commercially available sources of aluminum and iron. Aluminum, calcium or iron based coagulants added to poultry litter reduced soluble phosphorus concentrations (Moore and Miller 1994). Gypsum (calcium sulfate) is a source of calcium, however, it is only suggested for use in low alkalinity waters (Boyd 1990). Boyd (1990) observed that alum treatment caused almost immediate flocculation of clay particles, and a great reduction in turbidity within 2 hours in all the treated ponds. However, application of alum produces a strong acidic reaction in water and its use should be limited. Boyd (1998 and 1995) suggested alum for pond treatment if alkalinity is 50 mg/L or above in the water. Alum can remove organic particles and clay colloids in association with phosphorus in water through coagulation and entrapment. Generally, aluminum precipitates with inorganic phosphorus as aluminum phosphate compounds (Gensemer and Playle 1999). Welch and Cooke (1999) applied alum to the surface of eutrophic lakes at rates ranging from 5.5 to 10.9 g Al/m^3. They found a 50% decrease in total phosphorus and chlorophyll a concentrations.Cooke et al. (1993) investigated phosphorus removal in order to control algae blooms by using salts of iron, aluminum or calcium (ferric chloride, aluminum sulfate or calcium hydroxide). They reported aluminum salts as being most frequently used in lake restoration. Welch and Cooke (1999) observed decreased in cyanobacteria bio-volume after treatment with alum in unstratified lakes. Jacoby et al. (1994) found the magnitude and blooms of cyanobacteria reduced after 2 consecutive years of alum and sodium aluminate treatments in a polymictic lake. Phosphorus, total phytoplankton, and chlorophyll a concentrations in hypereutrophic lakes were reduced in 3 years following liquid alum, 10 mg Al/L treatment. However, Rowan (2001) found that application of alum at the rate of 50 mg/L immediately after seining resulted in a somewhat greater removal rate of some pollutants during the first hours of settling, but did not result in significantly improved water quality. She suggested using alum after the first hour of settling from seining, and higher application rates of alum would have been necessary to precipitate significant amounts of phytoplankton. Masuda and Boyd (1994) used alum as low as 20 mg/L concentration for catfish pond water, and found significant removal of Soluble Reactive Phosphorus (SRP) in the pond water. They reported no residual effects of alum treatment if used at low concentration (20 mg/L). Boyd (1995a) noted that to increase the amounts of solids removed from water for shrimp ponds utilizing alum would necessitate alum treatment of water in a settling basin. In most situations, settling ponds may be adequate to remove suspended solids (Boyd 1995a).

In ponds if acidity results from increased carbon dioxide and exchangeable aluminum in soil after chemical treatment, total alkalinity and total hardness concentrations are buffered by the applications of agricultural limestone, burnt lime, and hydrated lime (Boyd 1995a). Liming is applied simultaneously to neutralize H^+ ions, and eliminate or reduce the risks associated with alum toxicity.Masuda and Boyd (1994) used agricultural limestone and alum in catfish pond water in order to reduce nutrient concentrations. Twenty mg/L alum treatment reduced soluble reactive phosphorus, 80%; total phosphorus, 50%; and turbidity, 45% in their study. Precipitation of phosphorus after calcium hydroxide was rapid and higher than agricultural limestone. Calcium carbonate or calcium hydroxide treatments were also applied to hard water lakes by Prepas et al. (1990) and they reported significant decreases in the concentrations of total phosphorus and chlorophyll-a, resulting increased calcite precipitation and higher phosphorus binding affinity to the sediments. On the other hand, Salonen and Varjo (2000) applied gypsum to a hypereutrophic lake and observed that the treatment reduced the chlorophyll-a concentration. Masuda and Boyd (1994) suggest agricultural limestone or burnt lime in removal of Soluble Reactive Phosphorus (SRP) in the ponds. Schwartz and Boyd (1996) suggested that application of hydrated lime or quick lime to pond bottoms enhances ammonia volatilization, and kills pathogens, and they should not be used very frequently because of their inhibition on microbial activity.

Gypsum has a neutral reaction in water but it has been the least effective of the three (alum, ferric chloride, and gypsum) coagulants used in removing clay turbidity (Boyd 2000; 1990). However, Boyd (1995) suggested that gypsum treatment is better for use in low alkalinity waters because gypsum is a good source of calcium, and is more soluble than liming materials (agricultural limestone, burnt lime, and hydrated lime). Masuda and Boyd (1994) reported drastic decrease in SRP concentration, and lower phytoplankton concentration when calcium concentration was elevated with gypsum application. The effects of gypsum treatment on water quality in sunfish ponds with high alkalinity and low hardness conditions were studied and the gypsum treatment reduced phosphorus concentrations and phytoplankton abundance (Wu and Boyd 1990).

The effectiveness of several different compounds to immobilizing soluble reactive phosphorus found in soil from constructed wetlands was studied by Ann et al. (2000). They found that ferric chloride had immobilized the highest percentage of phosphorus in comparison to other amendments; alum, $Ca(OH)_2$, calcite, and dolomite. These amendments were only effective if applied at higher rates in their study.Cooke et al. (1993) reported that phosphorus inactivation with iron salts has shown only short term effectiveness, and subsequent failure was attributed to sediment anoxia because phosphorus precipitation with iron salts is possible if the water - sediment column is aerobic. However, Boyd (1995) reported that alum generally is cheaper than ferric chloride for pond treatment, and commercially available.

Ferric chloride is not suggested for frequent use in lake restoration because of the potential effects to redox reactions and relevant changes of pH on the solubility of iron-phosphate compounds. Under anaerobic condition, phosphorus bonded to the hydroxyl complexes of ferric iron is solubilized and released to the solution. Under anaerobic conditions, phosphorus from the sediments will be released to the water column therefore ferric chloride treatment for phosphorus precipitation can only be possible in aerobic condition (Rowan 2001).

Gutcho (1977) summarized the uses of polyelectrolytes and concluded that anionic, cationic, and nonionic polyelectrolytes are practical flocculating and clarifying agents in the clarification of water and sewage treatment. They are used in the removal of solids from various industrial wastes (mining, papermaking industries). Gutcho (1977) stated that polyelectrolytes if applied with ferric chloride are more effective in removal of phosphate and organic solids from municipal and wastewaters. Non-ionic polymer is generally applied to remove algae, diatoms, and organic contaminants in lakes and pond waters. Ozbay (2005) studied the effectiveness of gypsum, alum with agricultural limestone, ferric chloride, and ferric chloride with non-ionic polymer (polyacrylamide) removing excess nutrients and solids in the pond waters. She found alum with agricultural limestone treatment removed turbidity, suspended solids, and phosphorus during the sedimentation of pond effluents used in a laboratory set-up. Her research outcome was confirmed during her field application of alum with agricultural limestone and 1 hour was sufficient to remove most of the pollutants in the ponds. Figure 8 below shows the significant reduction in turbidity, TSS and ISS after alum with agriculture limestone application.

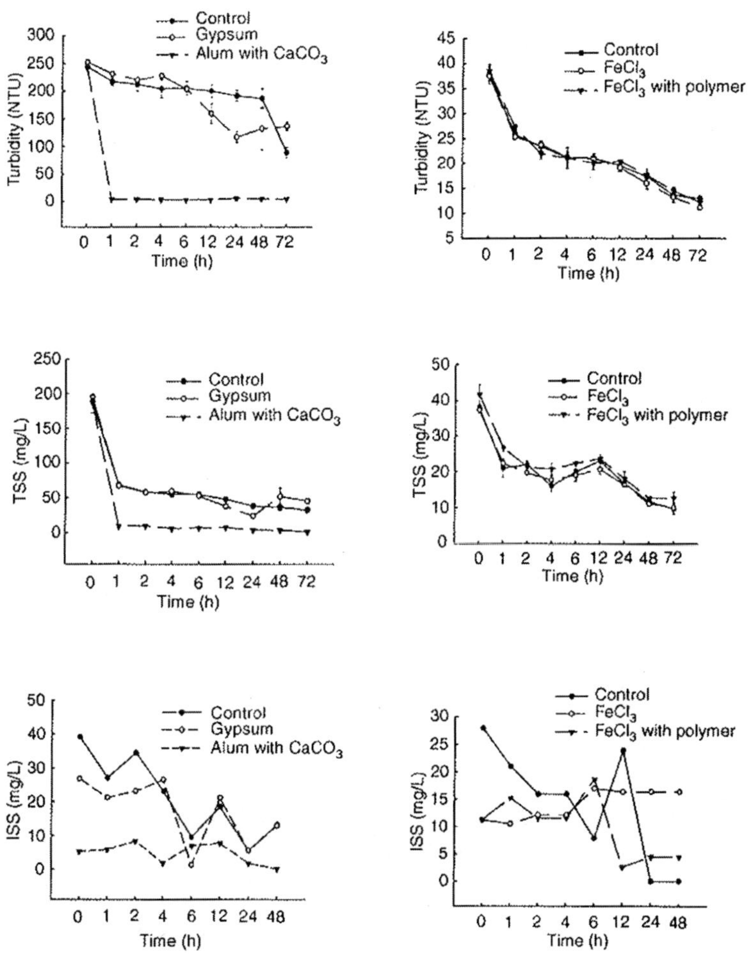

Figure 8. Means (±SE) of turbidity, total suspended solids (TSS), and inorganic suspended solids (ISS) concentrations in alum with agricultural limestone, gypsum, ferric chloride, and ferric chloride with polymer ((polyacrylamide) treated and control over a 72-72 hour sedimentation period, N=3 (Ozbay 2005).

Regardless of the chemicals selected for the aquaculture pond treatments, chemical treatment options should be minimized and carefully selected considering their effects on pond sediment pH and potential to increase solubility of various harmful metals.

3.5. Biological Amendments

Increased awareness concerning environmental issues has been intensified by the constant pressure propagating from agricultural activities and more specifically from aquaculture farming. Many aquaculture farms have invested in alternative methods to minimize their environmental foot prints and keep their operations profitable. Phytoplankton is an important component of estuaries and coastal waters, reaching high population densities and accounting for a large fraction of the particulate matter in these systems (Wright et al. 1982). Considering high organic loading and detritus inputs from aquaculture effluents, further increase in phytoplankton abundance may become detrimental to estuarine health. Fish reared semi-intensively and intensively on formulated diets generate wastes containing organic particles and soluble nutrients. As a result, un-utilized feed and feces generate additional sources of nutrients, which result in higher abundances of phytoplankton (Lin et al. 1998). Phytoplankton blooms have a drastic effect on the water quality in receiving waters of estuaries and rivers.

During the last few decades, many studies have focused on reducing the secondary effects of poor water quality by means of introducing chemicals (copper sulfate, herbicides...etc.) or introducing herbivorous fish species (tilapia, carp, etc.), or introducing filter-feeding bivalves in order to eliminate the problems associated with heavy phytoplankton blooms (Ozbay 2002).

An alternative method which consisted of rearing manila clams to treat the marine fish pond effluents was attempted in Israel by Shpigel and Fridman (1990). The effluents from gilthead sea bream ponds which contained potential edible organic loading were then passed through manila clams in the effluent pond. They improved water quality by using manila clam as the filter-feeding bivalve in their study, and were able to simultaneously produce a high value product, the manila clam itself. Using manila clams to remove the organic loading, primarily phytoplankton, was an effective method to minimize the nutrient loads in pond effluents, and also produce an additional product for the market. Shpigel and Fridman (1990) also found manila clam to be very adaptable to changes in temperature, salinity, and high organic loading (i.e. phytoplankton) making them an ideal candidate for treating intensive aquaculture effluents. They suggested that this type of operation might have the potential for improving water quality depending on the pond conditions and species cultured.

Newell et al. (1999) found that the eastern oyster can exert a top-down control on phytoplankton stocks and also reduce turbidity, thereby increasing light available to benthic communities. Rehabilitation of natural oyster stocks has the beneficial effect of removing phytoplankton from the water column without stimulating further phytoplankton production. Rensel et al. (2011) investigated nutrient and phytoplankton removal and shellfish growth near the salmon pens. They monitored the highest oyster growth near the salmon pens due to food availability caused by the nutrients in fish feces. Although they did not find substantial differences in water quality parameters, phytoplankton was constantly available and removed by the Pacific oysters (*Crassostrea gigas*). Chrzanowski et al. (1986) investigated the ability of an oyster reef community to remove suspended microbial biomass and observed significant reduction in the suspended microbial biomass. Toro et al. (1999) found a significant negative relationship between oyster growth and amount of particulate inorganic and organic matter in water. Higher organic matters increase oyster growth via their filtration of phytoplankton in the organic matter. Ulanowicz and Tuttle (1992) observed that oyster abundance decreased phytoplankton productivity as well as stocks of pelagic microbes, ctenophores, medusae, and particulate organic carbon. Reduction in turbidity resulted from the removal of suspended solids including inorganic particles and phytoplankton by the oysters that oyster filtration plays an important role for increasing light penetration in the water column (Leffler 2001). Miura and Yamshiro (1990) recorded that phytoplanktivorous freshwater bivalves reduce phytoplankton blooms in the water outflow from fish tanks. Lowe et al. (1991) used mussels to increase water transparency in a lake and also observed a shift toward increased densities of benthic algae and recorded an increase in the visibility of water.Senichieva (1990) observed that actively filtrating mussels transform algae and microorganisms into feces and pseudo feces. Santelices and Martines (1986) found that the production of fecal material by filter feeders function as a fertilizer, and stimulated macroalgae growth that provides a venue to the farmers to integrate filter feeders and macroalgae.

Filter feeding bivalves provide a strong venue for the marine finfish farmers to cope with excess nutrient issues that are a result of un-eaten fish feeds and feces. Through their filtration activities, those filter feeders remove phytoplankton which results from additional nutrients introduced to the system. Various commercially and ecologically important species are dependent on oyster reefs for feeding, reproduction, and shelter from predators, including the naked goby (*Gobiosoma bosc*), skilletfish (*Gobiesox strumosus*), striped blenny, (*Chasmodes bosquianus*), and oyster toadfish (*Opsanus tau*) (Marenghi and Ozbay 2010a,b). There is a unique feeding cycle as these resident fishes feed primarily on benthic invertebrates and fish eggs and they also prey on other benthic fishes and will also eat each other while mud crabs (*Panopeus herbstii*) prey upon their eggs (Harding and Mann 2000). Although not oyster reef obligate, there are many other species that utilize oyster reefs including: black sea bass (*Centropristis striata*), northern pipefish (*Syngathus fuscus*), and Atlantic spadefish (*Chaetodipterus faber*) (Harding and Mann 2000). Oyster shells create habitat and serve as spawning substrate for the Florida blenny (*Chasmodes saburrae*), feather blenny (*Hypsoblennius hentz*) and the frillfin goby (*Bathygobius soporator*) (Tolley and Volety 2005). The larger, more transient, bottom-feeding species such as striped bass (*Morone saxatilis*), juvenile summer flounder (*Paralichthys dentatus*), juvenile winter flounder (*Pleuronectes americanus*), spot (*Leiostomus xanthurus*), black drum (*Pogonias cromis*), American eel (*Anguilla rostrata*), and Atlantic silverside (*Menidia menidia*) also utilize oyster reefs for feeding (Breitburg 1999).

Oyster reefs provide nursery habitat for many economically important juvenile species. Posey (1999) discussed why these reefs become important habitat for those species when natural seagrass beds are limited or absent because of environmental degradation. It is important to mention that 10 m^2 of restored oyster reef in the southeast United States is estimated to yield an additional 2.6 kg per year of production of fish and large mobile crustaceans (Peterson et al. 2003). Various ways reefs enhance fish production include increased recruitment, providing refuge from predation, and providing reef-associated prey (Peterson et al. 2003). Because an average size oyster filters 76 liters of water per day, they play a significant role in maintaining natural habitats (The Nature Conservancy 2013). Although aquacultured oysters provide limited but similar services as the natural oyster reefs, they can still effectively remove nutrients and control phytoplankton as they do in nature (Ozbay et al. 2013).

Cultured oysters can serve as broodstock, contributing to enhance and promote naturally occurring populations in the bays. Consecutive research by Marenghi and Ozbay (2010a) and Reckenbeil (2013) found newly settled juvenile oysters within floating oyster gear in man-made, residential canal systems, and on riprap shorelines for the first time around the Delaware Inland Bays (DIB). It appears that the small scale oyster aquaculture for restoration yields hopeful results within the impaired estuarine conditions as more signs of natural recruitment were observed at several locations within the DIB (Marenghi and Ozbay 2010b).

Rice (2008) discussed how biodeposition of filter feeders, such as bivalves, transfer organic nitrogen in phytoplankton and suspended particles in the water to the sediment. Filter feeding bivalves cycle nitrogen and phosphorus which play an important role in maintaining aquatic productivity (Rice 2008). Similar to Rice (2008), Lin et al. (1993) stated that shrimp-bivalve integrated culture systems in Thailand served as a biological control on phytoplankton populations, thus relieving the nighttime BOD stress. Wright et al. (1982) observed bivalve filtration of natural marine bacterioplankton and their reduction in the presence of bivalves.

Boyd and Queiroz (1997) investigated the feasibility of using salt-tolerant plants (halophytes) that were used as crop plants to remove nutrients from the effluent wastewater stream. The plant-soil system sequestered inorganic nitrite and phosphorus, and removed over 94% and 97% of the applied inorganic nitrogen and phosphorus. Wilson et al. (2002)

summarized the use of plankton-feeding fish threadfin shad with channel catfish and monitored improved water quality conditions and enhanced catfish survival in the ponds with threadfin fish. Improvement in catfish production through the use of the Partitioned Aquaculture System (PAS) indicated that PAS offers the potential to eliminate blue-green algal dominance and associated off-flavor problems, while recovering wasted nitrogen and phosphorus discharges, which pose the threat of eutrophication to surface and groundwater supplies (Wilson et al. 2002).

In Yingbin Bay, China, the farmers set a large integrated aquaculture system that is capable of removing excess nutrients. By integrating seaweed and abalone into their main operation for shrimp culture, they were able to improve water quality. Pond bottoms are passed through seaweed and abalone to allow nutrients to be removed before using for shrimp grow-out ponds (Bennett et al. 2012). The authors found that farmers prefer seaweed farming because it reduces financial risks and leads to more consistent profits than shrimp farming however shrimp farming is more profitable for them.

Boyd and Tucker (1998) stated that the grass carp, the common carp, and certain species of tilapia have been evaluated for control of larger plant forms, including filamentous macroalgae. Plankton-feeding fish such as silver carp, bighead carp, tilapias, and gizzard shad are frequently employed in the ponds.Figure 9 shows the pictures of integrated aquaculture farm practices around the world and last picture is the illustration of the multi-trophic aquaculture system.

Tseng et al. (1991) reported that low concentrations of ammonia nitrogen and optimum algal density are better for controlling dissolved oxygen levels in tilapia ponds. Generally, microalgae stabilize pond water quality via either ammonia uptake or oxygen production. Burke and Bayne (1986) studied the effects of paddlefish on zooplankton, chlorophyll *a*, total ammonia nitrogen, and nitrite in a yearling paddlefish-catfish polyculture system. Higher seasonal mean chlorophyll *a* concentrations associated with lower zooplankton densities were measured in paddlefish treatment ponds. Smith (1985) found that filter feeders reduced algal biomass as much as 99%, increased phytoplankton diversity, and improved silver carp growth compared with other ponds without filter feeders, because filter feeders allowed high densities of zooplankters to remain and be available for fish. Fott et al. (1979) introduced carp in whitefish ponds and observed an increase in light penetration while primary production of phytoplankton and small zooplankton concentrations decrease substantially in the ponds.

Figure 9. Pictures of various integrated farm practices around the world; a. tilapia culture with hydroponics herbs culture (http:// land.allears.net/blogs/ jackspence); b. TamilNadu Agricultural University horticulture fish farming integrated system (agritech.tnau.ac.in); c. Malawi fish farm and fruit trees along the edges (http://www.afap.org/ africa_ masasa); d. shellfish macroalgae culture (E & T Magazine, eandt.theiet.org); and e. Integrated Multi-trophic Aquaculture Schema (Government of Canada, dfo-mpo.gc.ca).

During the last few decades integrated aquaculture practices have become a popular method to reduce the nutrient loads and pollutants entering natural waterways, and also increase profits by culturing more than one species of animal and/or plant. Canadian Aquaculture (2012) describes integrated multi-trophic aquaculture systems as the farming of various aquaculture species together where feces of one species serve as the feed to another, as demonstrated in fish /bivalve relationships. This system also increases profits for the farm and decreases its negative environmental footprint. An aquaculture operation consisting of blue mussels and kelps located near pre-established Atlantic salmon aquaculture sites could substantially increase water quality and profits for the farmers in question. We provided a detailed overview on recirculating aquaculture systems and associated aquaponics systems, and their applications for integrated farming practices which ideally will result in economic and ecological benefits in our next sections.

3.6. Feeding and Diet Manipulation

Discharge of unused nutrients in effluents impacts eutrophication and different ecological measures. Impacts from aquaculture feed derived wastes have been observed on the natural environments (Gowen 1991). Boyd and Queiroz (1997) reported that in channel fish ponds in the United States, pond water quality was correlated to stocking and feeding rates. Water quality rapidly deteriorates at feeding rates above 100 - 120 kg/ha per day. In ponds utilized for fertilization and feeding, water quality is related primarily to nutrient input rates. Boyd and Queiroz (1997) stated that part of the nutrients in feeds and fertilizers is recovered in the harvested product, but the remaining nutrients enter the pond ecosystem as inorganic nitrogen, phosphorus, and carbon, dissolved and particulate organic matter. Therefore, relatively small percentages of nitrogen, phosphorus, and organic carbon are recovered in the harvested product. Consequently, the concentrations of nutrients and organic matter in the pond waters and the amount of organic matter settling onto bottoms increase as fertilization and feeding rates increase.

High quality feeds improve feed conversion ratios and reduce quantities of metabolic wastes and uneaten feed (Schwartz and Boyd 1996). Conservative feeding practices, and lower stocking rates also reduces feed inputs and improves feed conversion ratios (Boyd et al. 2000). Feeds are the ultimate source of organic matter pollution in catfish pond effluents

(Boyd et al. 2000). The main types of wastes in aquaculture are residual feed particles, fecal matter, and metabolic by-products. Inefficient feed conversion results in poorer quality effluents and also decreases the concentrations of dissolved oxygen as shown in Figure 10. In a study by Filbrun and Culver (2013), dissolved oxygen levels in the ponds were increased by decreasing the feeding rates. The nitrogen in uneaten feed is transformed to ammonia by bacteria. Ammonia nitrogen tends to increase as feed application to a pond increases, and concentrations above 2 mg/L can be very harmful to aquaculture species at high pH (Gross et al. 1999). Ammonia is also added to ponds through fish excretions.

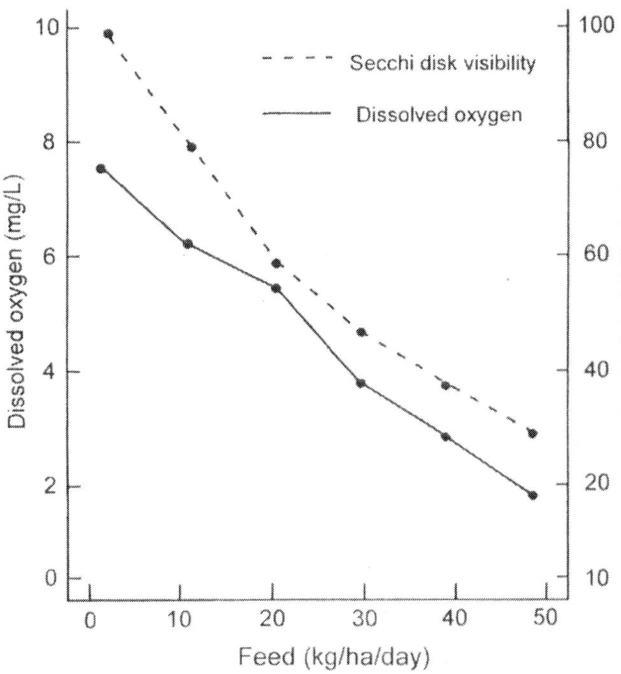

Figure 10. Effect of feeding rate on dissolved oxygen concentrations and Secchi disc visibilities at dawn (Boyd 2001).

Cripps (1995) stated that it is likely that ponds containing specific sized particles would have elevated nutrient concentrations, resulting from their origin in the diet. Abou et al. (2012) demonstrated that using fern (*Azolla spp.*) as a fish meal substitute for Nile Tilapia had tremendously limited phosphorous loss in the effluent and is considered environment friendly. Coloso et al. (2001) found that soluble phosphorus discharge in effluent water can be reduced in fish fed diets that contained little or no fishmeal, or in diets that were supplemented with a low level of dietary phosphorus. In their rainbow trout study, the dietary combination of low phosphorus and high vitamin D_3 decreased soluble and fecal phosphorus levels in the effluents, indicating a strategy whereby effluent phosphorus concentrations can be reduced by regulation of phosphorus metabolism. Increasing bioavailability of phosphorus will eliminate excess phosphorus in the effluent water.

According to Boyd and Queiroz (1997), increasing stocking density and feeding rates above assimilation capacity of pond water can harm the aquaculture pond in the long run. Heavy circulation, aeration, fish respiration and activities, plant abundance, feeding, fertilization, and stocking density in the ponds induced increases in concentrations of potential pollutants, which then require increased treatment efforts to reduce.

Nutrient manipulations were evaluated to promote more desirable phytoplankton communities by eliminating blue-green algae (Wilson et al. 2002). These methods include manipulating the ratio of nitrogen to phosphorus in the water, reducing the availability of phosphorus in bottom muds, enhancing the availability of inorganic carbon, increasing levels of salinity and potassium, and manipulating trace metal availability. Studies showed that various manipulations of waterborne plant nutrients have little promise for controlling phytoplankton community composition in catfish ponds with high feeding rates.Gross et al. (1999) found that the differences in phosphorus input among three feeds, containing 28, 32, and 36% crude protein, did not affect phosphorus concentrations in the effluents because most of the phosphorus from feed and fish excrement is absorbed by the soil. Gross et al. (1998) studied the phosphorus budgets for channel catfish ponds receiving one of five diets ranging from 0.60 to 1.03% phosphorus. They observed that low phosphorus diets did not decrease phytoplankton productivity or improve effluent quality. However, they suggested the use of low phosphorus diets in order to reduce the phosphorus load to bottom soils and to conserve the soils' ability to adsorb phosphorus.

Rangen Inc. (2013, Buhl, ID) provides important feeding tips of some of the commercially important aquaculture species. The tips that are species-specific would also minimize environmental impacts. Although recommended feeding practices differ by species, there are several common tips relevant to all species, including: 1) feeds should be broadcast well to allow all the fish to feed in the pond and minimize feed waste, 2) overfeeding should be avoided all costs to prevent from effluent pollution and gill damages, 3) feeding should be adjusted based on the percent body weight, 4) feed sizes should be adjusted as fish grows, 5) select the right feeding method for the species of interest, 6) feeding should be ceased before handling and shipping, 7) good record keeping is necessary to monitor fish growth and make the necessary adjustments, and 8) good storage and feeding management and feeding technique sanitation should be followed to avoid cross contamination and feed quality issues. In the last few decades, most fish farming has advanced from extensive rearing with few fish, to intensive rearing of high density populations in ponds and raceways. Cost effective good quality fish meals, proper feeding protocol, optimum growth and survival rates are the important goals of any fish farm operation.

3.7. Recirculating Aquaculture and Aquaponics

While many practices in aquaculture/mariculture (e.g., destruction of mangroves) have been criticized in years past for potential deleterious effects on the environment, the extent of any long term destruction due to aquaculture still remains debatable (Boyd and Schmittou 1990). Nevertheless in the United States and internationally, the most important environmental concern facing the aquaculture industry is the disposal of nutrient rich effluent water produced during the culture of aquatic animals (Goldburg and Triplett 1997; Frankic and Hershner 2003). Therefore, as aquaculture moves toward trying to feed the world, there is an inherent need to be stewards of the land, to protect, preserve and maintain conditions favorable to sustainability (Costa-Pierce 2002; United Nations FAO 2009). Recirculating aquaculture and its associated technology has largely developed out of concerns over water conservation and reducing environmental impacts (Martins et al. 2010). Besides growing fish, the purpose of a recirculating aquaculture system is to collect, concentrate, and process animal wastes rather than discharging wastes directly to the environment. Interestingly, an efficient recirculating aquaculture system is designed to reuse 90-99% of the water initially put into the system, while producing only a very small amount of waste or effluent (Chen et al. 2002, Hollingsworth et al. 2006;Badiola et al. 2012).

In their simplest form recirculating aquaculture systems (RAS) are similar to a home aquarium. Both a home aquarium and RAS have many of the same components including a tank or tanks, a pump to move water, some sort of filtration system, lighting, a heater or chiller, and fish. Also like a home aquarium, the RAS environment is very controlled to include lighting and room temperature good for the species of interest being cultured and other conditions in the facility. Fish are fed, water is added or taken out periodically, and water quality is monitored constantly and is often controlled through the addition of certain chemicals such as sodium bicarbonate. Unlike a home aquarium and relative to other types of aquaculture, RAS is very capital and energy intensive. It must rely on economic productivity for profitability, may require several additional components for processing water, and requires above average experienced personnel for successful management (Timmons et al. 2001; Timmons and Ebeling 2007; Ebeling and Timmons 2012).

With proper site selection, an advanced filtration capability, reduced water use and their small footprint, a recirculating aquaculture system lends itself to being a relatively environmentally friendly (Summerfelt and Vinci 2008; Ebeling and Timmons 2012, Losordo et al. 2013). Recirculating aquaculture systems do not rely on surface water for replenishment and with their ability to be located in close proximity to markets, they may be advantageous over other aquaculture systems (i.e., ponds, net pens, open ocean aquaculture) especially when comparing carbon footprint associated with food transport emissions (Martins et al. 2010). However, even with the positive attributes of a RAS, there is potential for it to be harmful to the environment and be considered unsustainable. Recirculating aquaculture systems are often described as the most effective way to grow large quantities of fish in a limited space. Furthermore, with their ability to control the environment they have the ability to grow fish year round (Hollingsworth et al. 2006; Ebeling and Timmons, 2012).

On a large production scale, recirculating aquaculture systems tend to be energy intensive and could be considered similar to other confined animal feeding operations, or CAFO's. In fact, under the 2004 United States federal aquaculture effluent limitation guidelines, recirculating aquaculture systems with an annual production exceeding 45,454 kg (100,000 lbs.) are classified as a concentrated aquatic animal production (CAAP) facility. Operations this large in scale are required to obtain a National Pollution Discharge Elimination System (NPDES) permit before discharging wastes into waters of the United States. Fortunately, the majority of recirculating aquaculture system operations in the United States choose alternatives to discharging effluent into natural waters, and instead either discharge into public municipal treatment works, collect the waste and apply it to nutrient deficient crops on land, or utilize treatment wetlands for processing effluent (Miller and Semmens 2002; Hollingsworth 2006;Summerfelt and Vinci 2008). A NPDES permit can be granted when the development of a facility specific Best Management Practices (BMP) plan specifies how discharges will be reduced of potential pollutants (Summerfelt and Vinci 2008).

The United States has a great deal of infrastructure that allows for regulation of discharge and more specifically, discharge into municipal treatment works, unfortunately the remainder of the world does not have this benefit. If recirculating aquaculture is to be adopted worldwide to raise fish in an environmentally and sustainable fashion, specific infrastructure is required. As previously mentioned, there is a plethora of literature available that describes RAS components and their efficiency at waste removal for large-scale fish culture. However, there is little information on dealing with the actual collected and concentrated solids that are generated from a large scale RAS. This is especially true when looking at recirculating aquaculture on an international capacity. Wetland ponds are often used in the United States and have been suggested on an international level, but wetland ponds have a limited lifetime and this is often a costly option. Another option for RAS effluent management that has been explored in other geographical areas is land application. Valencia et al. (2001) conducted a study to determine if effluent from a tilapia tank system could be used to replace nitrogen on guineagrass (*Panicum maximum*) managed as hay in a water limited area of the United States Virgin Islands. Interestingly, their results indicated that the tilapia tank system effluent could in fact serve as an adequate nitrogen and water replacement for guineagrass pasture, and hay production without changes in soil pH and phosphorous. Moreover, because this study used grass rather than row crops, it acts as a sink (similar to a wetland) with less risk of nutrient loss or leaching to the environment. The use of grass crops for assessing environmentally friendly ways to manage RAS effluent is but one step in the many ways research can explore repurposing and/or disposal of RAS effluent on and international level.

Overwhelmingly due to environmental concerns, but also to increase production efficiency, in recent years Best Management Practices have evolved across a number of industries from car manufacturing to food processing. Plain and simple, BMP's make sense and are a way of reducing multiple levels of risk. Within aquaculture, several entities including the Global Aquaculture Alliance (GAA) have created their own version of BMP's or Best Aquaculture Practices (BAP's). According to the GAA (2011), BAP's address environmental and social responsibility, animal welfare, food safety and traceability all on a national and international level. Through their BAP's the GAA further provides a certification program where they define elements of responsible aquaculture and evaluate adherence to these practices for each type of facility whether it be a hatchery, feed mill, farm or processing plant (www.gaalliance.org).

There are a number of BMP's that recirculating aquaculture system managers can use to make their facilities more environmentally friendly. Best management practices for recirculating aquaculture systems range in scope from choosing the right manager for the facility to using the most efficient types of filtration. Ensuring the use of high quality feeds with fewer fines will reduce nutrient levels and feed conversion ratios.

Incorporating hybrid technology such as bioflocs which help to reduce feed costs and enhancing energy efficiency by using less and reusing energy where possible will all help the economic and environmental sustainability of recirculating aquaculture systems (Miller and Semmens 2002; Summerfelt and Vinci 2008; Hanna et al. 2010; Martins et al. 2010; Badiola et al. 2012). Summerfelt and Vinci (2008) have presented a thorough explanation of RAS BMP's beginning with site selection, working through solids storage, treatment and disposal, and complete facility operation and maintenance. Interestingly, Summerfelt and Vinci (2008) consider the point source waste stream to be the major facility level environmental issue (see also Hollingworth 2006).

Ultimately, for RAS to truly be environmentally friendly BMP's must be incorporated into their everyday function. Agriculture and its water counterpart aquaculture have been scrutinized due to various practices that have been employed over the many years of operation. In recent years, the colloquial buzzword has been "sustainable". You can't speak to anyone, go anywhere, or do anything anymore in any area of agriculture and natural resources without the mention of "sustainable". But with regard to agriculture what does sustainable mean? According to the United States Department of Agriculture, National Institute of Food and Agriculture (USDA-NIFA 2013), sustainable agriculture is an integrated system of plant and animal production practices having site specific application that over the long term will be able to (1) satisfy human food and fiber needs, (2) enhance environmental quality and the natural resource base upon which the agriculture economy depends, (3) it should further make the most efficient use of nonrenewable resources and on-farm resources and integrate, where appropriate, natural biological cycles and controls, (4) sustain the economic viability of farm operations, and (5) enhance the quality of life for farmers and society as a whole.

Aquaponics developed from hydroponics, or the culture of plants with little to no soil. In hydroponics, plants are raised in some sort of structure where the roots are submerged in either water or some type of media base where they are fed via a solution containing all the nutrients (fertilizer, trace minerals, etc.) that they need (McMurtry et al. 1990). Aquaponics, however are virtually the same as any other RAS, except that they use the metabolic byproducts of one crop (i.e., finfish) to produce a secondary crop (plants), thereby adding value by producing two crops instead of one (Rakocy 2012). In turn, the plants in an aquaponics system filter potentially lethal nutrients (nitrite, nitrate) from the water and return it back to the fish culture tank (Losordo et al. 2013; Rakocy et al. 1992). With the mention of sustainability, increasing pressure from environmental groups, governmental regulations and the fact that aquaculture continues to play an ever increasing role in supplying the worlds' food supply it is not surprising that interest in aquaponics has begun to take a foothold with regard to RAS, especially in urban area of the US.

A typical aquaponics system is set up so that water flows from the fish culture tanks (Figure 12a) to a settling chamber, or clarifier where solids are removed from the waste stream, water then enters a biofilter where ammonia in the water from the fish and excess feed is converted to nitrite nitrogen and then nitrate nitrogen. Water then exits the biofilter and proceeds toward the plant component where there may or may not be several other components included (i.e., base addition, degasser). The plant component is either a raceway with floating rafts, or could be what is called an NFT (nutrient film technique). In general, this is where the plants feed off of the nitrate nitrogen before the water returns to the fish culture tank relatively free of nitrogen. In this system the plants receive trace minerals via the fish food; however, there is often the need to supplement with things like iron, calcium, and potassium. (For a complete description of an aquaponics system, refer to Rakocy 2012).

Aquaponics is beneficial for a number of other reasons including that the cycling of the fish water to the plant component in an aquaponics system reduces the amount of concentrated discharge coming from this system relative to other RAS. Also important is that while fertilizers, herbicides, and pesticides may be utilized in or around greenhouses housing hydroponics, these are highly discouraged around aquaponics systems because of the deleterious effects they would have on the fish. Similarly, because an aquaponics system is a form of RAS, the use of antibiotics is discouraged within the system so as not to kill beneficial bacteria that are involved in the natural nitrification process. For these and other reasons, aquaponics systems are considered to be broaching the realm of organic. Organic farming is often considered to be environmentally friendly and sustainable. Unfortunately, the United States Department of Agriculture (USDA) has yet to provide aquaculture with an organic certification. While there are other private agencies worldwide that provide an organic certification for fish, the stringent guidelines provided by the USDA elevate this title to a higher level. Fortunately, in recent years there has been an ever increasing attempt at creating sustainable 'fishmeal reduced' and 'fishmeal free' diets for a growing number of fish species, especially with regard to highly prized carnivorous, saltwater species (see also Rhodes et al. 2013; Watson et al. 2013). The continued development of these diets may ultimately lead to a USDA organic certification for United States aquaculture.

While aquaponics systems are perhaps the most environmentally sustainable form of RAS to date, it does have drawbacks. Like any aquaculture venture, costs associated with initial investment, system components, their availability, construction and operation can have a significant impact on the economic sustainability of a system (Rackocy 2012). Hanna et al. (2010), for example have shown how different managers' management practices can affect the operation of identical RAS. It therefore becomes extremely important for aquaponics/RAS managers to follow Best management practices that will allow for a system to be profitable and sustainable. Best management practices for recirculating aquaculture and aquaponics systems have been described extensively in the literature (Chen et al. 2002; Summerfelt and Vinci 2008,). However, as we move toward trying to feed the world and keep RAS as environmentally friendly as possible, there are many important factors in operating a RAS. With regard to profitability and sustainability, perhaps offsetting initial investment, component and/or construction costs can be achieved by targeting highly sought after plant and fish species. Again, as fish meal free fish diets are developed for highly sought after marine species this reality becomes closer. One farmer at a recent national aquaponics conference suggested dealing directly with restaurants and "setting your price" rather than letting someone tell you how much something is worth (personal communication, National Aquaponics Conference, Tucson, AZ 2013). Similarly, it may also be advantageous for the owner/operator of a small scale RAS or aquaponics system to maximize profit and sustainability through raising high dollar plant and fish species as long as they have an established market (Frankic and Hershner 2003). With the recent surge in "farm to table" interest, it is very apparent that this concept can be profitable and environmentally friendly in the United States.

Ultimately, in trying to keep up with the worlds' population growth and food needs, RAS and aquaponics will continue to play a major role. Costa-Pierce (2002) and others suggest there has to be a behavioral shift in humans rather than technology in order for aquaculture to become truly sustainable. Many individuals only seem to see aquaculture in the sense that we need to produce as much fish as possible in as small an area as possible, however perhaps instead of trying to create RAS that are on the same level as a CAFO we instead look to systems that are sized according to the supporting the local community (Frankic and Hershner 2003). Again, one of the major advantages to a RAS is its' small footprint. By building a RAS with community size in mind it can be sized to feed the community on an ongoing basis. Having the RAS near or within a community would also reduce the carbon footprint by a reduction in fossil fuels needed for shipping etc. Figures 11 and 12 show a recirculating aquaculture system and an airlift aquaponics raft system in the Aquaculture Research and Demonstration Facility at Delaware State University, Dover, DE, USA.

Figure 11. a. An indoor multi-tank recirculating aquaculture system (RAS); b. A parabolic screen filter in a RAS (Photo courtesy of Blank).

Figure 12. a. Construction of an airlift aquaponics system; b. Aquaponics raft raceway showing aeration for plant roots; c. Planting herbs in a floating Styrofoam sheet (Photos courtesy of Blank).

4. BEST MANAGEMENT PRACTICES (BMPS) AND SUSTAINABLE AQUACULTURE

Best Management Practices (BMPs) have been used as an important management tool for various aquaculture practices and management may vary based on the species cultured, type of aquaculture practice, location, surrounding habitat, economy and policy conditions of the area. Although there are differences in the application and level of engagement with BMPs, there are common issues through which the application of BMPs can be applied to all. While there are many types of different aquaculture operations, because worldwide ponds are the most prevalent the recommendation and management practices in this section will focus on the pond management.

BMPs reflect the most technically practical and economically feasible methods to reduce environmental impacts and limit costs at aquaculture facilities. One primary goal is to develop simple effluent treatment systems that reduce organic matter loads, suspended solids, and nutrients in effluents to prevent polluting receiving waters. The best methods to prevent soil and water quality problems include selecting a site with appropriate soils and an adequate water supply, and maintaining moderate fish densities and feeding rates (Hargreaves et al. 2003). Secondary management techniques to prevent soil and water quality imbalances include liming, fertilization, and aeration (Boyd 1998). Agricultural irrigation, created wetlands, settling basins, and biological filters are also practical methods for reducing and improving the quality of effluent from ponds (Setty 2013; Tucker 2009; Ozbay and Jackson 2006).

Countries with regulations on farm practices and effluent standards follow guidelines and permit processes. However, countries without aquaculture regulations can also apply BMPs to minimize off-site water pollution and related environmental impacts. The code of conduct and codes of practice become useful tools for the farmers to adopt and serve as the guiding tool. As described by Boyd (2003) the code of conduct is a system of rules on how aquaculture should be conducted. The guiding principles for responsible aquaculture by the Global Aquaculture Alliance (GAA) serve as the code of conduct. In order to avoid social and environmental problems, codes of practice are used to solve the problems through management activities. Overall the goal of these codes is to minimize, or remove the negative impacts of an aquaculture operation (Boyd 2003). Boyd (2003) describes BMPs as management practices that are the most effective at reducing pollution levels and other environmental impacts which meet water quality or resource management goals.

Although not comprehensive, Boyd (2003) highlights some BMPs for pond aquaculture: a. use stocking and feeding rates that do not exceed the assimilative capacity of ponds; b. avoid overfeeding and apply a strict feeding plan; c. do not use fertilizers unless it is absolutely necessary to promote healthy phytoplankton growth; d. reduce water exchange; e. reuse water as much as possible; f. use a settling basin if available to treat pond effluents before final discharge. Application of these recommendations is based on farm operation and design, species cultured, and culture methods used (Boyd 2003). As Boyd states, selection of BMPs is case specific. One such example includes 10 codes of practice established for responsible shrimp farming by the GAA including mangroves, site evaluation, design and construction, feeds and feed use, shrimp health management, effluents, solids waste and few others. Although one BMP may be sufficient for one small farm, multiple BMPs may be necessary for others. ALEARN (www.alearn.info) listed over 20 best management practices for ponds, raceways, cages, effluents, safety, and others.

Shrimp aquaculture is a rapidly expanding field and is being closely scrutinized by environmentalists and government agencies. Due to the need for saltwater, discharge from shrimp facilities often flows into fragile coastal ecosystems. Problems associated with discharge include eutrophication due to nitrogen loading and detritus, low dissolved oxygen levels, sedimentation, along with other problems (Villalon 1991). These problems however are not without solutions; treatment of effluent should be regarded as an opportunity rather than just an obligation. Our goal is to provide broader perspectives on how basic principles and natural solutions can make shrimp aquaculture longer lived and be more sustainable. We discuss some of the management strategies current shrimp aquaculture operations along with best management practices for reducing potential impacts of shrimp aquaculture.

Although this may not be the ultimate solution, one particular recommendation discussed in this section would be to improve shrimp management practices. The mariculture of shrimp may provide one of the best opportunities for polyculture and integrated systems. Shrimp require higher water quality standards than many other cultured species and, thus, would benefit from a more stable ecosystem. The ability of shrimp to utilize a broad spectrum of the food web would allow them to be cultured with a number of other species. Feed and fertilization management can be geared toward supporting the food web to produce food items which shrimp prefer, rather than relying on the direct consumption of pelleted feed. In addition, Hopkins et al. (1991) discussed how dissolved oxygen levels were higher in polyculture ponds presumably due to a healthier phytoplankton community. Thus, polyculture may actually reduce aeration costs.

Hollingsworth et al. (2006) suggests growers may develop a farm-specific Standard Operating Procedure (SOP) manual and apply the code of conduct for significant farm practices in their SOP manuals. Although not required for all farms, the development of farm specific procedures will promote efficient management decisions including trouble shooting problems, training employees, planning future expansions or developing biosecurity and emergency procedure plans.

Many states in the United States have adopted BMPs over the years and some states and countries have implemented further policies and regulations based on scientific knowledge to sustain the environment and aquaculture industry. One specific example is that of Louisiana's aquaculture producers (Lutz et al. 2011). By implementation and application of best management practices, producers minimize potential pollutants (i.e., mainly excess nutrients) to the state's water resources and by doing so they reduce the cost that would be incurred to treat water quality problems, potential disease outbreaks and wild fish stock mortality related costs. Lutz et al. (2011) suggested that sediment runoff reduction should be one of the most important practices a pond aquaculture farmer must adopt to save money and reduce the environmental foot print of their operation. As an example, in Thailand shrimp aquaculture, scientists and policy-makers have developed new ways to improve the quality of the culture system, ecosystem, as well as the efficiency of regulations. It is critical that advances such as this and many other practices are discussed and maintained with integrity and strong regulations to improve the quality of our shared water resources for future generations. The key is to make aquaculture an asset to the environment while continuing to food production simultaneously.

Initial efforts and guidance on BMPs have been developed by Hargreaves et al. (2003). He has provided guide sheets on various topics and issues of concerns including reducing storm runoff into ponds, managing ponds to reduce effluent volume, erosion control on watersheds and embankments, pond management to minimize erosion, control of erosion by effluents, settling basins and wetlands, feed management, fertilization of catfish ponds, water quality protection to improve effluents, water quality enhancers, therapeutic agents, fish carcasses, general operations and worker safety, emergency response and management, and a few others added as the technology advanced in recirculating, bioflocs and aquaponics systems and other aquaculture operations.

The Best Aquaculture Practices (BAP) standards developed by the Global Aquaculture Alliance (GAA 2011) “address environmental and social responsibility, animal welfare, food safety and traceability in a certification program voluntary for aquaculture facilities.” Certification for BAP ensures aquaculture operation is responsible and operates by the quantitative guidelines by which the farm operation is evaluated based on those practices. There are various standards developed in aquaculture sector including fish farm, hatchery, feed mill, and processing plant. The standard for the multi-species farming opens whole new area of attention with the new aquatic species used in integrated culture condition. Species BAP Standards used include channel catfish, shrimp, tilapia, and *Pangasius* initially and seabass, sea bream, cobia, seriola, trout, grouper, barramundi, perch, carp, flounder, turbot, striped bass, crabs, freshwater prawns, mussels, crawfish recent. According to GAA (2011), the new multi-species farm standards apply to all types of culture systems for finfish and crustaceans but not including cage-raised salmonids since this operation requires separate BAP standards. Seven of the most recent BAP standards listed in the GAA website (http://www.gaalliance. org/bap/standards.php) include Seafood processing/repacking plant, seafood processing plant, finfish and crustacean farm, salmon farm, mussel farm, shrimp hatchery, and feed mill.

The United States Environmental Protection Agency (USEPA 2004) initiated a new rule called the “effluent limitations guidelines (ELGs)’ for concentrated aquatic animal production facilities including aquaculture facilities. This rule is applied to all commercial aquaculture facilities, with the below mentioned specifications, that discharge their wastewater from their farms directly into waters of the United States. According to the final rule, aquaculture facilities that “produce at least 45,360kg a year in flow-through and recirculating systems that discharge wastewater at least 30 days a year (used primarily to raise trout, salmon, hybrid striped bass and tilapia); at least 45,360 kg a year in net pens or submerged cage systems (used primarily to raise salmon).” The whole expectation with implementation of this rule is that the ELGs will help reduce discharges of conventional pollutants, primarily total suspended solids. As the solids are removed, it is expected that non-conventional pollutants such as nutrients will also be reduced. Other contaminants not discussed in this chapter include heavy metals, drug residues and other hormonal chemicals used in facilities to manage fish health and chemicals and better growth and this regulation is expected to be effective for reducing those contaminants in discharges of the facilities. With the implementation of this rule through National Pollution Discharge Elimination System (NPDES) permits the discharge of total suspended solids are expected to be reduced more than 226,796 kg per year, and the biochemical oxygen demand and nutrients in discharge is to be reduced by about 136,078 kg per year. With the application of this rule it is expected that water quality conditions will be improved and provide increased opportunity for other users, swimmers, fisherman and environmentalists concerned about keeping biodiversity in the streams, rivers and estuaries.

There are many definitions for sustainability and sustainability with regard to a catfish farm may not be sustainable for a shrimp farm. For aquaculture, sustainable aquaculture is an ultimate goal with the application of all the best aquaculture standards and management practices. Sustainability is described by the Northwest Earth Institute (2012) simply as meeting the needs of the present without compromising the ability of future generations to meet their own needs" (taken from UN World Commission on Environment and Development, Our Common Future). According to the Monterey Bay Aquarium (2013), environmental impact of fish farming varies depending on the species cultured, location of the farm, life stage of the organism, methods of culture and culture technique. Creating a sustainable farm should ensure species cultured will last long and habitat damage be minimal. The key factor with aquaculture sustainability is to operate with sound environmental management practices in place (FAO FOCUS 2013). There are tremendous efforts being made to use integrated aquaculture-agriculture farming systems to sustain both aquaculture operations and maintain the healthy environmental conditions for aquatic life in rivers, streams and estuaries. Environmentally friendly methods are also beneficial to the species cultured and farm operation (FAO FOCUS 2013). Sustainable aquaculture should utilize the most readily available technology to produce high protein food diets while applying the same exact principle to reduce its environmental impact using similar technology. Sustainability is not a practice, it is a life style and condition we must grasp.

Although it is beyond the scope of this chapter, the application of innovative technology in sustainable aquaculture such as Geographic Information Systems (GIS) would be an effective tool for selecting sites for bivalve culture and farm management. Coupled with ecosystem models, this technology can assist in predicting the carrying capacity of estuaries (Newell et al. 2013). Similar to shellfish site selection and farm management, Clearwater Seafoods has utilized GIS to take an informed approach to harvesting which minimizes the impact of fishing activities and promotes sustainability both at sea and on land. By investing in GIS, this company saved and minimizes their impacts on ocean ecosystems and promotes a sustainable approach to fishing (ESRI News 2013).

5. CASE STUDY: SUSTAINABLE AQUACULTURE CULTURE IN THAILAND

Fisheries have long been integral to the Thai way of life. Management of fisheries in Thailand began in 1901 with the establishment of the Thailand Department of Fisheries (DOF). In 1901, the ministry of interior issued 3 guidelines to manage fisheries resources: 1) produce fisheries production for population in country, 2) produce fisheries production for country income, and 3) taxation for capture fisheries. However, only taxation was implemented because there was no fisheries biologist at that moment. In 1923, Dr. Hugh M. Smith, MD., LL.D (Commissioner of Fisheries USA) was invited as an advisor in fisheries to His Siamese Majesty's Government. After finishing his survey research, he published a book called "A Review of the Aquatic Resources and Fisheries of Siam, with Plans and Recommendation for the Administration, Conservation and Development". In 1926, the Department of Fisheries was established and Dr. Smith was appointed to be head official. Under his guidance, management systems were implemented and continue to be conducted. Fisheries production in the beginning relied mainly on wild capture since Thailand has many natural freshwater resources scattered all over the country, such as rivers, swamps, and reservoirs. Thailand also has a 3,500 kilometer-long coast line including both the Gulf of Thailand and Andaman Sea, including more than 900 islands. However, drastic changes to these habitats and overfishing have negatively impacted the wild capture fisheries. Therefore, production from aquaculture has gradually begun to play a more important role in maintaining total fisheries production (Figure 13).

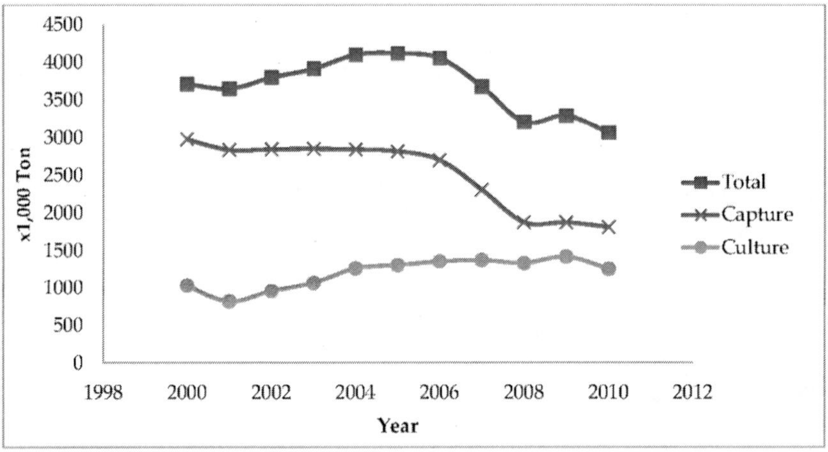

Figure 13. Thailand fisheries production from 2000-2010 (Thailand DOF Information System Center 2011).

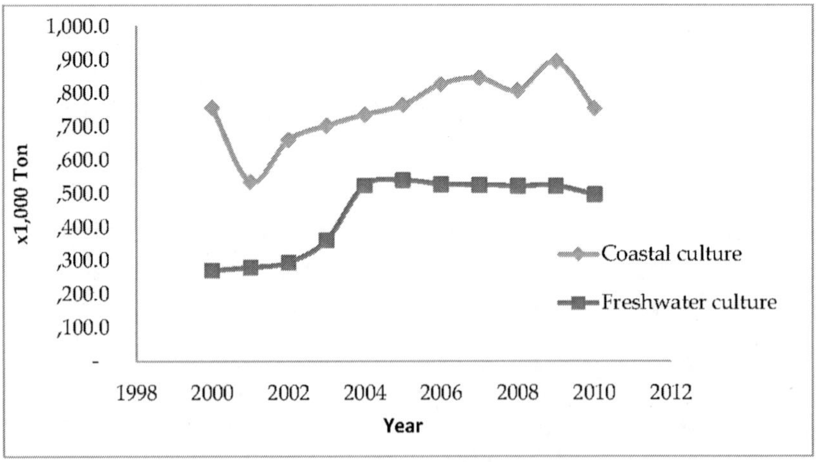

Figure 14. Freshwater and coastal aquaculture production from 2000-2010 (Thailand DOF 2011).

Aquaculture in Thailand evolved from traditional practice to modern science-based practices as aquaculture technology and innovations developed. Therefore, aquaculture in coastal areas, which contribute more national economics and provide more benefit to farmers, tended to increase and contribute more when compared to freshwater species, especially brackish water-cultured shrimp and prawn. However, aquaculture in freshwater areas has also increased due to population growth and market demands in the country (Figure 14) (Thailand DOF Information System Center 2011).

Of the three main groups of brackish water aquaculture – fish, shrimp, and shellfish – shrimp culture has increased dramatically while fish culture has decreased and become steady from 2000 – 2010, while shellfish culture production dropped because of shellfish diseases and natural changes (Thailand DOF Information System Center 2011) (Figure 15). However, shrimp culture is the most cultured species within brackish water aquaculture, globally.

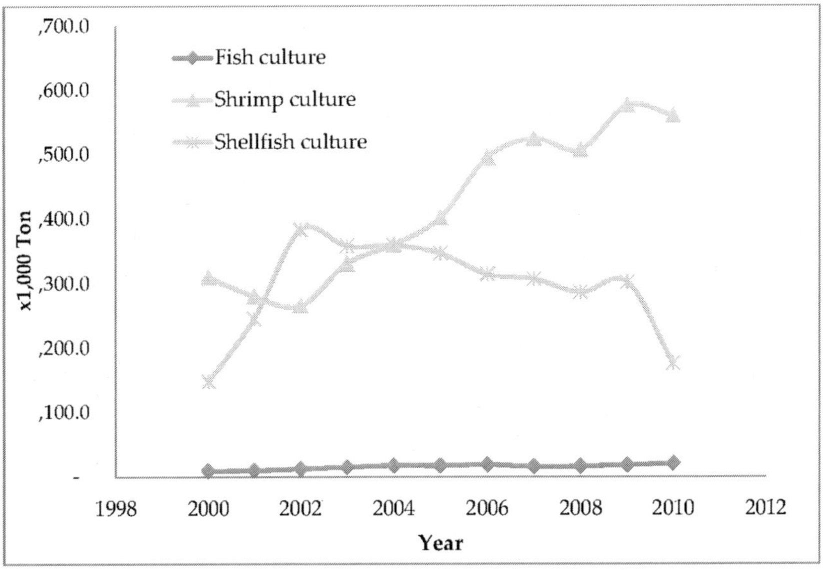

Figure 15. Brackish water aquaculture production by group from 2000-2010 (Thailand DOF 2011).

Shrimp culture in Thailand started more than 50 years ago. Production has been greatly increased within the last three decades as intensive farming techniques were developed and applied. In the beginning of brackish water shrimp culture the major shrimp species used was tiger prawn (*Peneaeus monodon*), which was substituted by white shrimp (*Peneaeus vannamei*) by more than 95%, by the year 2000 (Bureau of Agricultural Economic Research, 2011). Thailand now exports shrimp products at about 500,000 tons annually in the year 2010 (Center for Agricultural Information 2011), making Thailand one of the largest shrimp exporters in the world.

Cultured shrimp, in particular, has come under threat in key export markets due to adverse publicity concerning the environmental and social impact of some aquaculture activities (Kongkeo, 2001). Some of this publicity has served to highlight some of the negative environmental and social impacts that have occurred in the development of aquaculture. These include the destruction of wetlands and mangrove forests, water pollution, reduction of biodiversity, waste of natural resources, and loss of access to fishing grounds by artisanal fisherman (Boyd and Tucker, 1998). Therefore, management inputs that remedy these problems that occur between culturing period and after harvesting such as chemicals used, water management, water discharge, etc. must be considered in order to encourage sustainable growth and practices within the shrimp aquaculture sector.

Shrimp culture in Thailand is performed under intensive culture conditions which consumes heavy feed, water supply, and aeration; therefore, management must play a key role in helping to reduce problems and the impacts of effluent from culturing systems. Sustainable shrimp culture, economically speaking, is less about increasing production but more about the ability to maintain steady production, customer satisfaction and reliability, and mitigate social and environmental impact concerns.

Long before Good Aquaculture Practices (GAPs) and Food and Agriculture Organization (FAO) Code of Conduct for Responsible Fisheries (CoC) were integrated in Thailand, best management practices were implemented by the Department of Fisheries to increase farm productivity while minimizing the environmental impacts. The environmental issues caused by the effluent discharges of the shrimp culture facilities have been one major concern of the aquaculture operation. Some of these management practices facilitated by the Department of Fisheries at the National and local level have focused on increasing feed conversion ratios, better water exchange, aeration, and pond management, and, if available, applying integrated multi-trophic aquaculture technology to reduce nutrient loads from farm effluents and increase profits. Specifically, aquaponics using commercial crop systems become popular and practical to the farmers while the operation is sustained and become environmental friendly.

The Thailand DOF integrated GAP and CoC in 1998 with support from the World Bank. Under GAP standard, the requirement are farm registration;, farm management; use of veterinary drugs chemical, hazardous substances, and probiotics used in aquaculture; farm sanitation; harvest and post-harvest prior to distribution; effluent and sediment management; energy source and fuel use; social and environmental responsibility; and record keeping. Code of Conduct for shrimp culture has two components: operational guidelines for hatcheries and farms, and guidelines for harvesting and transport. These practices - GAP and CoC - are standard for shrimp culture to ensure that shrimp culture has minimal to no chemical residues which protects consumers and applies environmental friendly practices. Three year GAP or CoC certificates are issued by the Thailand DOF to shrimp farms after they meet qualifications and comply with annual surveillance.

Shrimp culture systems in Thailand also have a traceability system from farm to product which initially started as a form of hatchery management. Some hatcheries complied with bio-secure systems to ensure that larvae produced are healthy and viable before selling to grow-out farmers. Hatcheries must provide Fry Movement Documents (FMD) to their customers to indicate the number of fry that a farmer purchases in addition to other hatchery information since this document is checked if any problems occur during grow-out. Shrimp farmers must also provide Movement Documents (MD) which indicate the weight of shrimp in the shipment in addition to other farm information to processing plants or their customers. A DOF officer checks MD, Hazard Analysis and Critical Control

Point (HACCP), GMP, and product quality at the processing plant before issuing product health certificates.

Moreover, many policies and projects are established for sustainable aquaculture support. The agencies involved at the national level include, the Department of Fisheries, Pollution Control Department and Department of Marine and Costal Resources, while there are Provincial government and Local Administrative Organization involve at local level. Water quality testing program pond and discharge water responsible by DOF and water quality in natural water responsible by Pollution Control Department, etc. Mangrove re-habitation projects are established by the national and local government sectors, private sector, and Non-Governmental Organizations (NGOs) to increase mangrove forest area along the costal zones. Thailand also supports Non Illegal Unreported Unregulated Fishing (Non-IUU) and issued several programs for fish resource conservation, which include combatting IUU fishing, prohibiting certain fishing gears within spawning season, and expanding fish conservation areas. Although wild fisheries are not a part of the discussion in this chapter, it does play a part serving as a source of fish meal industry which is used in aquaculture feed. Therefore, control of IUU and certain fishing gear, will support sustainable aquaculture.

In conclusion, fisheries production in Thailand has decreased while aquaculture production has increased and plays a vital role for providing a high protein food source for economic development in the future. Aquaculture in Thailand evolved from traditional practice to science-based due to a number of policies and regulations put in the place to sustain both the aquaculture industry and the environment. Shrimp aquaculture in Thailand is an excellent example of why sustainable aquaculture practices are necessary and how they have become implemented. With the establishment of GAP and CoC, Thailand has ensured that shrimp farming results in production of a high quality product, safe from chemical residues, that is environmental friendly. However, truly sustainable aquaculture will only be attainable when the balance between food security, economic benefit, social benefit, and a reduction of environmental impact is achieved. Figure 16 below shows the steps involved in best management practices in shrimp aquaculture farm.

Figure 16. a. Clean and dried pond before start; b. Water quality monitoring; c. Water drained into treatment ponds; d. Aeration in culturing period using paddle wheel aerators; e. Shrimp health sampling with a cast net; f. Shrimp monitoring with a lift net; and g. Collected specimens are assessed for growth and survival rate.

6. FINAL REMARKS

In recent years aquaculture has gone through the "blue revolution" in which there has been rapid growth worldwide in aquaculture production of both fresh and saltwater fish and shellfish species. In part this is due to the fact that the natural fisheries are close to their maximum sustainable yield. However, this rapid growth in aquaculture may also be attributed to the ever increasing world population and an increase in demand for high protein sources of seafood. In the past, aquaculture has been demonized for destruction of mangroves worldwide for shrimp production as well as causing potential eutrophication through unwelcomed discharges of nutrient rich effluents.

With increased environmental awareness and the general populations increased concern over its food sourcing, aquaculture has stepped up to the proverbial plate to try and fulfill the worlds seafood demands through increased production while trying to maintain more environmentally friendly practices of culturing fish through many technological advancements. Unfortunately, aquaculture has not yet truly reached its sustainability goal. However, in addition to much technological advancement, aquaculture has begun to incorporate best management practices to create a more environmentally friendly way of producing fish. In this chapter we have gone over several areas associated with BMPs and described them with regard to how their incorporation can impact or reduce the impacts of aquaculture on the environment. Many of the BMPs discussed are simple and rely on common sense approaches to nutrient problems. Others are more technologically advanced and require additional components and or descriptions that are beyond the scope of this chapter.

In the end we are all trying to get to a point at which aquaculture can be considered a sustainable farming entity so that its impacts are minimal at best to the surrounding environment, there is a continuous supply of food, and it is profitable for all of those who are involved. As we consider moving ahead we must continue to remember that the world's resources are there for all and we want to maintain them for future generations to come. Again, we quote Costa-Pierce (2002) in suggesting there has to be a behavioral shift in humans rather than technology in order for aquaculture to become truly sustainable.

7. TECHNICAL SUMMARY

Tremendous efforts have been made to improve aquaculture farm practices through disease prevention and treatment, planning and management of facilities, feeding, and advances in aquaculture technology and sustainable practices. However, the industry is not without its issues and faults. Although significant accomplishments have been made in minimizing the negative impacts of aquaculture operations on the environment, it is not reached sustainability worldwide. Further efforts ranging from husbandry practices to policy and regulations are essential to ensure the sustainability of aquaculture on a global scale. As aquaculture moves from feeding millions to billions of people in the last century, intensive culture practices have become common and require better management and monitoring efforts. Intensive production of fish farming requires significant inputs of nutrients in the form of inorganic fertilizers or feeds. Of these inputs, typically only 25% of the chemical constituents of the feed are assimilated into fish biomass while the rest is released into the water as metabolic wastes. In pond culture, fish are usually harvested after draining the pond partially or fully. The waste water expelled from these ponds into watershed, laden with organic matter and nutrients, concerns regulatory agencies as a point-source of pollution. In addition, in most countries, including the United States, statistics are lacking on the amounts of chemicals used, and as a result, regulations cannot truly be effective. Furthermore, it is difficult to make correlations of aquaculture effluents to environmental impacts without accurate records. The main goals of effluent management or, more often referred to as best management practices, are to minimize impacts to the environment while maintaining productivity. Fortunately, most of these strategies are as beneficial to the aquaculturists/ farmers as they are to the environment. Both production costs and effluent can be reduced by using stock-specific feeds applied in smaller quantities several times a day, good aeration, improved husbandry practices, and paying good attention not to exceed to the carrying capacity of the system. By lowering concentrations of phytoplankton, savings on herbicides and aeration are inevitable. Limited water exchange, integrated aquaculture, and good monitoring are further best management practices measures.

ACKNOWLEDGEMENTS

The authors would like to thank Dr. Karuna Chintapenta, Brian Reckenbeil, Kris Roeske, and Cory Janiak for their editorial assistance and review of the chapter. The authors would also like to thank Dr. Claude Boyd for the resources and guidance he provided both authors for their research and professional development. Some of the projects discussed in this chapter were funded by SRAC, USDA-NIFA, USDA Evans-Allen, NSF-EPSCOR, AU, DSU, and Government of Thailand.

REFERENCES

1. United Nations Food and Agriculture Organization. UNFAO: 2009. How to Feed the World 2050. High-Level Expert Forum. Global agriculture towards 2050. http://www.fao.org/fileadmin/templates/wsfs/ docs/ Issues (accessed 10 September 2013).
2. University of Maryland Medical Center. UMD: 2013. Omega-3 fatty acids. Overview. http://umm.edu/health/medical/altmed/ supplement/omega3-fatty-acids (accessed 3 October 2013).
3. United Nations Food and Agriculture Organization. UNFAO: 2013. FOCUS. Fisheries and food security. Sustainable aquaculture development. http://www.fao.org/focus/e/fisheries/sustaq.htm (accessed 5 September 2013).
4. Ewart JW. 2013. Shellfish Aquaculture in Delaware's Inland Bays: Status, Opportunities, and Constraints.In Delaware Sea Grant Program; 2013. p 43.Available from http://darc.cms.udel.edu/ibsa/Inland (accessed 7 November 2013).
5. Bush SR, Belton B, Hall D, Vandergeest P, Murray FJ, Ponte S, Oosterveer P, Islam MS, Mol APJ, Hatanaka M, Kruijssen F, Ha TTT, Little DC, and Kusumawati R. Global Food Supply. Certify Sustainable Aquaculture? Science 2013;341(6150) 1067-1068.
6. National Oceanic and Atmospheric Administration. NOAA: 2013. 10 Myths about Marine Aquaculture. http://www.nmfs.noaa.gov/ aquaculture (accessed 7 November 2013).

7. United States Food and Drug Administration. USFDA: 2008. Taken from Seafood Health Facts: Making Smart Choices. Balancing the Benefits and Risks of Seafood Consumption. http://seafoodhealthfacts. org/ seafood_nutrition/practitioners/composition_chart.php (accessed 3 September 2013).

8. Wilson JL, Stone NM, Bayne D, Popma TJ, Boyd C, Brune DA, Collier JA, Schwedler TE, Burtle GJ, Lewis GW, Styer EL, Romaire RR, Hargeaves JA, Kingsbury SK, Nagle DG, Daniels HV, Hodson RG, Shrader KK, Zimba PV, Grimm CC. Control of Blue-green Algae in Aquaculture Ponds. SRAC Fifteenth Annual Progress Report 2002;30.

9. United Nations Food and Agriculture Organization. FAO: 2011. FAO STAT. http://faostat.fao.org/site/629/default.aspx (accessed 11 December 2013).

10. Dealteris JT, Kilpatrick BD, Rheault RB. A comparative evaluation of the habitat value of shellfish aquaculture gear, submerged aquatic vegetation and a non-vegetated seabed. Journal of Shellfish Research 2004;23(3) 867-874.

11. Erbland PJ, Ozbay G. A Comparison of the macrofaunal communities inhabiting a Crassostrea virginica oyster reef and oyster aquaculture gear in the Indian River Bay, Delaware. Journal of Shellfish Research 2008;27 757-768.

12. Shumway SE, Davis CR, Downey R, Karney J, Kraeuter J, Parsons R, Rheault, Wikfors G. Shellfish aquaculture: in praise of sustainable economies and environments. World Aquaculture 2003;34 15-18.

13. O'Beirn F X, Ross PG, Luckenbach MW. Organisms associated with oysters cultured in floating systems in Virginia, USA. Journal of Shellfish Research 2004;23(3) 825-829.

14. Tallman JC, Forrester GE. Oyster grow-out cages function as artificial reefs for temperate fishes. Transactions of the American Fisheries Society 2007;136(3) 790-799.

15. D'Amours O, Archambault P, McKindsey CW, Johnson LE. Local enhancement of epibenthic macrofauna by aquaculture activities. Marine Ecological Progress Series 2008; 371 73-84.

16. Taylor J, Bushek D. An Assessment of Habitat Value of Constructed Intertidal Oyster Reefs and Oyster Aquaculture Systems in Delaware Bay, USA. Journal of Shellfish Research 2008;27(4) 1056-1056.

17. Emerson C. Aquaculture impacts on the environment. Cambridge Scientific Abstracts 1999.

18. World Wildlife Foundation. WWF: Unsustainable fishing. Our oceans are being plundered. http://wwf.panda.org/about_our_earth/ blue_ planet/problems/problems_fishing/ (accessed 2 November 2013).

19. National Oceanic and Atmospheric Administration. NOAA: Fish watch U.S. Seafood Facts: Outside the U.S. 2011 http://www. fishwatch. gov/farmed_seafood/outside_the_us.htm (accessed 6 October 2013).

20. National Oceanic and Atmospheric Administration. NOAA: 2010. U.S. Seafood Consumption Declines Slightly in 2009. http://www.noaanews. noaa.gov/stories2010/20100909_consumption.html (accessed 8 December 2013).

21. Monterey Bay Aquarium Foundation. MBAF: Fishing and Farming Methods. http://www.montereybayaquarium.org/cr/ cr_ seafoodwatch/sfw_gear.aspx (accessed 2 December 2013).

22. Briggs MRP, Funge-Smith S. Unsustainable shrimp culture- causes and potential solutions from experience in Thailand. In: Development of Strategies Shrimp Farming. In report to the Overseas Development Administration, Research Project R4751;1994. p31.

23. Dalsgaard JPT, Lightfoot C, Christensen V. Towards quantification of ecological sustainability in farming systems analysis. Ecological Engineering 1995;4 181-189.

24. Food and Drug Administration. FDA: Food and Drug Administration Compliance Program Guidance Manual, 2008. http://www.fda.gov/ downloads/Food/ComplianceEnforcement/ucm073192.pdf (accesses 10 October 2013).

25. Virtanen JK, Mozaffarian D, Chiuve SE, Rimm EB. Fish consumption and risk of major chronic disease in men. *American Journal of Clinical Nutrition* 2008;88(6) 1618–1625.

26. Soft Bottom. In Coastal Habitat Protection Plan; 2005. p365-448. Available from http://ncfisheries.net/habitat/chppdocs/ G_Soft% 20Bottom.pdf (accessed 3 August 2013).

27. Garland Science. In: Microbes in the Marine Environment; 2011. p23. Available from http://www.fisheries.go.th/library/app/ (accessed 3 November 2013).

28. Friend M. Disease Emergence and Resurgence: The Wildlife-Human Connection: Reston, Va., U.S. Geological Survey. 2006;400.

29. Yanong RPE, Erlacher-Reid C. Biosecurity in Aquaculture, Part 1: An Overview. Southern Regional Aquaculture Center 2012;(4707) 16.

30. Chatterjee S, Haldar S. Vibrio Related Diseases in Aquaculture and Development of Rapid Accurate Identification Methods. Journal of Marine Science 2012; 7.

31. Mori K. Effects of artificial eutrophication on the metabolism of the Japanese oyster Crassostrea gigas. Marine Biology 1979;53(4) 361-369.

32. Bastin L. International Aquafeed. Incorporating Fish Farming Technology 2013;16(1) 26.

33. Boyd CE. Water quality management and aeration in shrimp farming 1989: Auburn University, Department of Fisheries and Allied Aquacultures, Auburn, AL, Series No. 2. p183.

34. Ryther JH, Goldman JC, Gifford JE, Huguenin AS, Wing JP, Clarner LD, Williams, Lapontie BE. Physical models of integrated waste recycling of "Marine Polyculture Systems." Aquaculture 1975;5 163-177.

35. Neori A, Ragg NLC, Shpigel M. The integrated culture of seaweed, abalone, fish and clams in modular intensive land-based systems: II. Performance and nitrogen partitioning within an abalone (Haliotis tuberculata) and macroalgae culture system. Aquacultural Engineering 1998;17(4) 215-239.

36. Avnimelech Y., editor. Biofloc Technology - A Practical Guide Book. The World Aquaculture Society, Baton Rouge, Louisiana; 2012. p272.

37. The University of Virgin Islands Aquaculture Program. 2013. Biofloc Systems. http://www.uvi.edu/research/agricultural-experiment-station/aquaculture-home/biofloc-systems/default.aspx.

38. The Aquaponics Association. AA: What is aquaponics? http://www.aquaponicsassociation.org/ (accessed 8 November 2013).

39. Bennett L, Diana J, Qiuming L. Learning From Tradition. In: Integrated Aquaculture Practices in Yingbin Bay;2012. http://pdf.gaalliance.org/pdf/GAA-Bennett-Nov12.pdf (accessed 10 October 2013).

40. Nutreco. Feeding the Future? How can aquaculture contribute to feeding 9 billion people in 2050 in a Sustainable way? In Nutreco Prins Frederiklaan; 2011. http://www.nutreco.com. (accessed 8 November 2013).

41. Scottish Association for Marine Science. SAMS: Aquaculture Impacts and Coastal Eutrophication. http://www.sams.ac.uk/ microbial/aquaculture-impacts-and-coastal-eutrophication#sthash.pZzJ4ask.dpf (accessed 9 December 2013)

42. Science Daily. In Science Reference; 2013. Eutrophication. http://www.sciencedaily.com/articles/e/eutrophication.htm (accessed 8 November 2013).

43. Boyd CE. Soil and Water Quality Considerations in Shrimp Farming. In: Haws MC. and Boyd CE. (ed.) Methods for Improving Shrimp Farming in Central America. In Central American University Press; 2001. p17-20.

44. Boyd CE, Queiroz J. Aquaculture pond effluent management. Aquaculture Asia 1997;2 (2) 43-46.

45. Shireman JV, Cichra CE. Water quality from Florida catfish and fingerling culture ponds. In: Joseph St. (eds) 1996: national livestock, poultry, and aquaculture waste management, Proceedings of the National Workshop, American Society of Agricultural Engineers, St. Joseph, MI. p206-210.

46. Schwartz MF, Boyd CE.Channel catfish pond effluents. The Progressive Fish-Culturist 1994a;56(4) 273-281.

47. Boyd CE. Effluents from catfish ponds during fish harvest. Journal of Environmental Quality 1978; 7 59-63.

48. Schwartz MF, Boyd CE. Effluent quality during harvest of channel catfish from watershed ponds. The Progressive Fish-Culturist 1994b;56(1) 25-32.

49. Seok K, Leonard S, Boyd CE, Schwartz ME. Communications: Water Quality in Annually Drained and Undrained Channel Catfish Ponds over a Three-Year Period. The Progressive Fish-Culturist 1995;57(1) 52-58.

50. Boyd CE, Queiroz J, Lee J, Rowan M, Whitis GN, Gross A. Environmental assessment of channel catfish, Ictalurus punctatus, farming in Alabama. Journal of the World Aquaculture Society 2000;31(4) 511-544.

51. Hariyadi S, Tucker C, Steeby J, Van der Ploeg M, Boyd CE. Environmental conditions and channel catfish (Ictalurus punctatus) production under similar pond management regimes in Alabama and Mississippi. Journal of the World Aquaculture Society 1994;25 236-249.

52. Boyd CE, Gross A, Rowan M. Laboratory study of sedimentation for improving quality of pond effluents. Journal of Applied Aquaculture 1998;839-48.

53. Boyd CE, Tucker CS. Pond Aquaculture Water Quality Management. In Kluwer Academic Publishers; 1998. p541-575.

54. Wheaton FW. Aquaculture Engineering. In Robert E. Kreiger; 1977. Malabar, Florida, USA.1977. p414-462.

55. Coloso RM, Basantes SP, King K, Hendrix MA, Fletcher JW, Weis P, and Ferraris RP. Effect of dietary phosphorus and vitamin D3 on phosphorus levels in effluent from the experimental culture of rainbow trout (*Oncorhynchus mykiss*). Aquaculture 2001;202(1) 145-161.

56. Queiroz JF, Boyd CE, Gross A. Evaluation of a bio-organic catalyst in channel catfish, Ictalurus punctatus, ponds. Journal of Applied Aquaculture 1998;8(2) 49-61.

57. Tucker CS, Kingsbury SK, Pole JW, Wax CL. Effects of water management practices on discharge of nutrients and organic matter from channel catfish (Ictalurus punctatus) ponds. Aquaculture 1996;147(1) 57-69.

58. Thompson JR, Polz MF. Dynamics of Vibrio populations and their role in environmental nutrient cycling. The biology of Vibrios. In ASM; 2006. p90-203.

59. ICAP Bio-Organic. 2013. Technology. http://www.icapbio-organic.com/technology.html (accessed 8 November 2013).

60. Dame RF, Spurrier JD, Wolaver TG. Carbon, nitrogen and phosphorus processing by an oyster reef. Marine ecology progress series 1989;54(3) 249-256.

61. Seaweed Energy Solution. SES: Seaweed is simple. http://www.seaweedenergysolutions.com/seaweed/ (accessed 8 November 2013).

62. Becker V, Huszar VLM, Crossetti LO. Responses of phytoplankton functional groups to the mixing regime in a deep subtropical reservoir. *Hydrobiologia*; 2008; 628(1) 137-151.

63. National Oceanic and Atmospheric Administration. NOAA: State of the Coastal Environment. Oxygen Depletion in Coastal Waters. http://oceanservice.noaa.gov/websites/retiredsites/sotc_pdf/HYP.PDF (accessed 8 December 2013).

64. Boesch D, Brinsfield R, Magnien R. Chesapeake Bay eutrophication: Scientific understanding, ecosystem restoration, and challenges for agriculture. Journal of Environmental Quality 2001;30 303-320.

65. Boyd CE. Water quality for pond aquaculture. Alabama Research and Development Series 1998; 43 37.

66. Boyd CE. Water Quality in Ponds for Aquaculture: Auburn University; 1990; p482.

67. Boyd CE, Musig Y. Shrimp pond effluents: observations of the nature of the problem on commercial farms. In: Proceedings of the Special Session on Shrimp Farming. World Aquaculture Society 1992, Baton Rouge, Louisiana 2006; 195-197.

68. Boyd CE., editor. Bottom Soils, Sediment, and Pond Aquaculture. Chapman and Hall; 1995a. p348.

69. Ozbay G. Studies of Channel Catfish *Ictalurus punctatus* Pond Effluent Characteristics and Treatment Options. PhD thesis. Auburn University Auburn; 2002; p109.

70. Ebeling JM, Vinci B. Solids Capture. Recirculating Aquaculture Systems Short Course; 2013. http://ag.arizona.edu/azaqua/ista/ISTA7/RecircWorkshop/Workshop %20PP%20%20&%20Misc%20Papers% 20 Adobe%202006/5%20Solids%20Capture/Solids%20Control.pdf (accessed 15 November 2013).

71. Bowser PR. General Fish Health Management. Northeastern Regional Aquaculture Center 2012;111 p16.

72. Pietrak M, Leavitt D, Walsh M. Biosecurity on the Farm-Guidelines & Resources for Developing a Biosecurity Plan. Northeastern Regional Aquaculture Center 2010;208 p6.

73. Figueras MJ, Borrego JJ, Pike EB, Robertson W, Ashbolt N. Sanitary Inspection and Microbiological Water Quality. In: Bartram J. and Rees G. (eds.) Monitoring Bathing Waters-A Practical Guide to the Design and Implementation of Assessments and Monitoring Programmes. In E and FN Spon; 2000. p311.

74. Thompson FL, Iida T, Swings J. Biodiversity of Vibrios. Microbiology and Molecular Biology Review 2004;68(3) 403-431.

75. Austin. Vibrios as causal agents of zoonoses. Veterinary Microbiology 2010;140(3-4) 310-317.

76. Urakawa H., Rivera ING. Aquatic Environment. In: Thompson FL, Austin B, Swings J. (eds.) *The Biology of Vibrios*, D.C. ASM Press; 2006. p175–189.

77. Austin, B. Miscellaneous animal pathogens. In: Thompson, F.B., Austin, B. Swings, J. (eds.) The Biology of Vibrios In ASM Press; 2006. p423.

78. Grimes DJ. Ecology of estuarine bacteria capable of causing human disease: a review. Estuaries 1991;14(4) 345-360.

79. Baker-Austin CL, Stockely R, Rangdale, Martinez-Urtaza J. Environmental occurrence and clinical impact of *Vibrio vulnificus* and *Vibrio parahaemolyticus*: A European perspective. Environmental Microbiology Reports 2010;2 7-18.

80. Lipp EK, Huq A, Colwell RC. Effects of Global Climate on Infectious Disease: the Cholera Model. Clinical Microbiology Reviews 2002;15(4) 757-770.

81. Moyle PB, Cech JJ. 2004. Fishes: an introduction to ichthyology. Prentice Hall; 2004. p 465, 615.

82. Boyd CE, Boyd C, Chainark S. From The Shrimp Book: Shrimp Pond Soil and Water Quality Management. Global Aquaculture Advocate 2012;15(1):28-31.

83. Kalff, J. Limnology: inland water ecosystems. Prentice Hall; 2002. p592.

84. Shine C, de Klemm C. 1999. Wetlands, Water and the Law. Using law to advance wetland conservation and wise use. IUCN Environmental Policy and Law Paper 1999; p330.

85. United States Environmental Protection Agency. USEPA: Economic Benefits of wetlands. http://www.epa.gov/owow/wetlands (accessed 8 October 2013).

86. United States Environmental Protection Agency. USEPA: Nutrient Criteria Technical Guidance Manual- Wetlands. http://www2.epa.gov/sites/production/files/documents/wetlands_full.pdf (accessed 6 November 2013).

87. Rückauf U, Augustin J, Russow R, Merbach, W. Nitrate removal from drained and reflooded fen soils affected by soil N transformation processes and plant uptake. Soil Biology and Biochemistry 2004;36(1) 77-90.

88. Reddy KR, Delaune RD. Biogeochemistry of Wetlands: Science and Applications. CRC Press; 2008. http://www.crcpress.com. (accessed 15 October 2013).

89. Mitsch WJ, Gosselink JG. Editors. Wetlands. In John Wiley and Sons, Inc., New York, NY. USA. 2007. p600.

90. Moshiri S. Identification of resistance source in barley to certain fungal pathogen in Khorasan 1995: conference proceedings, September 2-7, 1995, Karadj.

91. Schwartz MF, Boyd CE. Constructed wetlands for treatment of channel catfish pond effluents. The progressive fish-culturist 1995;57(4) 255-266.

92. Boyd CE. Hydrology and pond construction, In: Tucker CS. (ed.) Channel Catfish Culture. Elsevier; 1985. p107-134.

93. Shpigel M, Ben-Ezra D, Shauli L, Sagi M, Ventura Y, Samocha T, Lee JJ. Constructed wetland with Salicornia as a biofilter for mariculture effluents. Aquaculture 2013;412 52-63.

94. Lymbery AJ, Kay GD, Doupé RG, Partridge GJ, Norman HC. The potential of a salt-tolerant plant (Distichlis spicata) to treat effluent from inland

saline aquaculture and provide livestock feed on salt-affected farmland. Science of The Total Environment 2013;445 192-201.

95. Serpa D, Pousão-Ferreira P, Caetano M, Cancela da Fonseca L, Dinis MT, Duarte P. A coupled biogeochemical-Dynamic Energy Budget model as a tool for managing fish production ponds. Science of the Total Environment 2013;463 861-874.

96. Boyd CE. Potential of sodium nitrate to improve environmental conditions in aquaculture ponds. Journal of the World Aquaculture Society 1995b;26 38-40.

97. Cathcart TP, Pote JW, Rutherford DW. Reduction of effluent discharge and groundwater use in catfish ponds. Aquacultural Engineering 1999;20(3) 163-174.

98. USDA/Natural Resources Conservation Service. [DVD]: Alabama Aquaculture Best Management Practices (BMP). Settling Basins and Wetlands 2002;BMP No. 6.

99. Tucker CS. Aquaculture Effluents and the Environment. Aquaculture Meeting; 2009 NC USA p61.

100. United States Environmental Protection Agency. USEPA: NPDES. Final Permit Fact Sheet. http://www.epa.gov/region9/water/ npdes/ pdf/ca/GoletaSdNpdesPermitFinalOrder.pdf (accessed 11 October 2013).

101. Teichert-Coddington DR, Rouse DB, Potts A and Boyd CE. Treatment of harvest discharge from intensive shrimp ponds by settling. Aquacultural Engineering 1999;19 147-161.

102. Ozbay G, Boyd CE. Treatment of Channel Catfish Pond Effluents in Sedimentation Basins. The World Aquaculture Society Magazine 2004;35(3)10-13.

103. Ozbay G, Boyd CE. Particle Size Fractions in Pond Effluents. The World Aquaculture Society Magazine 2003a;34(4) 56-59.

104. Ozbay G, Boyd CE. U.S. Study Finds Turbidity Useful in TSS, ISS Estimates. Global Aquaculture Advocate 2003b;6(4) 58-60.

105. Boyd CE. Aquaculture sustainability and environmental issues. World Aquaculture 1999;30(June) 10-15.

106. Hargreaves J, Cathcart T, Boyd C, Engle C, Romaire R, McClain R, Boardman G, Browdy C. Management of Aquacultural Effluents from Ponds. SRAC Sixteenth Annual Progress Report 2003; p18.

107. Boyd CE. Farm effluent during draining for harvest. Global Aquaculture Advocate 2000;3(4) 26-27.

108. Schwartz MF, Boyd CE. Suggested management to improve quality and reduce quantity of channel catfish pond effluents. Alabama Agriculture Experiment Station; 1996. Auburn University, Alabama, USA. Leaflet 108.

109. Lutz G, LeBlanc BD, Sheffield R, Nix KE. Aquaculture Environmental Best Management Practices. BMPs. LSU Ag Center Communications; 2011. Louisiana State University, LA, USA. p28.

110. Tucker CS, Hargreaves JA. (eds.). Environmental Best Management practices for aquaculture. John Wiley and Sons: Blackwell Publishing; 2008. p582.

111. Setty K. Design Manual: Retention Basin. Bren School of Environmental Science and Management; 2013. University of California, Santa Barbara, CA USA. p7.

112. Boyd CE, Tucker CS. Sustainability of channel catfish farming. World Aquaculture 1995;2645-53.

113. Cripps SJ. The application of suspended particle characterization techniques to aquaculture systems. In: Wang, J. (ed.) Techniques for Modern Aquaculture; 1993. p26–34.

114. Cripps SJ. Serial particle size fractionation and characterization of an aquacultural effluent. Aquaculture 1995;133 323–339.

115. Ackefors H, Enell M. The release of nutrients and organic matter from aquaculture systems in Nordic countries. Journal Applied Ichthyology 1994;10(4) 225-241.

116. Cripps SJ. 1992. Characterization of an aquaculture effluent based on water quality and particle size distribution data 1992: conference proceedings, May 21-25, 1992. Orlando, FL, USA.

117. Bergheim A, Tyvold T, Seymour EA. 1991. Effluent loadings and sludge removal from land based salmon farming tanks. In: Bredene EAS

(ed.) Europe'91: Aquaculture and the Environment International Aquaculture Conference: proceedings of Aquaculture, Dublin, Ireland, 10–12 June, 1991. Belgium: EAS Special Publication.

118. Cripps SJ. Minimizing outputs: treatment. Journal of Applied Ichthyology 1994;10 284-294.

119. Cripps SJ., Kelly L. Reduction in wastes from aquaculture. In: Baird DG, Beveridge M, Kelly GA, Muir BF. (eds.) Aquaculture and Water Resource Management. Blackwell Scientific; 1996. p219.

120. Pepper VA, Sutterlin AM, Nicholls T, Collier C. Newfoundland experience with development of all-female and non-reproductive Atlantic salmon for marine aquaculture. Bulletin-Aquaculture Association of Canada 1996;14-23.

121. Moore PA, Miller DM. Decreasing phosphorus solubility in poultry litter with aluminum, calcium, and iron amendments. Journal of Environmental Quality 1994;23(2) 325-330.

122. Gensemer RW, Playle RC. The bioavailability and toxicity of aluminum in aquatic environments. Critical reviews in environmental science and technology 1999;29(4) 315-450.

123. Welch EB, Cooke GD. Effectiveness and longevity of phosphorus inactivation with alum. Lake and Reservoir Management 1999;15(1) 5-27.

124. Cooke GD., Welch EB., Peterson SA., Newroth PR. editors. Restoration and Management of Lakes and reservoirs. Lewis Publishers: CRC Press; 1993. p548.

125. Jacoby JM, Gibbons HL, Hanowell R and Bouchard DD. Wintertime blue-green algal toxicity in a mesotrophic lake. Journal of Freshwater Ecology 1994;9(3) 241-251.

126. Rowan M. Chemical phosphorus removal from aquaculture pond water and effluent. PhD thesis. Auburn University Auburn; 2001. p180.

127. Masuda K, Boyd CE. Effects of aeration, alum treatment, liming, and organic matter application on phosphorus exchange between soil and water in aquaculture ponds at Auburn, Alabama. Journal of the World Aquaculture Society 1994;25(3) 405-416.

128. Prepas EE, Murphy TP, Crosby JM, Walty DT, Lim JT, Babin J, Chambers PA. Reduction of phosphorus and chlorophyll a concentrations following calcium carbonate and calcium hydroxide additions to hypereutrophic Figure Eight Lake, Alberta. Environmental Science and Technology 1990;24(8) 1252-1258.

129. Salonen VP, Varjo E. Gypsum treatment as a restoration method for sediments of eutrophied lakes–experiments from southern Finland. Environmental Geology 2000;39(3-4) 353-359.

130. Wu R, Boyd CE. Evaluation of calcium sulfate for use in aquaculture ponds. The Progressive Fish-Culturist 1990;52(1) 26-31.

131. Ann Y, Reddy R, Delfi JJ. Influence of chemical amendments on phosphorus immobilization in soils from a constructed wetland. Ecological Engineering 2000;14 157–167.

132. Gutcho SJ. Waste Treatment with Polyelectrolytes and Other Flocculants No. 31. Noyes Data Corp 1977; Technology & Engineering. University of Minnesota, p24.

133. Ozbay G. Effects of Coagulant Treatments on Aquaculture Effluent Quality. Journal of Applied Aquaculture 2005;17(4) 1-23.

134. Wright RT, Coffin RB, Ersing CP, Pearson D. Field and laboratory measurements of bivalve filtration of natural marine bacterioplankton. Limnology and Oceanography 1982;27(1) 91-98.

135. Lin CK., Shrestha MK., Thakur DP., Diana JS. Management to minimize the environmental impacts of pond draining. In: Burke D., Baker J., Goetze B., Clair D. and Egna H. (eds.) Fifteenth Annual Technical Report; 1998. p132-139.

136. Shpigel M, Fridman R. Propigation of the Manila clam, Tapes semidecussatus, in the effluent of fish aquaculture ponds in Eliat, Israel. Aquaculture 1990;90 113-122.

137. Newell RIE, Cornwell JC, Owens M, Tuttle J. Role of oysters in maintaining estuarine water quality. National Shellfisheries Association. Journal of Shellfish Research 1999;18(1) 300-301.

138. Rensel J, Bright K, Siegrist Z. Integrated Fish-Shellfish Mariculture in Puget Sound. Final Report; 2011. p82.

139. Chrzanowski TH, Spurrier JD, Dame RF, Zingmark RG. Processing of microbial biomass by an intertidal reef community. Marine Ecology-Progress Series 1986;30 181-189.

140. Toro JE, Paredes PI, Villiagra JD, Senn CM. Seasonal variation in the planktonic community, seston and environmental variables during a 2-year period and oyster growth at two mariculture sites, southern Chile. Marine Ecology 1999;20(1) 63-89.

141. Ulanowicz RE, Tuttle JH. The trophic consequences of oyster stock rehabilitation in Chesapeake Bay. Estuaries 1992;15(3) 298-306.

142. Leffler M. Improving Bay Water Quality: The role of oyster reefs. Maryland Sea Grant. Maryland Aquafarmer 2001. http:// www. mdsg.umd.edu/Extension/Aquafarmer/ Winter01.html (accessed 10 October 2013).

143. Miura T, Yamashiro T. Size selective feeding of Anodonta calipygos, a phytoplanktivorous freshwater bivalve, and visibility of egested algae. Japanese Journal of Limnology 1990;51(2) 73-78.

144. Lowe R, Reich LA, Sierra J. Algal herbivory by the zebra mussel: Fate of algae in feces and pseudofeces. Journal of Shellfish Research 1991;10(1) 250-251.

145. Senichieva MI. Dynamics of phytoplankton community in the region with mussel mari-farming. Hydroores 1990;7(8) 86-89.

146. Santelices B, Martinez. The effect of filter feeders on macroalgae spores. 2. Indirect effects and landscape determination. Biota 1986;1:102.

147. Marenghi FP, Ozbay G. Floating Oyster, *Crassostrea virginica* Gmelin 1791, Aquaculture as habitat for fishes and macroinvertebrates in Delaware Inland Bays: The comparative value of oyster clusters and loose shell. Journal of Shellfish Research 2010a; 29(4) 889-904.

148. Marenghi FP, Ozbay G. Preliminary habitat assessment of floating oyster (Crassostrea virginica) gardens (Delaware). Ecological Restoration 2010b;28(3) 254-257.

149. Harding JM, Mann R. Estimates of naked goby (*Gobiosoma bosc*), striped blenny (*Chasmodes bosquianus*) and eastern oyster

(*Crassostrea virginica*) larval production around a restored Chesapeake Bay oyster reef. Bulletin of Marine Science 2000;66(1) 29-45.

150. Tolley SG, Volety AK. The role of oysters in habitat use of oyster reefs by resident fishes and decapod crustaceans. Journal of Shellfish Research 2005;24(4) 1007-1012.

151. Breitburg DL. Are three-dimensional structure and healthy oyster populations the keys to an ecologically interesting and important fish community. Oyster reef habitat restoration: a synopsis and synthesis of approaches; 1999. Virginia Institute of Marine Science Press, Gloucester Point, Virginia. p239-250.

152. Posey DA. Developing sui generis options for the protection of living aquatic resources of indigenous and local communities. In: ICLARM Conference Proceedings, 1999.http://www.fao.org/ag/againfo/programmes/en/genetics/documents/ITWG-AnGR4/MontpellierAnGR-Papers.pdf (accessed 8 November 2013).

153. Peterson CH, Grabowski JH, Powers SP. Estimated enhancement of fish production resulting from restoring oyster reef habitat: quantitative valuation. Marine Ecology-Progress Series 2003;264 249–264.

154. The Nature Conservancy. 2013. Oyster Restoration Reaches New Depth. New Hampshire. Do you know how many gallons of water an oyster can filter in a day? http://www.nature. org/ ourinitiatives/regions/northamerica/unitedstates/newhampshire/explore/oyster-restoration-reaches-new-depths.xml.

155. Ozbay G, Reckenbeil B, Marenghi F, Erbland P. Can't Go Wrong With Eastern Oysters (*Crassostrea virginica*): Enhancing Biodiversity of the Delaware Inland Bays. Journal of Biodiversity and Endangered Species 2013;1 2.

156. Reckenbeil BA. Assessment of oyster restoration along human altered shorelines in the Delaware Inland Bays: An examination of riprap stocked with the eastern oyster (Crassostrea virginica). Master's thesis Delaware State University; 2013. p.174.

157. Rice MA. Environmental effects of shellfish aquaculture in the Northeast. NRAC Publication Fact Sheet 105; 2008.

158. Lin CK, Ruamthaveesub P, Wanuchsoontorn P. Integrated culture of the green mussel in wastewater from an intensive shrimp pond: concept and practice. World Aquaculture 1993;24 68-72.

159. Tseng KF, Huang JS, Liao I. Species control of microalgae in an aquaculture pond. Water Research 1991;25(11) 1431-1437.

160. Burke JS, Bayne DR. Impact of paddlefish on plankton and water quality of catfish ponds. The Progressive Fish Culturist 1986;48 177-183.

161. Smith NP. The suitability of routine weather data for estimating local estuarine heat energy fluxes. Estuaries 1985;8 270-278.

162. Fott J., L. Pechar and M. Prazakova. Fish as a factor controlling water quality in ponds. In: Barica J., Mur LR. (eds.) SIL Workshop on Hypertrophic Ecosystems, Vaxjo, Sweden; 1979. Sept. 10-14.

163. Canadian Aquaculture. Integrated Multi-trophic Aquaculture. What is it? Canadian Aquaculture Industry Alliance; 2012. http://www.aquaculture.ca/files/species-multi-trophic.php. (accessed 11 October 2013).

164. Gowen RJ. Aquaculture and the environment. In: De Pauw N., Joyce J. (eds.) Aquaculture and the Environment. European Aquaculture Society Special Publication 1991;16 23 – 48.

165. Filbrun JE, Culver DA. Can Reduced Provision of Manufactured Feed Improve Fish Production Efficiency in Ponds? North American Journal of Aquaculture 2013; 1 64-76.

166. Gross A, Boyd CE, Wood CW. Ammonia volatilization from freshwater fish ponds. Journal of Environmental Quality 1999;28(3) 793-797.

167. Abou Y, Saidou A, Mama D, Fiogbé ED, Micha JC. Evaluation of Nitrogen and Phosphorus Wastes Produced by Nile Tilapia (Oreochromis niloticus L.) Fed Azolla-Diets in Earthen Ponds. Journal of Environmental Protection 2012;3(6) 502-507.

168. Gross A, Boyd CE, Lovell RT, Eya JC. Phosphorus budgets for channel catfish ponds receiving diets with different phosphorus

concentrations. Journal of the World Aquaculture Society 1998;29(1) 31-39.

169. Rangen. Salmon Feeds/ Trout Feeds/ Tilapia Feeds/ Quality Feed for Catfish. Rangen Inc; 2013 Buhl, ID, USA. http://www.rangen.com.

170. Boyd CE, Schmittou HR. Achievement of sustainable aquaculture through environmental management. Aquaculture Economics and Management 1999;3 59-69.

171. Goldburg R. Triplett T. Murky water: environmental effects of aquaculture in the United States. Environmental Defense Fund, New York, USA; 1999.

172. Frankic A, Hershner C. Sustainable aquaculture: developing the promise of aquaculture. Aquaculture International 2003;11 517-530.

173. Costa-Pierce, B. A. Ecology as the paradigm for future of aquaculture. In: Barry AC., (ed.) Ecological Aquaculture: The evolution of the blue revolution. Blackwell; 2002 Science, MA USA. p. 339-372.

174. Martins CIM, Eding EH, Verdegem MCJ, Heinsbroek LTN, Schneider O, Blancheton JP, Roque D'Orbcastel E, Verreth JAJ. New developments in recirculating aquaculture systems in Europe: a perspective on environmental sustainability. Aquaculture Engineering 2010;43 83-93.

175. Chen S., Summerfelt S., Losordo T., Malone R. Recirculating systems, effluents, and treatments. In: Aquaculture and the environment in the United States; 2002. p29.

176. Hollingsworth CS, Baldwin R, Wilda K, Ellis R, Soares S. Best management practices for finfish aquaculture in Massachusetts. Publication AG-BPFA; 2006. Western Massachusetts Center for Sustainable Aquaculture. University of Massachusetts Extension.

177. Badiola M, Mendiola D, Bostock J. 2012. Recirculating aquaculture systems (RAS) analysis: main issues on management and future challenges. Aquaculture Engineering 2012;51 26-35.

178. Timmons MB, Ebeling JM, Wheaton FW, Summerfelt ST, Vinci BJ. Recirculating Aquaculture Systems. Northeastern Regional Aquaculture Center. In Cayuga Aqua Ventures; 2001. Ithaca, NY 14850.

179. Timmons MB, Ebeling JM. Recirculating Aquaculture Systems. In Northeastern Regional Aquaculture Center; 2007. 2nd Ed. Cayuga, AquaVentures LLC. Ithaca, NY 14850.

180. Ebeling JM, Timmons MB. Recirculating Aquaculture Systems. In: James HT. (ed.) Aquaculture Production Systems. John Wiley and Sons; 2012. p245-277.

181. Summerfelt ST., Vinci. Better Management Practices for Recirculating Aquaculture Systems. In: Craig ST., John AH. (eds.) In. Environmental Best Management Practices for Aquaculture. Blackwell Publishing; 2008. p389-426.

182. Losordo T, Haider D, Tran H. Recirculation meets aquaponics: Engineering flexibility into fish/plant production systems. The Global Aquaculture Advocate 2013;16 88-90.

183. Miller D, Semmens K. Waste management in aquaculture. Publication # AQ02-1. Aquaculture information series: 2002. Agriculture and Resource Economics Program, Division of Resource Management. College of Agriculture, Forestry, and Consumer Sciences. West Virginia University, Morgantown, WV.

184. Valencia E, Adjei M, Martin J. Aquaculture effluent as a water and nutrient source for hay production in the seasonally dry tropics. Community Soil Science Plant Annals 2001;32 1293-1301.

185. Global Aquaculture Alliance. GAA: BAP Standards. www. galliance. org. (accessed 5 October 2013).

186. Hanna K, Wheaton F, Lazur A, Vankeuren S. Water quality in identical recirculating systems managed by different aquaculturists. International Journal of Recirculating Aquaculture 2012;11 55-74.

187. United States Department of Agriculture, National Institute of Food and Agriculture. USDA-NIFA: Legal definition of sustainable agriculture (US code Title 7, Section 3103). www.csrees.usda.gov/ nea/ag_ systems/in_focus/sustain_ag_if_legal.html (accessed 8 November 2013).

188. McMurtry MR, Nelson PV, Sanders DC, Hodges L. Sand culture of vegetables using recirculated aquaculture effluents. Applied Agriculture Research 1990;5 280-284.

189. Rakocy JE., Aquaponics-Integrating Fish and Plant Culture. In: James HT. (ed.) Aquaculture Production Systems. John Wiley and Sons; 2012. p343-386.

190. Rakocy JE, Losordo TM and Masser MP. Recirculating aquaculture tank production systems: integrating fish and plant culture. SRAC Publication 1992: Sourthern Region Aquaculture Center, Mississippi State University, Stoneville, MS. p1-11.

191. Rhodes MA, Hanson T, Alavi S, Davis DA. High-soy, fishmeal-free diets support Florida pompano growth. The Global Aquaculture Advocate 2013;16 100-103.

192. Watson AM, Barrows FT, Price AR. Taurine supplementation of plant derived protein and n-3 fatty acids are critical for optimal growth and development of cobia, *Rachycentron canadum*. Lipids 2013;48 819-913.

193. Ozbay G, Jackson AR. Aquaculture Effluents: Perspectives on Best Management Practices. Global Aquaculture Advocate 2006;9(5) 68-70.

194. Boyd CE. Guidelines for aquaculture effluent management at the farm-level. Aquaculture 2003;226(1) 101-112.

195. Villalón JR. Practical Manual for Semi-intensive Commercial Production of Marine Shrimp. Texas A&M University Sea Grant College Program; 1991. TAMU-SG-91-501. College Station, Texas, USA. p1-104.

196. Hopkins JS, Stokes AD, Browdy CL, Sandifer PA. The relationship between feeding rate, paddlewheel aeration rate and expected dawn dissolved oxygen in intensive shrimp ponds. Aquacultural Engineering 1991;10(4) 281-290.

197. United States Environmental Protection Agency. USEPA: Effluent Guidelines: Aquatic Animal Production Industry. http://water. epa. gov/scitech/wastetech/guide/aquaculture/index.cfm (accessed 8 November 2013).

198. Northwest Earth Institute. 2012. Discussion Course on Choices for Sustainable Living. Northwest Earth Institute, Portland, OR USA. p138.

199. Newell CR, Hawkins AJS, Morris K, Richardson J, Davis C, Getchis T. ShellGIS: A Dynamic Tool for Shellfish Farm Site Selection. World Aquaculture 2013;44(3) 50-53.

200. ESRI News. 2013. Esri News for Agriculture. Clearwater Seafoods Achieves Sustainable Operations through GIS. Courtesy of Esri Canada Limited. Spring 2013.

201. Thailand DOF Information System Center. Fisheries Statistics of Thailand 2010; 2011. Department of Fisheries. Bangkok, Thailand. http://www.fisheries.go.th/dof/en/ (accessed 10 October 2013).

202. Bureau of Agricultural Economic Research.Situation of Agricultural products and Trend in the Year 2011. Office of Agricultural Economics, Bangkok, Thailand.

203. Center for Agricultural Information. Thailand Foreign Agricultural Trade Statistics 2011; 2011. Office of Agricultural Economics, Bangkok, Thailand.

204. Kongkeo H. Current Status and Development Trends of Aquaculture in the Asian Region. In: Subasinghe RP, Bueno P, Phillips MJ, Hough C, McGladdery SE, Arthur JR. (eds.) 2001: Aquaculture in the Third Millennium. Technical Proceedings of the Conference on Aquaculture in the Millennium, 20–25 February 2000, NACA, Bangkok and FAO, Rome, Italy.

Chapter 6

Using stable isotope analysis to determine the contribution of naturally occurring pond biota and supplementary feed to the diet of farmed Australian freshwater crayfish, redclaw (*Cherax quadricarinatus*)

Melissa K. Joyce , Igor Pirozzi

Centre for Sustainable Tropical Fisheries and Aquaculture and College of Marine and Environmental Sciences, James Cook University, Townsville, Australia

ABSTRACT

Nutritional requirements of redclaw (*Cherax quadricarinatus*) farmed in Australia are poorly understood and little is known on what is actively being consumed in semi-intensive pond culture. In this study the isotopic signatures of δ13C and δ15N of potential food sources were analysed with a multi-source mixing model to determine the extent of their contribution to the diet of farmed redclaw. Four commercial redclaw farms in North Queensland, Australia (Mareeba, Biboorha, Wondecla and Topaz) were sampled for naturally occurring pond organisms and commonly used supplemental feed such as raw corn, soybean, lupin, commercial redclaw, and chicken feed. Both naturally occurring pond biota and supplemental feed contribute to the tissue composition of redclaw to some degree. However, the contribution varies with the type of feed and availability of natural sources, for example plant material at Topaz Farm was a greater contributor with 43.9 ± 19.5 % compared to supplementary sources raw corn 8.20 ± 3.10 and lupin 1.60 ± 1.70 %. Moreover, some supplemental feeds provided a direct nutrient source for primary pond productivity; contribution of the redclaw pellet to zoo- and phytoplankton at Wondecla Farm was 83.1 ± 6.50 and 50.0 ± 9.50 %, respectively, with similarly high values for chicken feed at Biboorha Farm of 72.6 ± 4.70 and 83.4 ± 6.90 %. The cost effectiveness of such feeds needs to be questioned if these are not being consumed and utilised directly. Providing a species-specific formulated feed with improved water stability would enhance production reliability and facilitate growth within the industry.

KEYWORDS

Crayfish, Nutrition, Pond biota, Supplemental feed, Stable isotope analysis, Semi-intensive aquaculture

1. INTRODUCTION

Redclaw (*Cherax quadricarinatus*) are a tropical species of freshwater crayfish native to Northern Queensland, Australia. Physically robust, tolerant of low dissolved oxygen concentrations, varied water quality conditions, relatively high stocking densities and with no planktonic larval stage production techniques are relatively straightforward (Jones 2002; Saoud et al. 2012; Thompson et al. 2003). Redclaw can grow to 200 g in 6–9 months under optimal conditions (Jones et al.2002; Thompson et al. 2004) which is superior to many overseas commercially important species such as the Red Swamp Crayfish (*Procambarus clarkii)* which reaches up to 50 g in 7–10 months (Ackefors 2000; McClain et al. 2007).

Further, approximately 30 % of the total body weight of redclaw is edible tail meat which is one of the highest yields compared to other farmed crustaceans (Thompson et al.2005). These characteristics have seen redclaw exported to countries such as Argentina, Mexico, Spain, The United States, South-East Asia, and Central/South America (Food and Agriculture Organisation 2012; Rodríguez-Canto et al. 2002; Saoud et al. 2013). However, the industry is relatively small with total global production totalling 61.5 tonne worth USD$1.3million in 2011 (FAO 2012). In Australia, commercial aquaculture of redclaw is limited to a small number of farms in Queensland where production has continued to decline over the past decade. Only 41 tonnes worth AUD$862,000 was produced in 2010–2011 down from a peak of 100 tonnes in 2004–2007 (Heidenreich 2013). In comparison the global production (primarily China and the United States) of *P. clarkii* was 539,764 tonnes worth over USD$2.5billion (FAO 2012). There is clearly a market for freshwater crayfish that Australian farmers have the potential to capitalise on if production can meet demand. However, limited information on fundamental nutritional requirements and the lack of an industry feeding standard has led to the stagnation of the redclaw industry and limits the ability of Australian farmers to increase production and compete on a global scale.

Current redclaw farming practices have remained fairly similar to when the industry began in the late 1980s. Farmed in earthen ponds, growth of redclaw is largely attributed to the consumption of naturally occurring pond biota. Farmers commonly promote growth of primary productivity with the use of fertilisers and detrital forage such as hay (Jones et al. 1996). These ponds are essentially small ecosystems that display many of the processes that occur in natural environments. Naturally occurring pond organisms such as zooplankton including cladocerans, copepods, chironomid larvae, rotifers, and eggs of other organisms, associated periphyton and other natural feedstuff are often thought to explain differences in growth rates and survival (Hernández-Vergara et al. 2003; Viau et al. 2012), providing adequate nutrition and inducing strong feeding responses (Duffy et al. 2011).

While the importance of primary productivity is clearly understood, supplemental feeding is a common component of intensive and semi-intensive farming practices. Currently there is no feed available that is formulated based on the nutritional requirements for redclaw farmed in Australia. Rather, a generic commercial feed modified from other aquatic species, particularly marine shrimp, is used (Saoud et al. 2012). Commercial chicken pellets, unprocessed legumes and/or food scraps are also often used as supplemental feed; however, this varies from farm-to-farm with the assumption that if a good crop is produced then the feeding regime must be adequate (Jones 1989; Saoud et al. 2012). Efforts to determine the nutritional requirements of crayfish species have increased in recent years with many studies looking at the suitability and digestibility of plant-based and other alternative protein sources to formulate cheaper and more nutritionally complete feed for culture (Campaña-Torres et al. 2005; Duffy et al. 2011; Pavasovic et al. 2007; Thompson et al. 2003, 2004, 2005). However, there has been little research conducted on what is actively being consumed by redclaw in commercial ponds under semi-intensive farming conditions. Understanding the contribution of naturally occurring organisms to the diet of crayfish and the role of supplemental feed is critical to advancing feed management strategies that will improve production efficiencies in pond aquaculture of redclaw.

Stable isotope analysis (SIA) is a method frequently used by ecologists to identify what food sources are assimilated by the organism and provides a comprehensive understanding of an organism's diet. Ingested material has a measurable ratio of heavy to light isotopes defining its isotopic signature (Fry 2007). The elements most often used in food web studies are carbon (13C and 12C) and nitrogen (15N and 14N) with both having two stable isotopes (Stenroth et al. 2006). Metabolic processing of ingested organic matter causes changes in the ratio of heavy to light isotopes (isotopic fractionation) of 13C:12C and 15N:14N which leads to distinct isotopic compositions (Jacob et al. 2005). The ratio of 13C:12C changes very little with increasing trophic level making carbon stable isotope analysis useful in indicating which primary producers are the important nutritional sources for consumers (Whitledge and Rabeni 1997). Similarly the 15N:14N ratio is a reflection of the consumers' food source with the enrichment of 15N consistent with increasing trophic level (Whitledge and Rabeni 1997). Thus, sources that contribute to growth can be identified by measuring the carbon and nitrogen stable isotope ratios of both the consumer and any of its potential food sources. The aim of this study was to use SIA to determine the extent to which naturally occurring pond organisms and supplemental feed contribute to the diet of *C. quadricarinatus* in commercial pond culture.

2. MATERIALS AND METHODS

2.1 Study sites

Field work was conducted across four farms located in the Atherton Tablelands region (17°16'S, 145°53'E) of North Queensland over 3 days. All four farms had different farming and feed management practices including the use of a variety of hides (e.g. old car tyres or milk crates, etc.), time and frequency of feeding, pond fertilisation methods, water source, and supplemental feed provided.

Mareeba Farm (17°01'37"S, 145°23'46"E) was situated in the semi-rural area of Mareeba. Ponds were all of the same size (approx. 45 m × 18 m × 0.5 m) and distribution across the farm area. Hides were old milk crates and hessian-type bags strung from ropes that ran across the pond. Unhulled raw soybeans were the only supplemental feed provided. Hay was used as a forage material at the beginning of the season.

Biboorha Farm (16°48'51"S, 145°31'33"E) is situated alongside the Barron River and in the middle of rugged bushland. Ponds were staggered along the hillslope making use of gravity fed water. Size and depth of the ponds varied slightly (approx. 53 m × 20 m × 1.5 m) with old tyres used as hides. Diamonnium phosphate (DAP) and hay were used as fertilisers. Supplemental feed was a combination of unhulled soybean and a commercial chicken feed.

Wondelca (17°18'34"S, 145°27'37"E) utilised natural spring and rain water to supply the farm. Pond size varied with the larger measuring approx. 53 m × 40 m and the average being around 40 m × 10 m. Depth was similar at 1.2 m. Supplemental feed was a combination of corn and a commercial redclaw pellet. Hay was also used as forage material at the start of the season.

Topaz Farm (17°24′18″S, 145°43′46″E) is located next to Mount Bartle Frere. Lush vegetation surrounded the ponds more so than was present at the other farms. Ponds were gravity fed with an approximate size of between 30 and 60 m (L) × 20 m (W) × 1.2 m (D). Supplemental feed was a combination of corn, unhulled lupin and the commercial redclaw pellet. A mixture of dolomite and lime was used as a fertiliser and was often combined with the supplemental feed during feeding.

2.2 Redclaw sample collection

Traps were set by the farmer the evening prior to sampling. On arrival at each site traps were raised and the contents collected. Berried females were released back into the pond. The remaining crayfish from each pond were placed in a plastic bag and euthanized in an ice-water slurry. (Mareeba Farm $n = 22$, Biboorha Farm$n = 26$, Wondecla Farm $n = 24$ and Topaz Farm $n = 23$).

2.3 Collection of feed samples

Two ponds at each farm ($n = 2$) were sampled for zooplankton, phytoplankton and any macro-invertebrates by dragging a 50 μm plankton net through the water column. The filtered material was then rinsed through a 500 μm mesh to separate any larger particulates, such as leaves and macro-invertebrates, before being stored in polyethylene containers (2 L per pond); sediment samples were collected by scooping approximately 2 cm of the topmost substrate layer into polyethylene containers (70 g per pond); any abundant plant matter seen growing in or around the pond margins was collected (100 g); hides were sampled for periphyton by scraping surfaces (30 g). Supplemental feed, fertiliser and forage material used by the farmers were also sampled (50 g of each feed source per farm).

All material requiring cold storage were kept in a 250 L esky filled with ice-water while in the field and then frozen to −20 °C on return to the lab while awaiting further preparation.

2.4 Sample preparation

Redclaw were removed from the freezer and slightly defrosted at room temperature to remove excess ice. Crayfish were then individually weighed to the nearest 0.01 g with an average of: Mareeba Farm = 54.29 ± 14.36 g; Biboorha Farm = 33.09 ± 8.93 g; Wondecla Farm = 43.99 ± 23.94 g; and Topaz Farm = 42.83 ± 12.11 g.

Stable isotope analysis was determined for whole-body samples as well as tail-muscle samples. The first segment of muscle (closest to the abdomen, exoskeleton removed) was sampled and weighed to the nearest 0.01 g, with the remaining tail muscle material included in the whole-body sample. Whole-body and tail-muscle sub-sample were homogenised with a small sample (1 g) taken to analyse moisture content. Material for moisture content analysis was oven-dried (105 °C) to a constant weight.

Pond water was defrosted and filtered through a sequence of mesh sizes (>500 μm, 500–300 μm, 300–60 μm, and 60–20 μm) to separate into zoo- and phytoplankton. Particles 60–300 μm were considered to be zooplankton with filtration to 20 μm being phytoplankton (Duffy et al. 2011; Meakin et al. 2009) Where there was enough material from 300 to 500 μm this was treated as another individual sample. As disintegration or lysis comprises cell integrity samples were inspected under a light microscope prior to freeze-drying. No damaged cells were observed and samples were considered to be suitable for further analysis. Sediment samples were later thawed and sifted to remove rocks and larger particulates.

2.5 Preparation for stable isotope analysis

Samples with high carbonate content such as the whole-body sample underwent additional treatment in preparation for carbon SIA (Feuchtmayr and Grey 2003). As the process to reduce the carbonate content can influence $\delta^{15}N$ values freeze dried whole-body, tail-muscle, and zoo- and phytoplankton samples were allocated into two groups. Those for nitrogen analysis were left untreated whilst those for carbon analysis were acidified by adding 2 M HCl drop-by-drop to 1 g material (with cessation of bubbling used as the indicator to determine the amount of acid to add (Bunn et al. 1995)). Acid treated samples were not washed with distilled water, instead HCl was removed by decanting after centrifuging the samples at 3000 rpm, a process which was repeated until removal of HCl was sufficient. This prevented a loss of $\delta^{13}C$ which can occur if washing in distilled water (after Carabel et al. 2006).

All samples were placed into −80 °C freezer 48 h prior to being freeze-dried for 24 h (Virtis benchtop 2 K, VWR, Australia). Dried samples were milled to a fine powder consistency. $\delta^{13}C$ and $\delta^{15}N$ values, carbon (%) and nitrogen (%) weight, and C:N ratios were determined using a Costech Elemental Analyzer fitted with a zero-blank auto-sampler coupled via a ConFloIV to a ThermoFinnigan DeltaV^{PLUS} using Continuous-Flow Isotope Ratio Mass Spectrometry (EA-IRMS) at James Cook University's Advanced Analytical Centre (AAC), Cairns. Stable isotope results are reported as per mil (‰) deviations from the VPDB and AIR reference standard scale for $\delta^{13}C$ and $\delta^{15}N$ values, respectively. Precisions (S.D.) on internal standards were better than ±0.1 and 0.2 ‰ for carbon and nitrogen, respectively.

2.6 Lipid extraction

Lipids are known to be depleted in $\delta^{13}C$, preferentially deposited as $\delta^{12}C$ (Bodin et al. 2007) with the removal of lipids routinely conducted, making samples isotopically heavier to better reflect the $\delta^{13}C$ of the diet in natural systems (Bunn et al. 1995; Stenroth et al. 2006). However, the resulting isotopic shift may be of little relevance if the effect on the estimated variable is negligible (Anthony et al. 2007). Further, diet estimates in a complex natural system (several sources) of a generalist consumer will likely be less impacted by isotopic variation compared to a specialist consumer in a simple (few sources) system (Tarroux et al. 2010). This will likely also be the case in an "artificial" food web of a semi-intensive aquaculture production system. To test this duplicate lipid and non-lipid extracted redclaw and source samples obtained from Mareeba Farm were compared. To extract lipid content samples were immersed initially in 100 % methanol and sonicated for 15 min, then in successive 2:1 Dichloromethane/Methanol solutions (each sonicated for 15 min), centrifuged and supernatant removed until clear. SIA was conducted as described above. No significant differences were noted for $\delta^{13}C$ between lipid and non-lipid extracted samples ($P > 0.05$). All subsequent SIA results are presented on non-lipid extracted samples.

2.7 Stable isotope data analysis

2.7.1 Multi-source mixing model

A multi-source mixing model estimating the relative contribution of dietary nitrogen and carbon to source material was used to calculate the range of all possible source contributions in systems with two or more sources. (Isosource; Phillips and Gregg2001, 2003). The multi-source mixing model iteratively creates each possible combination of source proportions (that sum to 100 %) in small increments (in this case 1 %). The predicted isotopic signatures for the mixture are compared with the observed signatures and if they fit within a small tolerance (i.e. 'mass balance tolerance'; Beatty 2006) it is therefore considered feasible and recorded. The resulting dataset represents all feasible solutions and distribution of proportions (Phillips and Gregg 2003). A mass balance tolerance of 0.1 % was initially used, however, as there were large numbers of potential food sources for each farm it was common for no feasible solutions to occur (indicated by zero observations being generated). As such, the minimum mass balance tolerance that resulted in a feasible solution was used. In the present study tolerances were between 0.2 and 2 % which are within the range used in previous studies with no bias introduced with larger tolerance levels (Benstead et al. 2006; Cole et al. 2011; Duffy et al. 2011; Stenroth et al. 2006)

Reporting the distribution of feasible solutions is recommended by Phillips et al. (2005) however, sources with large distributions tend to not reflect the number of observations (frequency) of the source contributions. To adequately represent the source contribution the distribution is supported by the mean ± SD. The distribution range is presented 1–99th percentiles as the full range is sensitive to the small observations present in the tails of the distribution (Phillips and Gregg 2003). Mean SI values of ponds ($n = 2$) were pooled together with the average representing the farm overall. The isotopic composition of potential food sources was examined in relation to whole animal and tail muscle. Additionally, the contribution of supplementary feed was examined in relation to zooplankton and phytoplankton.

A one-way ANOVA followed by Tukey's test on significant terms was conducted on the mean stable isotope values of whole animal and tail-muscle among farms. Significance level was accepted at $P < 0.05$.

3. RESULTS

3.1 Stable isotope data analysis

There was an overall significant difference between farms for whole-body $\delta^{13}C$ ($F = 27.23$, d.f. = 7, $P < 0.05$), and $\delta^{15}N$ ($F = 42.65$, d.f. = 7, $P < 0.05$), and for tail-muscle $\delta^{13}C$ ($F = 117.30$, d.f. = 7, $P < 0.01$) and $\delta^{15}N$ ($F = 54.025$, d.f. = 7, $P < 0.01$) (Fig. 1).

Tukey post hoc comparisons of the farms indicate that there were no significant differences in whole-body $\delta^{13}C$ and $\delta^{15}N$ between Wondecla Farm and Topaz Farm or Mareeba Farm and Biboorha Farm. There were significant differences for $\delta^{13}C$ between Mareeba Farm and Wondecla Farm ($P < 0.01$) and Mareeba Farm and Topaz Farm ($P < 0.01$) and between Biboorha Farm and Wondecla Farm ($P < 0.01$). Whole-body $\delta^{15}N$ differed significantly between Mareeba Farm and Wondecla Farm ($P < 0.05$), Mareeba Farm and Topaz Farm ($P < 0.01$), Biboorah Farm and Wondecla Farm ($P < 0.01$) and for Biboorah Farm and Topaz Farm ($P < 0.01$). Tail-muscle isotopic values were significantly different ($P < 0.01$) except for $\delta^{13}C$ between Wondecla Farm and Topaz Farm ($P < 0.05$) and $\delta^{15}N$ between Mareeba Farm and Biboorha Farm ($P < 0.05$).

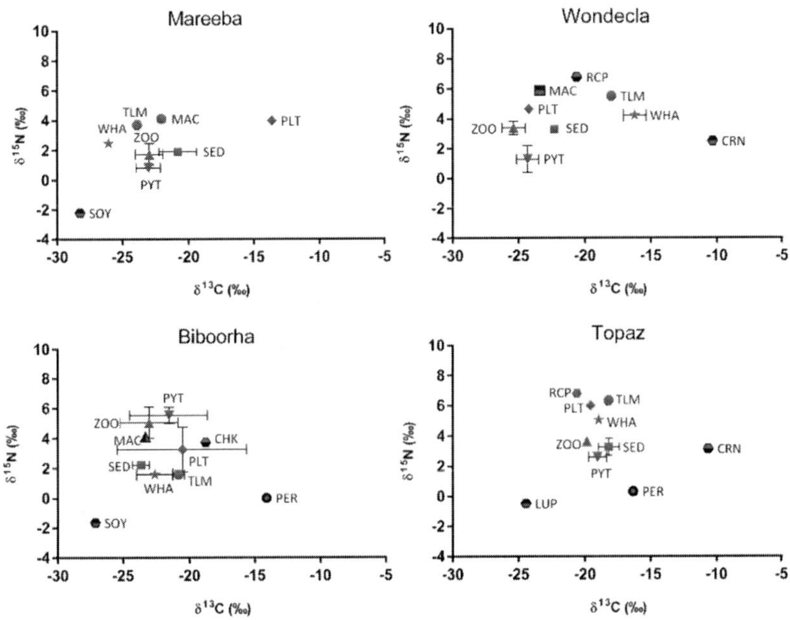

Figure 1. Isotopic signature of *C. quadricarinatus* and potential food sources sampled at four commercial redclaw farms. Plotted values are means, *error bars* represent ±SD. *WHA* whole animal, *PYT* phytoplankton, *ZOO* zooplankton, *SED* sediment, *PLT* plant material, *PER* periphyton, *MAC* macroinvertebrates, *LUP* lupin, *RCP* commercial redclaw pellet, *SOY* soybean, *CHK* chicken feed

3.2 Multi-source mixing model

Proportions of the different food sources that contribute to the isotopic signature of the crayfish were estimated using the multi-source mixing model. As no organic carbon values are present in diammonium phosphate

(Biboorha Farm) or dolomite and lime (Topaz Farm) these were excluded from the model. Results show the majority of supplemental feed provided is assimilated by redclaw. Similarly plant material and periphyton predominantly contribute when present in relation to supplemental feed and/or natural pond biota. Source contributions are presented in Table 1. Refer to Online Resources 1 (Fig. S1–4) for histograms generated of the feasible source contributions for each farm.

3.3 Whole-body: naturally occurring pond organisms

The highest contribution of naturally occurring pond biota occurred at the Mareeba Farm with zooplankton contributing between 0–57 % (mean ± SD; 17.4 ± 14.3 %) and 0–44 % (11.9 ± 10.7 %) for phytoplankton. Additionally, the contribution of macro-invertebrates was the highest for all farms ranging between 1 and 49 % (22.6 ± 10.7 %). Contribution of naturally occurring pond biota was lower at the remaining farms particularly at Wondecla Farm where zooplankton had a contribution of only 0–11 % (2.90 ± 2.80 %) and phytoplankton 0–9 % (2.10 ± 2.20 %).

3.4 Whole-body: supplementary feed

Supplemental feed was readily consumed at most of the farms. Corn used at Mareeba Farm had a contribution of between 25 and 52 % (mean ± SD; 37.7 ± 6.20 %). The contribution of supplemental feed was quite high in relation to other sources at Wondecla Farm with the commercial redclaw pellet of 11–41 % (29.9 ± 6.60 %) and corn ranging between 45 and 53 % (49.0 ± 1.70 %). While the redclaw pellet had a contribution at the Topaz Farm of 0 to 82 % (26.1 ± 19.1 %) plant material was greater with a feasible contribution of between 0 and 86 % (43.9 ± 19.5 %) and lupin contributing very little 0–12 % (1.60 ± 1.70 %). Conversely, chicken feed at the Biboorha Farm did not contribute greatly to the diet with a range of 0–37 % (10.0 ± 8.80 %) compared with soybean which ranged between 5 and 46 % (25.8 ± 8.80 %). The contribution of the other food sources sampled (naturally occurring pond biota, periphyton, plant material, etc.) were less easily distinguished at this farm.

Table 1. Summary of the average source contributions (%; ±SD) to the diet of farmed *C. quadricarinatus*

Source (Relative % contribution)	Farm							
	Mareeba		Biboorha		Wondelca		Topaz	
	Whole animal	Tail muscle	Whole animal	Tail muscle	Whole animal	Tail muscle	Whole animal	Tail muscle
Naturally occurring pond biota								
Zooplankton	17.4 ± 14.3	6.90 ± 6.20	9.00 ± 7.50	13.0 ± 11.0	2.90 ± 2.80	1.20 ± 1.40	6.10 ± 5.60	2.90 ± 2.90
Phytoplankton	11.9 ± 10.4	4.20 ± 4.00	8.30 ± 7.50	14.8 ± 12.2	2.10 ± 2.20	0.80 ± 1.10	4.50 ± 4.20	2.00 ± 2.20
Macro-invertebrates	22.6 ± 10.7	70.40 ± 4.40	10.2 ± 8.90	12.3 ± 10.47	8.40 ± 7.40	3.80 ± 3.60	–	–
Supplemental feed								
Corn	–	–	–	–	49.0 ± 1.70	27.6 ± 1.10	8.20 ± 3.10	8.30 ± 3.10
Redclaw Pellet	–	–	–	–	29.9 ± 6.60	63.0 ± 3.70	26.1 ± 19.1	18.2 ± 16.1
Soybean	37.7 ± 6.20	15.0 ± 2.40	25.8 ± 8.80	6.00 ± 4.90	–	–	–	–
Chicken feed	–	–	10.0 ± 8.80	16.40 ± 12.90	–	–	–	–
Lupin	–	–	–	–	–	–	1.60 ± 1.70	0.60 ± 0.90
Additional sources								
Sediment	7.70 ± 6.90	2.60 ± 2.60	15.6 ± 13.7	10.70 ± 9.40	3.90 ± 3.70	1.70 ± 1.80	6.50 ± 5.90	3.00 ± 3.10
Periphyton	–	–	9.40 ± 6.00	11.50 ± 7.20	–	–	3.00 ± 2.90	1.30 ± 1.50
Plant material	–	–	11.6 ± 10.3	15.20 ± 13.20	4.60 ± 4.20	2.00 ± 2.10	43.9 ± 19.5	63.6 ± 16.4

Blank cells indicate source material that was not used at a particular site

3.5 Assimilation of sources in tail-muscle

When the mixing model is applied to the tail-muscle the potential food source that is being assimilated by the crayfish for growth become more identifiable as there is a distinct difference in the contribution value to that of the whole-body (Table 1). At the Mareeba Farm where zoo- and phytoplankton had high contributions for whole-body this fell to 0–26 % (mean ± SD; 6.90 ± 6.20 %) and 0–16 % (4.20 ± 4.00 %), respectively. Macro-invertebrates increased to a contribution of between 60 and 81 % (70.4 ± 4.40 %). Soybean was also readily assimilated in the tail muscle at a range of 10–20 % (15.0 ± 2.40 %). At Wondecla Farm the contribution of the redclaw pellet was 55–72 % (63.0 ± 3.70 %), and corn the second highest source contribution at 25–30 % (27.6 ± 1.10 %). Naturally occurring pond biota and macro-invertebrates contributed less than 25 %. Plant material at the Topaz Farm remained high contributing between 13 and 86 % (63.6 ± 16.4 %) with the redclaw pellet falling slightly to 0–69 % (18.2 ± 16.1 %). Sources utilised for growth at Biboorha Farm are less distinguishable, however, soybean contribution was 0–20 % (6.0 ± 4.90 %) and chicken feed 0–53 % (16.4 ± 12.9 %). Overall, the contribution of zooplankton and phytoplankton in relation to tail-muscle was less than that observed in the whole-body analysis.

3.6 Contribution of supplementary feed to primary productivity

Stable isotope values for supplementary feed were also analysed in relation to their contribution to zooplankton and phytoplankton productivity (Table 2). As the Mareeba Farm only had one supplemental feed source (soybean) the minimum of three isotope mixtures required by the mixing model were not met and the farm was therefore excluded from this analysis (Phillips and Gregg 2003). For the remaining farms the results clearly show that supplemental feed acts as an additional nutrient source for primary pond productivity. The contribution of commercial chicken feed at Biboorha Farm ranged between 64–83 % (mean ± SD; 72.6 ± 4.70 %) for zooplankton and 71–99 % (83.4 ± 6.90 %) for phytoplankton, similarly, results for the redclaw pellet at Wondecla Farm were also high, 72–98 % (83.1 ± 6.50 %) and 58–97 % (50.0 ± 9.50 %), respectively. At the Topaz Farm, the redclaw pellet ranged from 34 to 74 % (54.8 ± 11.2 %) and 10–52 % (31.0 ± 11.2 %) with lupin comparable with ranges of 12–42 % (27.3 ± 8.10 %) for zooplankton and 23–54 % (38.7 ± 8.10 %) for phytoplankton.

Table 2. Feasible source contributions (%) of supplementary feed to primary productivity*

Source	Farm					
	Biboorha		Wondecla		Topaz	
	Phytoplankton (%)	Zooplankton (%)	Phytoplankton (%)	Zooplankton (%)	Phytoplankton (%)	Zooplankton (%)
Corn	–		0–29	0–20	18–42	6–30
Redclaw pellet	–	–	58–97	72–98	10–52	34–76
Soybean	0–20	16–30	–	–	–	–
Chicken pellet	71–99	64–83	–	–	–	–
Lupin	–	–	–	–	23–54	12–42
Hay	0–17	0–11	0–40	0–27	–	–

Blank cells indicate source material that was not used at a particular site

* Minimum of three sources required to calculate feasible contributions with Mareeba Farm not meeting this requirement

4. DISCUSSION

This study identified for the first time the feed sources contributing to the diet of redclaw grown in semi-intensive commercial pond aquaculture. SIA demonstrated that while supplemental feed is a primary nutrient source contributing to the diet of redclaw; the availability of naturally occurring sources appears to be a factor in the relative contribution of the supplemental feed to the diet. For, e.g. natural vegetation contributed a significant proportion of redclaw diet on some farms even though there was a number of supplementary sources provided (lupin, corn and commercial redclaw pellet). Further, raw lupin, a commonly used feed, had the lowest contribution for all supplemental feed sources sampled. More importantly, a clear contribution of the redclaw pellet, chicken feed, lupin, and soybean to both zoo- and phytoplankton production was expressed when the multi-source mixing model was applied to supplementary feed in relation to primary productivity. Chicken feed and lupin which were shown to provide little direct contribution to the diet of the redclaw, rather acting as direct C and N sources for primary pond productivity. These results have direct implication on the effectiveness of current feeding practices in Australian redclaw pond aquaculture, highlighting the benefit of providing a water stable formulated feed suited to the requirements of redclaw.

Current intensive and semi-intensive crayfish farming methods rely on the availability of natural food items as an important source of nutrition for freshwater crayfish species (Jones 1989; Meakin et al. 2009; Momot 1995). Using SIA Duffy et al. (2011) found that *Cherax destructor* consumed natural food items even in the presence of formulated feed. Similarly, zooplankton and biofilm (comprised of organisms such as flagellates, ciliates, rotifers, and nematodes) improved survival and growth of juvenile redclaw (Viau et al. 2012) and juvenile *C. destructor* (Verhoef et al. 1998). Production in ponds is initially stimulated with chemical fertilizers and/or organic matter (e.g. hay) (Jones et al. 1996) and some farms will routinely add additional fertiliser, e.g. dolomite and lime, during the culture period. Stimulation of primary pond productivity is important in semi-intensive culture, particularly when supplementary feeding doesn't provide adequate nutrition, and also because of their role in nutrient cycling and the influence on water quality (Moriarty 1997). Results from the current study corroborate this understanding as zoo- and phytoplankton, associated periphyton and plant material were also found to contribute to a similar or greater degree as supplemental feed. Conversely, the role of primary productivity in a system where adequate supplemental feed is not available is also demonstrated, particularly at the Mareeba Farm. Here, SIA and the multi-source mixing model show that the provision of soybean was secondary to the contribution of naturally occurring pond biota, in particular macro-invertebrates. The Mareeba Farm had little vegetation in and around the ponds (restricted by a steep pond edge) and the depth was quite shallow at ~0.6 m in comparison to the depth of the other farms

(~1.2 m). As crayfish are benthic in nature it is possible that the shallower ponds resulted in increased consumption of naturally occurring pond organisms due to their accessibility within the water column (Duffy et al. 2011; Meakin et al. 2009). Migration of *Daphina* spp. in shallow lakes has shown to result in very high densities of zooplankton in vegetated littoral zones (Burks et al. 2002; Burns 2000) and could allow for increased consumption by crayfish (Meakin et al. 2009). Indeed, redclaw are known to switch to a more herbivorous diet as they grow larger (Giling et al.2009; Momot 1995) and *C. destructor* of up to 45 g have been shown to actively respond to, and, prefer live zooplankton (Meakin et al. 2009). Crayfish from the Mareeba Farm were the largest by weight (54.29 ± 14.36 g (n = 22) and unlike other farms where multiple feed sources was supplied soybean was the only alternative food source provided. The contribution of macro-invertebrates to the diet is thought to be a reflection of the pond depth, lack of vegetation and alternative food sources.

Plant-based sources are palatable and readily consumed by redclaw (Cronin et al.2002) therefore the use of raw plant-based feeds would be expected to be reflected in the isotopic signatures of redclaw. However, this was not the case for a number of the supplementary sources (i.e. chicken feed, lupin, corn). While the commercial redclaw pellet is readily consumed and digested (Conrado, Pichette & Pirozzi; unpub. data 2014) it has an inherent low water stability (Gamble & Pirozzi, unpub data, 2014). Instead of the intact feed being directly consumed the disintegration of pellets and decomposing raw feed may form a nutrient-source and/or substrate for biofilm to accumulate (Loya-Javellana et al. 1993; Salame and Rouse 2000; Viau et al. 2012; Wood et al. 2012). Sources such as sediment, plant material and tyre scrapings (periphyton) were also found to contribute to a small degree to the diet of redclaw. Additionally benthic detritus and associated micro-organisms (biofilm) have been suggested to be a source of nutrition for juvenile redclaw (Jones 1989),*Procambarus clarkii* (Cronin et al. 2002), *C. destructor* (Giling et al. 2009) and*Pacifastacus leniusculus* (Bondar et al. 2005). It is unknown whether biofilm is improving the nutritional profile of the decomposing feed, or is an important dietary source for crayfish and is a concept that should be investigated further. Similarly, it is unknown if sources such as pond sediment and plant material have any nutritional benefit to the redclaw sampled at this farm, or whether accumulation of biofilm is improving the nutritional profile and palatability of such sources and therefore increasing intake.

In natural systems C. *destructor* and *C. cainii* are known to consume predominantly animal matter, such as small fishes and gastropods in summer; however, during winter, diets shift towards a more herbivorous predominance with crayfish consuming greater amounts of *Melaleuca spp.* when the abundance of high-protein food sources became limited (Beatty 2006). There are no studies that have documented this dynamic in extensive or semi-intensive aquaculture pond systems. Species richness and abundance of naturally occurring pond biota will likely vary seasonally, particularly at farms where water is sourced from natural river systems such as Biboorha Farm, which obtains water directly from the Barron River, and at farms with little to no filtration of incoming water. However, Stenroth et al. (2006) reported that there were no significant changes in isotopic signatures in the months or years of their study which investigated the use of stable isotopes as an indicator of the diet for the signal crayfish (*Pacifastacus leniusculus*) in the lakes of southern Sweden. The authors suggested that the crayfish either had a consistent diet throughout the seasons or that crayfish muscle tissue had a turnover rate slow enough to mask any differences of the diet during those months in-between sampling. Given the variability of pond productivity and performance of supplementary feed a nutritionally complete feed that is water stable, digestible by redclaw and therefore utilised more efficiently will result in improved food conversion ratios (FCRs), reduce nutrient loading in pond aquaculture systems and decrease harvest times, all of which have direct cost benefits to the farmer. Implementing more efficient feeds and feed practices for redclaw aquaculture will help to promote the growth of the industry.

When studying diet and trophic interactions the $\delta^{15}N$ enrichment of sources needs to be considered. In crustaceans $\delta^{15}N$ is reported to be approximately 2 ‰ (Vanderklift and Ponsard 2003) with values from Mareeba Farm and Biboorha Farm fitting this standard. Enrichment values for Wondecla Farm and Topaz Farm are higher at 4.21 and 5.04 ‰, respectively, however, $\delta^{15}N$ values of 5.7 ‰ (Duffy et al. 2011) and 6 ‰ (Rudnick and Resh 2005) have previously been reported for crayfish. The

$\delta^{15}N$ of the redclaw may be a reflection of the contribution of the redclaw pellet and plant material to the diet as they had $\delta^{15}N$ values of 6.78 ‰ and, respectively, 5.99 ‰. As the isotopic signatures identified for redclaw in the current study were determined from whole-body and tail-muscle samples, which are generally tissues that integrate isotopic signatures from the diet over a longer time period (Bodin et al. 2007), short term changes in feedstuffs should therefore have little influence on stable isotope signatures. The contribution of feed sources generally decreased slightly for tail-muscle compared to whole-body. There were some exceptions, however, particularly the contribution of the redclaw pellet at Wondecla Farm (63.0 ± 3.70 vs 29.9 ± 6.60 %), plant matter at Topaz Farm (63.6 ± 16.4 vs 43.9 ± 19.5 %) and macro-invertebrates at Mareeba Farm (70.40 ± 4.40 vs 22.6 ± 10.7 %). The reason for these increases are unknown and it is speculated that the utilisation efficiency and assimilation of these sources is different in muscle tissue compared to Whole animal (e.g. Stenroth et al. 2006). Regardless, those sources with the greatest contribution at each farm were the same whether looking at whole-body or tail muscle (Table 1).

Redclaw are known to have cannibalistic tendencies, particularly when individuals are most vulnerable when moulting (Barki et al. 1997). Jones and Ruscoe (2000) commented that high stocking densities (maximum density recommended to be between 9 and 15 m^2), insufficient supplemental feeding, and limited shelter options are other factors that can influence such behaviour. The farms in this study initially stock according to the industry recommendation of five redclaw per m^2 in a standard 1000 m^2 pond with sufficient shelter provided (Stevenson and Keast 2014). This is a relatively low stocking density but it is possible that some cannibalism still occurs. Regardless, stable isotope analysis cannot differentiate nutrient contribution through cannibalistic events and in such circumstances radio labelling techniques in a controlled trial would be more appropriate.

5. CONCLUSION

Both naturally occurring pond biota and supplemental feed contributed to the tissue composition of redclaw indicating the omnivorous nature of this species. In general, supplemental feed had a greater contribution to the diet than zooplankton, phytoplankton or macro-invertebrates. However, the contribution varied with the type of feed provided and the availability of alternative sources. Given the reliance and predominant use of relatively cheap and readily available raw feeds there is surprisingly little information available in relation to their suitability for redclaw. SIA techniques demonstrated that raw feeds such as soybean and corn were consumed to some degree but others like lupin were not. Further research into the use of different processing of raw feed could be beneficial (e.g. dehulling) give current feed practices; however, ultimately a move towards a nutritionally appropriate formulated feed is required to significantly improve redclaw production efficiencies. Some supplemental feeds sampled in this study such as commercial chicken and lesser extent redclaw pellets were shown to indirectly provide a nutrient supply for redclaw, instead providing a direct nutrient source for primary pond productivity. While maintaining healthy and productive natural pond biota is important to optimise growth of redclaw, in this case, the use of cheaper fertilisers would likely be more cost effective. The development of a formulated feed of an appropriate nutrient specification that meets the nutritional requirements of redclaw and has good water stability is vital for the industry to grow, particularly with the increase towards more intensive culture systems. Such a move would reduce the reliance on primary productivity and supplemental feed ultimately improving current feed management strategies and ensuring production security.

ACKNOWLEDGMENTS

The authors would like to thank John Stevenson, President of Queensland Crayfish Farmers Association (QCFA), J. Whan of Advanced Analytical Centre, Cairns and A. Pichette from the Centre for Sustainable Tropical Fisheries and Aquaculture for their assistance. This research was supported by James Cook University, the Queensland Crayfish Farmers Association and the Rural Industries Research and Develop Corporation (RIRDC; project no. 8536).

REFERENCES

1. Ackefors HE (2000) Freshwater crayfish farming technology in the 1990s: a European and global perspective. Fish Fish 1:337–359

2. Anthony RG, Deng X, Hung SSO, Miles AK, Ricca MA (2007) Effect of lipid extraction on analyses of stable carbon and stable nitrogen isotopes in coastal organisms of the Aleutian archipelago. Can J Zool 85:40

3. Barki A, Levi T, Shrem A, Karplus I (1997) Ration and spatial distribution of feed affect survival, growth, and competition in juvenile red-claw crayfish, *Cherax quadricarinatus*, reared in the laboratory. Aquaculture 148:169–177

4. Beatty SJ (2006) The diet and trophic positions of translocated, sympatric populations of *Cherax destructor* and*Cherax cainii* in the Hutt River, Western Australia: evidence of resource overlap. Mar Freshw Res 57:825–835

5. Benstead JP, March JG, Fry B, Ewel KC, Pringle CM (2006) Testing IsoSource: stable isotope analysis of a tropical fishery with diverse organic matter sources. Ecology 87:326–333

6. Bodin N, Le Loc'h F, Hily C (2007) Effect of lipid removal on carbon and nitrogen stable isotope ratios in crustacean tissues. J Exp Mar Biol Ecol 341:168–175

7. Bondar CA, Bottriell K, Zeron K, Richardson JS (2005) Does trophic position of the omnivorous signal crayfish (Pacifastacus leniusculus) in a stream food web vary with life history stage or density? Can J Fish Aquat Sci 62:2632–2639

8. Bunn S, Loneragan N, Kempster M (1995) Effects of acid washing on stable isotope ratios of C and N in penaeid shrimp and seagrass: implications for food-web studies using multiple stable isotopes. Limnol Oceanogr 40:622–625

9. Burks R, Lodge D, Jeppesen E, Lauridsen T (2002) Diel horizontal migration of zooplankton: costs and benefits of inhabiting the littoral. Freshw Biol 47:343–365

10. Burns CW (2000) Crowding-induced changes in growth, reproduction and morphology of Daphnia. Freshw Biol 43:19–29

11. Campaña-Torres A, Martinez-Cordova LR, Villarreal-Colmenares H, Civera-Cerecedo R (2005) In vivo dry matter and protein digestibility of three plant-derived and four animal-derived feedstuffs and diets for juvenile Australian redclaw, *Cherax quadricarinatus*. Aquaculture 250: 748–754

12. Carabel S, Godínez-Domínguez E, Verísimo P, Fernández L, Freire J (2006) An assessment of sample processing methods for stable isotope analyses of marine food webs. J Exp Mar Biol Ecol 336:254–261

13. Cole JJ, Carpenter SR, Kitchell J, Pace ML, Solomon CT, Weidel B (2011) Strong evidence for terrestrial support of zooplankton in small lakes based on stable isotopes of carbon, nitrogen, and hydrogen. Proc Natl Acad Sci 108:1975–1980

14. Cronin G et al (2002) Crayfish feeding preferences for freshwater macrophytes: the influence of plant structure and chemistry. J Crustacean Biol 22:708–718

15. Duffy RE, Godwin I, Nolan J, Purvis I (2011) The contribution of naturally occurring food items to the diet of*Cherax destructor* when fed formulated diets of differing protein levels. Aquaculture 313:107–114

16. Feuchtmayr H, Grey J (2003) Effect of preparation and preservation procedures on carbon and nitrogen stable isotope determinations from zooplankton. Rapid Commun Mass Spectrom 17:2605–2610

17. Food and Agriculture Organisation of the United Nations (FAO) (2012) The State of World Fisheries and Aquaculture. Rome. Retrieved from http://www.fao.org/docrep/016/i2727e/i2727e00.htm Accessed 20 August 2015

18. Fry B (2007) Stable isotope ecology. Springer Science & Business Media, New York

19. Giling D, Reich P, Thompson RM (2009) Loss of riparian vegetation alters the ecosystem role of a freshwater crayfish (*Cherax destructor*) in an Australian intermittent lowland stream. J N Am Benthol Soc 28:626–637

20. Heidenreich M (2013) Ross Lobegeiger report to farmers: aquaculture production summary for Queensland 2011–2012 fisheries Queensland. Department of Agriculture, Fisheries and Forestry

21. Hernández-Vergara MP, Rouse DB, Olvera-Novoa MA, Davis DA (2003) Effects of dietary lipid level and source on growth and proximate composition of juvenile redclaw (*Cherax quadricarinatus*) reared under semi-intensive culture conditions. Aquaculture 223:107–115

22. Jacob U, Mintenbeck K, Brey T, Knust R, Beyer K (2005) Stable isotope food web studies: a case for standardized sample treatment. Mar Ecol Prog Ser 287:251–253

23. Jones CM (1989) The biology and aquaculture potential of *Cherax quadricarinatus*, final report submitted by the Queensland Department of Primary Industries to the Reserve Bank of Australia Rural Credits Development Project No QDPI/8860

24. Jones CM (2002) Redclaw crayfish information pacakge: introductory information for prospective redclaw crayfish farmers. Northern Fisheries Centre, Department of Primary Industries, Cairns, Australia

25. Jones CM, Ruscoe IM (2000) Assessment of stocking size and density in the production of redclaw crayfish,*Cherax quadricarinatus* (von

Martens)(Decapoda: Parastacidae), cultured under earthen pond conditions. Aquaculture 189:63–71

26. Jones CM, Ruscoe IM, Fisheries F (1996) Production technology for redclaw crayfish (*Cherax quadricarinatus*). Fisheries Research & Development Corporation, Walkamin, Australia

27. Jones PL, Chavez JR, Mitchell BD (2002) Production of Australian freshwater crayfish in earthen-based systems using pelleted diets and forage crops as food. Aquac Int 10:157–175

28. Loya-Javellana G, Fielder D, Thorne M (1993) Food choice by free-living stages of the tropical freshwater crayfish, *Cherax quadricarinatus* (Parastacidae: Decapoda). Aquaculture 118:299–308

29. McClain W, Romaire R, Lutz CG, Shirley M (2007) Louisiana crawfish production manual. Louisiana State University Agricultural Center Publication 2637. Louisiana State University Agricultural Center, Baton Rouge

30. Meakin CA, Qin JG, Mair GC (2009) Zooplankton predation and consumption by freshwater crayfish, *Cherax tenuimanus*, at different sizes and light conditions. J Freshw Ecol 24:159–167

31. Momot WT (1995) Redefining the role of crayfish in aquatic ecosystems. Rev Fish Sci 3:33–63

32. Moriarty DJ (1997) The role of microorganisms in aquaculture ponds. Aquaculture 151:333–349

33. Pavasovic A, Anderson AJ, Mather PB, Richardson NA (2007) Effect of a variety of animal, plant and single cell-based feed ingredients on diet digestibility and digestive enzyme activity in redclaw crayfish, *Cherax quadricarinatus* (Von Martens 1868). Aquaculture 272:564–572

34. Phillips DL, Gregg JW (2001) Uncertainty in source partitioning using stable isotopes. Oecologia 127:171–179

35. Phillips DL, Gregg JW (2003) Source partitioning using stable isotopes: coping with too many sources. Oecologia 136:261–269

36. Phillips DL, Newsome SD, Gregg JW (2005) Combining sources in stable isotope mixing models: alternative methods. Oecologia 144:520–527

37. Rodríguez-Canto A, Arredondo-Figueroa JL, Ponce-Palafox JT, Rouse DB (2002) Growth characteristics of the Australian redclaw crayfish, Cherax quadricarinatus, cultured in an indoor recirculating system. J Appl Aquac 12:59–64

38. Rudnick D, Resh V (2005) Stable isotopes, mesocosms and gut content analysis demonstrate trophic differences in two invasive decapod crustacea. Freshw Biol 50:1323–1336

39. Salame MJ, Rouse DB (2000) Forage-based feeding in commercial red claw ponds in Ecuador. J Appl Aquac 10:83–90

40. Saoud I, Garza De Yta A, Ghanawi J (2012) A review of nutritional biology and dietary requirements of redclaw crayfish *Cherax quadricarinatus* (von Martens 1868). Aquac Nutr 18:349–368

41. Saoud IP, Ghanawi J, Thompson KR, Webster CD (2013) A review of the culture and diseases of redclaw crayfish *Cherax quadricarinatus* (von Martens 1868). J World Aquac Soc 44:1–29

42. Stenroth P, Holmqvist N, Nyström P, Berglund O, Larsson P, Granéli W (2006) Stable isotopes as an indicator of diet in omnivorous crayfish (*Pacifastacus leniusculus*): the influence of tissue, sample treatment, and season. Can J Fish Aquat Sci 63:821–831

43. Stevenson J, Keast B (2014) In: Asher D, Curtis M (eds) Redclaw crayfish farming: getting started. Masa Services, Sunshine Coast, Australia

44. Tarroux A, Ehrich D, Lecomte N, Jardine TD, Bêty J, Berteaux D (2010) Sensitivity of stable isotope mixing models to variation in isotopic ratios: evaluating consequences of lipid extraction. Methods Ecol Evol 1:231–241.

45. Thompson KR, Muzinic LA, Christian TD, Webster CD, Rouse DB, Manomaitis L (2003) Effect on growth, survival, and fatty acid composition of Australian red claw crayfish *Cherax quadricarinatus* fed practical diets with and without supplemental lecithin and/or cholesterol. J World Aquac Soc 34:1–10

46. Thompson KR, Muzinic LA, Engler LS, Morton SR, Webster CD (2004) Effects of feeding practical diets containing various protein levels on growth, survival, body composition, and processing traits of Australian red claw crayfish (*Cherax quadricarinatus*) and on pond water quality. Aquac Res 35:659–668

47. Thompson KR, Muzinic LA, Engler LS, Webster CD (2005) Evaluation of practical diets containing different protein levels, with or without fish meal, for juvenile Australian red claw crayfish (*Cherax quadricarinatus*). Aquaculture 244:241–249

48. Vanderklift M, Ponsard S (2003) Sources of variation in consumer-diet δ15N enrichment: a meta-analysis. Oecologia 136:169–182.

49. Verhoef GD, Jones PL, Austin CM (1998) A comparison of natural and artificial diets for juveniles of the Australian freshwater crayfish *Cherax destructor*. J World Aquac Soc 29:243–248

50. Viau VE, Ostera JM, Tolivia A, Ballester EL, Abreu PC, Rodríguez EM (2012) Contribution of biofilm to water quality, survival and growth of juveniles of the freshwater crayfish *Cherax quadricarinatus* (Decapoda, Parastacidae). Aquaculture 324:70–78

51. Whitledge GW, Rabeni CF (1997) Energy sources and ecological role of crayfishes in an Ozark stream: insights from stable isotopes and gut analysis. Can J Fish Aquat Sci 54:2555–2563

52. Wood SA, Phillips NR, de Winton M, Gibbs M (2012) Consumption of benthic cyanobacterial mats and nodularin-R accumulation in freshwater crayfish (Paranephrops planifrons) in Lake Tikitapu (Rotorua, New Zealand). Harmful Algae 20:175–179

Chapter 7

Prokaryotic Horizontal Gene Transfer in Freshwater Lakes: Implications of Dynamic Biogeochemical Zonation

Christopher N. Drudge, Lesley A. Warren

School of Geography and Earth Sciences, McMaster University, Hamilton, Canada.

ABSTRACT

The highly adaptive nature of prokaryotic communities in the face of changing environmental conditions reflects in part their ability to share advantageous genetic information through horizontal gene transfer (HGT). Natural freshwater lacustrine (lake) systems are a vital and finite resource, and the influence of HGT on their quality (e.g. enabling the spread of antibiotic resistance and xenobiotic catabolism genes) is likely significant. Laboratory and in situ studies indicate that the dynamic physical, chemical, and biological conditions that structure freshwater systems can influence HGT within freshwater prokaryotic communities. Thus, understanding how biogeochemical parameters impact HGT in freshwater lakes is an emerging knowledge gap with potential implications for ecosystem and human health on a global scale. In this review, we provide a general synopsis of what is known about HGT in freshwater prokaryotic communities, followed by an integrated summary of current knowledge identifying how biogeochemical factors may influence prokaryotic HGT in freshwater lacustrine systems.

KEYWORDS

Gene Transfer; Freshwater; Water Quality; Prokaryotic; Biogeochemistry

1. INTRODUCTION

Intercellular horizontal gene transfer (HGT), the movement of genetic information between cells by mechanisms operating independently of cell division (i.e. vertical gene transfer), has contributed to the macroevolution of prokaryotes and their attainment of tremendous structural and functional diversity [1,2]. Further, the highly adaptive nature of prokaryotic communities in the face of changing environmental conditions reflects in part their ability to share advantageous genetic information through HGT. The horizontal spread of antibiotic resistance genes among pathogenic bacteria [3], catabolic genes among bacteria inhabiting soil contaminated with organic pollutants [4,5], and foreign genes carried by genetically engineered organisms into natural prokaryotic communities [6,7] highlights the breadth of the environmental impact of prokaryotic HGT.

Prokaryote-mediated biogeochemical processes are fundamental to aquatic ecosystem functioning [8], such that functional community changes resulting from HGT may have substantial repercussions. The dissemination of introduced genetic information in aquatic prokaryotic communities via HGT can potentially impact water quality by such diverse processes as altering their health hazard potential (e.g. acquisition of introduced pathogenicity-enhancing genes) or ability to carry out particular metabolic pathways (e.g. degradation of recalcitrant organic pollutants). Natural freshwater lacustrine (lake) systems are a vital and finite resource, collectively accounting for 94% of the world's fresh surface water by volume [9]. As water use increases and water quality decreases, it is increasingly clear that our freshwater resources have gone beyond a critical status on a global scale, where many countries are without adequate safe and secure water supplies. Reflecting the links between water quality and HGT, the growing spread of clinically derived antibiotic resistance genes, an emerging class of environmental contaminants [10,11], in freshwater prokaryotic communities by HGT has been the focus of recent attention [12-14]. Laboratory and in situ studies indicate that the dynamic physical, chemical, and biological conditions that structure freshwater systems can influence the nature, extent and rates of HGT within freshwater prokaryotic communities. However, mechanistic understanding of how biogeochemical parameters impact HGT in freshwater lakes is an emerging knowledge gap for both ecosystem and human health on a global scale. This review provides an integrated summary of current knowledge identifying how biogeochemical factors influence the horizontal flow of genetic information in freshwater lacustrine prokaryotic communities.

2. PROKARYOTIC HORIZONTAL GENE TRANSFER IN FRESHWATER SYSTEMS

Intercellular HGT consists of the acquisition, stable assimilation, and expression of foreign DNA via specific cellular mechanisms. Foreign DNA derived from a donor cell is introduced into a recipient cell predominantly via 1) direct intercellular exchange (conjugation); 2) virusmediated transport (transduction); or 3) uptake from the extracellular environment (transformation). As the emphasis of this review is on the regulation of HGT by variable biogeochemical parameters within freshwater lacustrine systems, the reader is referred to specific reviews covering conjugation [15], transformation [16], and transduction [17] for more detailed discussion of the mechanisms themselves.

Single-cell DNA sequencing of bacteria inhabiting freshwater lakes has revealed evidence for inter-phyla HGT [18] and the three principal mechanisms of intercellular HGT have been observed in situ in freshwater environments (e.g. [19-21]); however little is known about the magnitude and relative contributions of each mechanism, the indigenous participant prokaryotes, or the overall impact of freshwater HGT on water quality and prokaryotic microevolution. In the Baltic Sea, genomic sequence analysis of Shewanella baltica isolates from various depths within the water column indicated that genes enabling adaptation to particular environmental conditions (e.g. anaerobic metabolism genes) were horizontally exchanged to a greater extent at more similar depths (i.e. biogeochemical conditions) as well as over seasonal time frames [22]. This finding suggests that HGT rates may vary with water column depth-dependent biogeochemical zonation within freshwater zystems, and that the horizontal transfer of ecologically relevant genes occurs over timescales that are sufficiently rapid to have appreciable system impacts.

Most studies of aquatic HGT have been carried out in microcosms, which while capable of providing approximate in situ transfer rates [23] and permitting some control of individual environmental parameters, cannot entirely replicate the complexity of in situ conditions [24]. These studies have predominantly utilized culture-dependent approaches, although more recent investigations have utilized molecular microscopy approaches to directly assay HGT within freshwater prokaryotic communities, albeit under laboratory conditions. Shintani et al. [25] detected the conjugative transfer of a plasmid tagged with a fluorescent protein gene in river water samples. Kenzaka et al. [26] directly detected transduction in a freshwater river prokaryotic community mediated by three Escherichia coli-infecting phages, including one previously isolated from a freshwater river, using a DNA amplification-based in situ microscopy technique. A similar approach was used to detect the uptake and expression of an introduced plasmid tagged with a fluorescent protein gene by indigenous prokaryotic communities in two Japanese rivers up to three weeks following plasmid introduction [27,28]. A small number of field studies of freshwater HGT have been performed to date, and will be briefly described in the following section.

2.1. Conjugation

Conjugative plasmids have been isolated from bacterial communities in freshwater lakes [29] and rivers [30-33], indicating that conjugation is common in freshwater systems. O'Morchoe et al. [20] detected conjugative plasmid transfer in situ between strains of the freshwater bacterium Pseudomonas aeruginosa introduced into a semi-permeable chamber suspended in the upper pelagic waters of a freshwater lake. Conjugative plasmid transfer has also been demonstrated in situ in biofilms on the submerged surfaces of river stones (epilithon) between introduced strains of P. aeruginosa [30-32] and from an indigenous freshwater epilithic prokaryotic community to an introduced recipient Pseudomonas strain [32,34].

2.2. Transduction

Viruses are ubiquitous in freshwater environments, typically exceeding bacteria in concentration by an order of magnitude [35]. Phages (for the purposes of this review, viruses that infect prokaryotes) constitute the majority of freshwater viruses, exhibit high diversity, and consistently infect a substantial portion of freshwater prokaryotic cells, indicating that transduction is an important form of HGT in freshwater systems [36]; however to date this remains largely unconstrained. Most known species of bacteria can be infected by phages, some of which can infect more than one species or even genus [37]. Many prokaryotic cells in freshwater systems are lysogenic, meaning that phage DNA capable of being reactivated to produce viral particles has been integrated into their chromosomes as inducible prophage [19]. Induction of prophage in lysogenic cells can potentially stimulate both transduction and transformation by increasing the availability of phage particles and extracellular DNA, respectively, which are released following cell lysis. Phage particles may be rapidly inactivated in aquatic environments [36], so the frequency and induction of lysogeny in prokaryotic cells are important determinants of transduction potential in aquatic environments [19].

Transduction of chromosomal and plasmid DNA has been demonstrated in situ between strains of Pseudomonas aeruginosa intentionally infected with species-specific phages and introduced into a semi-permeable chamber suspended in the upper pelagic waters of a freshwater lake

[19,38-40]. Transduction between introduced lysogenic bacterial strains following natural induction of lysogeny has also been demonstrated in situ in river epilithic biofilms [41].

2.3. Transformation

Transformation in aquatic environments is contingent upon the availability of genetically competent (able to uptake extracellular DNA) prokaryotes and extracellular DNA sharing homology with the intracellular DNA of recipient competent cells. The reader is referred to Nielsen et al. [42] for a detailed review of the release and fate of extracellular DNA in aquatic environments. Dissolved (<0.2 µm) extracellular DNA including fragments of sufficient size to encode prokaryotic genes has been measured at concentrations on the order of 10-1 to 101 µg/l in bulk water from freshwater lakes, reservoirs, ponds, and rivers [43-45], with transformable free DNA molecules (as opposed to viral or colloid-associated DNA) constituting a major form of dissolved DNA in aquatic systems [44,46]. Extracellular DNA is also a component of the extracellular polymeric substances that surround prokaryotic cells in biofilms [47]. Sources of prokaryotic extracellular DNA in aquatic systems include excretion by living cells and release from dead cells following grazing or lysis (e.g. phage-induced) [48,49]. Aquatic extracellular DNA is degraded by extracellular and cell-associated nucleases, genotoxic chemicals, as well as solar ultraviolet radiation [50]. In particular, the activity of extracellular nucleases can strongly influence the persistence of extracellular DNA in freshwater environments [51].

In situ transformation in freshwater systems remains poorly characterized. The transformation of an introduced recipient strain of Acinetobacter calcoaceticus by co-introduced extracellular DNA from cell lysates or intact donor A. calcoaceticus cells has been demonstrated in situ in river epilithic biofilms [21].

While these studies indicate the occurrence of prokaryotic HGT within freshwater systems, their relevance is limited by two important factors.

Firstly, they were restricted to the lighted and well-oxygenated temperate environments of either biofilms on the submerged surfaces of river stones or the upper offshore open waters of lakes, poorly representing the range of habitats and extensive biogeochemical variation found within freshwater systems. Secondly, they employed laboratory-grown strains of freshwater-inhabiting species of Acinetobacter or Pseudomonas bacteria (both belonging to the Pseudomonadales order of Gammaproteobacteria) as donors (cells, DNA, or phage particles from lysogens) and/or recipients. In stark contrast, environmental prokaryotic communities are well known to be highly diverse and remain largely uncharacterized [52,53] and Gammaproteobacteria are typically not major constituents of freshwater lake bacterial communities [54]. Of key significance, prokaryotes utilize a wide variety of metabolic pathways to obtain energy and organic carbon for growth. Autotrophic prokaryotes (e.g. photoautotrophs, sulfuroxidizing lithoautotrophs) are widespread in freshwater environments, and heterotrophs capable of using terminal electron acceptors other than oxygen (e.g. ironand sulphur-reducing bacteria) can dominate prokaryotic communities in oxygen-poor and anoxic regions (e.g. bottom sediments and bottom waters of productive lakes). These metabolic guilds are capable of substantially altering their local environments (i.e. generating biogeochemical zonation) such that they and their neighbours encounter distinct biogeochemical conditions [55-58] that may differentially influence HGT within freshwater systems. Further, these prokaryotes are often challenging to grow under laboratory conditions owing to their oxygen, nutrient and/or consortial requirements and so are likely underrepresented among HGT studies to date. The potential influence of the particular metabolism a prokaryote is carrying out on HGT remains essentially undefined. Thus, the characterization of HGT based principally on only two species of bacteria of the same order in a highly limited range of habitats and biogeochemical conditions means that the full context of freshwater HGT has yet to be determined.

3. BIOGEOCHEMICAL FACTORS INFLUENCING HGT IN FRESHWATER LAKES

The biogeochemical conditions found in freshwater lacustrine systems can potentially influence all mechanisms of HGT (**Figure 1**(a)). Many of the factors known to influence HGT typically exhibit spatial and temporal variation among and within freshwater lakes. While the impacts of most of these factors have been individually assessed with respect to HGT, most studies have assessed impacts in isolation, impeding understanding of the role of environmental interactions or spatial/temporal variability of these factors within freshwater lakes.

Field studies of freshwater HGT to date have been highly limited in their scope of system characterization, and with the exception of a small number of studies examining temperature [21,30-32] or suspended particulate matter concentrations [40], lack data linking biogeochemical parameters to observed HGT rates in the particular freshwater compartment investigated. Most studies of biogeochemical factors influencing HGT have been conducted in vitro, utilizing a limited number of easily culturable chemoorganoheterotrophic bacteria grown aerobically in nutrient rich media under highly constrained laboratory conditions. Such an approach is poorly representative of the myriad permutations of biogeochemical conditions to which freshwater prokaryotic communities are typically exposed, as well as their substantial phylogenetic and metabolic diversity. This has limited the generation of new insight into environmenttally linked HGT and reduced the relevance of any findings for environmental systems. Here we provide a summary of what has been shown for these factors in terms of their influence on HGT to date, providing context for what is known about their variability within freshwater lacustrine systems where possible.

3.1. Temperature

Conjugative plasmid transfer rates for laboratory cultures of Gram-negative bacteria have been observed to be temperature sensitive, potentially reflecting the limited expression of conjugative machinery within certain temperature ranges independent of optimal temperature ranges for growth [59,60]. The relationship between temperature, optimal growth, and HGT in natural freshwater prokaryotic communities is likely to be appreciably more complex. In stream epilithic biofilms, rates of conjugation and transformation between laboratory strains of bacteria have been positively correlated with temperature over seasonal timeframes, suggesting that HGT increases during summer months in temperate climates, particularly in surface waters warmed by solar radiation, and consistently occurs with greater frequency in warmer climates [21,30-32]. However, the observed influence of temperature on HGT in these studies may simply reflect the optimal temperature range for growth of the mesophilic bacterial strains that were used. Temperature was positively correlated with the abundance, production, and specific growth rate of marine heterotrophic bacteria throughout Chesapeake Bay during non-summer seasons [61], suggesting that it may similarly control the growth of prokaryotes in temperate freshwater lakes and their metabolic capacity for participating in HGT.

Transduction may be influenced by temperature, but the direction and magnitude of this relationship are currently poorly defined. Viral decay rates are generally reduced at lower temperatures in aquatic systems [36], suggesting that transduction rates are greater in colder water due to greater phage availability. However, viral abundance has been significantly positively correlated with temperature in a thermally stratified freshwater reservoir [62] and prophage induction was demonstrated in vitro to increase by several orders of magnitude with an increase in temperature from 22°C to 28°C in the presence of ultraviolet radiation [63], so the relationship between temperature and transduction in freshwater systems remains unclear.

Transformation rates may be greater in cooler waters as the reduced activity of nucleases at lower temperatures increases the availability of transformable extracellular DNA [64]. Using water collected from different regions within a thermally stratified freshwater lake during summer, Matsui et al. [65] found that degradation of introduced exogenous plasmid DNA with a concurrent decrease in transformation efficiency was observed over several days in epilimnetic water (18.0°C) but not in hypolimnetic water (7.8°C), both held at the same temperature as measured in the field. Taken together, these observations indicate that rates of transformation may be higher in summer hypolimnetic regions of temperate lakes. However, what is not known is whether these findings are directly translatable to rates of transformation within the naturally occurring prokaryotic communities within these regions.

3.2. Dissolved Oxygen

As depth-related variation in the concentration of dissolved oxygen can often mirror thermal stratification in freshwater lakes (**Figure 1**(b)), it is inherently difficult to tease out the separate effects of temperature and oxygen concentration on freshwater HGT. Further, no in situ studies examining the influence of dissolved oxygen concentration on HGT in freshwater systems have been published to date. As discussed previously, under low oxygen and anoxic conditions prokaryotes often utilize terminal electron acceptors other than oxygen, and so may distinctly change their local environments, in turn influencing HGT. In a laboratory setting, the efficiency of competence development and frequency of transformation were both reduced in the facultative anaerobe Streptococcus pneumoniae under anaerobic conditions [66]. Similarly, lower rates of conjugation have been observed under anaerobic conditions relative to aerobic conditions in laboratory studies utilizing the facultative anaerobe Escherichia coli [67,68]. Conjugation rates were observed to increase 10-fold following increased aeration of a laboratory culture of E. coli [69]. Notably, these organisms are not typically dominant or key players in freshwater prokaryotic communities observed within freshwater systems and thus the relevance of these findings to environmental prokaryotes, many of which are autotrophic and/or obligate anaerobes, is unknown. However, collectively, these findings indicate that conjugation and transformation may be inhibited in anoxic zones, both in bulk (e.g. hypolimnion of highly productive thermally stratified lakes) and microenvironments (e.g. within biofilms), among facultatively anaerobic members of inhabitant prokaryote communities. Laboratory research clearly indicates that HGT among bacteria capable of both aerobic and anaerobic growth is influenced by dissolved oxygen concentration; however in situ studies are required to characterize this regulation in freshwater lacustrine systems.

3.3. Chemical Agents

A suite of natural and anthropogenic chemicals can potentially influence HGT in freshwater lakes, although their in situ effects largely remain uncharacterized. Chemicals can damage prokaryotic cells, phage particles, and extracellular DNA, limiting their participation in HGT. Detergents, mutagens, and antibiotics have been observed to inhibit conjugation in vitro, either directly or indirectly via induction of plasmid degradation and/or selective toxicity targeting plasmid-containing cells [70, 71]. Conversely, in vitro exposure to chemicals has been demonstrated to stimulate HGT between bacterial species otherwise unlikely to exchange genetic information, likely by inactivating restriction enzymes that normally limit interspecific HGT [71-73].

Genotoxic chemicals including antibiotics and trace elements can promote all three principal mechanisms of HGT in vitro by activating bacterial stress responses (e.g. SOS response), leading to the induction of competence [74,75], prophage [76,77], and expression of conjugation machinery [78]. In particular, prophage induction via activation of the SOS response in lysogenic strains of both Gram-positive and Gram-negative bacteria has been demonstrated in vitro for a variety of chemicals at sublethal concentrations including antibiotics, antineoplastic drugs, halogenated organics, polycyclic aromatic hydrocarbons, and trace elements [79-83]. SOS response activation was observed in cultures of the aquatic bacterium Vibrio cholerae following exposure to subinhibitory concentrations of most major groups of clinically used antibiotics [84], indicating that antibiotic-mediated stimulation of transduction can occur in freshwater prokaryotic communities. Collectively, in vitro investigations of the influence of chemical agents on HGT via bacterial stress response activation suggest that freshwater lake compartments containing biologically accessible genotoxic chemicals at concentrations sufficient to induce stress responses are sites of higher HGT rates. Supporting this conclusion, a genotoxicity assay based on prophage induction in Escherichia coli produced positive results for bulk water samples collected at sites along a freshwater river most highly contaminated by trace elements [85].

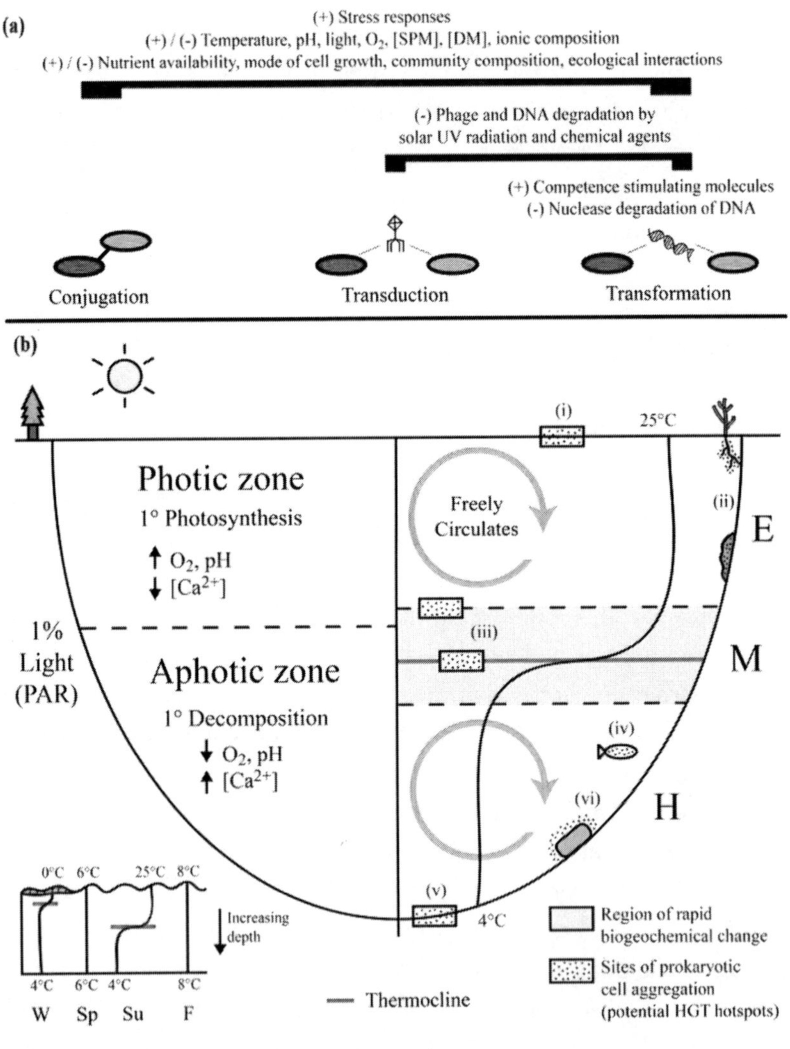

Figures 1. Physical, chemical, and biological parameters capable of regulating horizontal gene transfer (HGT) can vary appreciably over space and time within freshwater lacustrine systems. (a) Biogeochemical

parameters as a whole remain poorly characterized with respect to their in situ influence on HGT. Many parameters may alternatively stimulate (+) or inhibit (−) all three principal mechanisms of gene transfer (conjugation, transduction, and transformation) in freshwater systems based on their magnitude and interactions with other parameters. In general, activation of stress responses can stimulate HGT. Degradation of phage particles and extracellular DNA, the external vehicles for transduction and transformation, respectively, can inhibit these mechanisms. Competence stimulating molecules, produced by prokaryotes and algae, can promote transformation; (b) Biogeochemical parameters capable of influencing horizontal gene transfer can vary appreciably within freshwater lakes. In particular, the vertical physical separation of chemical and biological processes by depth-dependent changes in light availability and thermal density can lead to the formation of distinct local environmental conditions that likely have a specific cumulative influence on HGT. In general, HGT is enhanced in surficial aggregations of metabolically active cells (i.e. biofilms). Sites representing potential hotspots for HGT within freshwater lakes include 1) the air-water interface; 2) aquatic plants and algae; 3) vertical zones of rapid density and subsequent biogeochemical change within the metalimnion (e.g. thermocline); 4) aquatic animals; 5) bottom sediments and the sediment-water interface; and 6) submerged abiotic surfaces (e.g. rocks). SPM = suspended particulate matter, DM = dissolved matter; PAR = photosynthetically active radiation; W = winter, Sp = spring, Su = summer, F = fall; E = epilimnion, M = metalimnion, H = hypolimnion.

Horizontally mobile genetic elements including integrons, transposons, and plasmids were found to be more abundant in freshwater river sediments exposed to waste water containing relatively high concentrations of antibiotics [86]. Similarly, exposure to trace elements has been associated with increased abundance of IncP-1 plasmids in freshwater river sediments [87] and class 1 integrons in freshwater river epiphytic biofilms, bulk surface water, and bottom sediments [88,89]. The general extent of freshwater system contamination with chemicals has been positively correlated to the incidence of potentially mobile plasmids in bacterial isolates [90]. As the abundance of mobile genetic elements can be used as an indicator of relative HGT potential in prokaryotic communities [88], these findings support the notion that freshwater lake compartments with relatively high concentrations of antibiotics and trace elements may be hotspots for HGT.

Chemical agents capable of influencing HGT are principally introduced into freshwater lacustrine systems as a consequence of anthropogenic activities [91,92]. Generally speaking, the lowest concentrations of trace elements and antibiotics necessary to stimulate HGT in vitro are several orders of magnitude greater than concentrations typically found in freshwater environments [93,94], so it is unclear whether bacterial stress response activation by these chemicals is relevant to HGT regulation in most freshwater systems. However, chemicals may be sufficiently enriched in compartments within lakes that they can locally influence HGT. Trace elements and organic pollutants can accumulate within bottom sediments and at the sediment-water and air-water interfaces of aquatic systems [95-97], and aquatic biofilms can sorb a wide variety of inorganic and organic chemicals [47,98,99], suggesting that these compartments could be hotspots for HGT within freshwater lakes. Antibiotics are naturally produced by common freshwater microorganisms includeing Actinobacteria and filamentous fungi, and so may influence HGT in freshwater habitats in which these microorganisms are active and abundant. As discussed previously, prokaryotes are capable of a wide variety of metabolisms, and particular metabolisms may distinctly alter biogeochemical conditions, in turn influencing HGT, although this remains essentially undefined. For example, iron-reducing bacteria in anoxic lake bottom sediments can degrade recalcitrant organic pollutants and liberate trace elements through the dissolution of iron minerals [58], potentially resulting in their differential exposure to chemicals that can regulate HGT via activation of stress responses. Prokaryotes can also transform the redox state of trace elements such as arsenic, altering their mobility, bioavailability, and toxicity [100], and ultimately their influence on HGT.

3.4. Solar Ultraviolet Radiation

As with chemicals, solar ultraviolet radiation can potentially stimulate or inhibit all three principal mechanisms of HGT by activating bacterial stress responses [101] and damaging participating bacterial cells [102], phage particles [102-104], and DNA [105], respectively. In situ effects of solar ultraviolet radiation on HGT in freshwater lacustrine systems are not well characterized. Ultraviolet wavelengths are rapidly attenuated within water bodies and thus their impacts will largely be restricted to the upper 1 - 2 meters of a lake. Prophage induction in lysogenic prokaryotic cells following exposure to solar radiation has been observed in water samples collected from freshwater lakes [106,107]. Exposure of microcosms prepared in situ from seawater to solar radiation increased prokaryotic mortality and production of viral particles, suggesting an increase in transduction rates following prophage induction [37]. Consistent with the surficial limitation of ultraviolet radiation impacts, Tapper and Hicks [106] determined that the surface microlayer at the air-water interface of a freshwater lake was enriched in virus particles relative to the underlying epilimnion water, indicating the occurrence of appreciable prophage induction by the high intensity of solar ultraviolet radiation incident on the water surface. Ozonedriven increases in solar ultraviolet-B radiation over the last several decades with concurrent negative impacts on aquatic ecosystems [108] indicate that solar ultraviolet radiation will continue to increase in importance as a regulator of HGT in lacustrine systems.

3.5. Suspended Particulate Matter and Dissolved Matter

Suspended particulate matter and dissolved matter (typically operationally defined as materials suspended in the water column with diameters greater and less than 0.45 µm, respectively) can influence rates of HGT in freshwater lakes through a number of processes. Both contribute to the attenuation of solar radiation and thereby limit the vertical region in which it can directly influence HGT. In particular, dissolved organic matter, a major form of dissolved matter in most freshwater systems, can strongly reduce ultraviolet penetration [109]. Photo-oxidation mediated by solar ultraviolet-B radiation in humic (organic carbon rich) lakes was observed to be restricted to the top couple of centimeters [110], suggesting that the vertical region in which solar ultraviolet radiation can appreciably influence

HGT is further restricted in systems with higher dissolved organic matter concentrations. Concentrations of dissolved organic matter are generally higher in highly biologically productive systems, reflecting their in situ production. Dissolved organic matter is also higher in lake systems impacted by anthropogenic organic-rich wastewater inputs (e.g. sewage, agricultural runoff) or those receiving significant wetland inputs. The influence of dissolved organic matter on HGT may be increasingly pronounced in Boreal Forest lake systems impacted by climate change, as melting permafrost is driving significant exports of organic matter via surrounding organic-rich wetlands to downstream lakes.

Conversely, the presence of suspended particulate matter can potentially stimulate all three principal mechanisms of HGT by providing surfaces for the adsorption of prokaryotic cells, phage particles and extracellular DNA, which stabilizes and protects these HGT participants from damage by extracellular enzymes, toxic chemicals, and solar ultraviolet radiation while also increasing the probability of cell-cell, DNA-cell, and phage-cell encounters and resultant HGT [40,51,111-114]. Furthermore, particulate-associated prokaryotes are typically more metabolically active than planktonic (free-living) prokaryotes [115,116], suggesting that the former can participate to a greater extent in HGT. Ripp and Miller [40] measured a 100-fold increase in the number of transductants when sterilized freshwater microcosms were prepared using water collected near a lake bottom and enriched in suspended particulate matter compared to surface water with little particulate matter. The addition of clay mineral particles has also been shown to enhance in vitro transduction rates [40]. The affinity of a phage particle for adsorption to a surface is dependent on the type of phage and surface involved [117]. Kernegger et al. [118] found that the abundances of adsorbed bacteria and viruses, as well as the ratio of adsorbed virus to adsorbed bacteria, were significantly higher for suspended organic particles compared to suspended mineral (inorganic) particles, suggesting that organic particles may be sites of higher HGT rates. Extracellular DNA adsorbs to clay minerals, sand particles, and natural organic matter [119] and in this form is available for transformation [111, 112,120,121]. The high affinity of DNA for organic and mineral phrases present in bulk in aquatic environments and the protection that adsorption confers against degradation by nucleases and toxic chemicals [119] inndicate that adsorbed DNA represents an important reservoir of transformable DNA in freshwater systems. Transformation frequencies of adsorbed DNA have been observed to be alternatively higher [111], lower [112], or similar [122] relative to dissolved DNA under laboratory conditions, and the type of DNA sorbent can influence transformation efficiency [122,123], so the net effect of DNA adsorption on transformation in aquatic systems is likely complex. Variability in the amount of different mineral and organic phases available for adsorption of cells, phages, and DNA within freshwater lacustrine systems may result in system and system compartment specific control of transformation rates (e.g. nutrient poor subglacial lakes with high suspended clay mineral concentrations compared to nutrient rich lakes with high concentrations of organic matter). Finer (smaller grain size) suspended matter possesses relatively greater surface area to volume ratios and typically stay in suspendsion longer, and so may enhance contact between HGT participants to a greater extent than more coarse particles due to a greater collective capacity for HGT participant adsorption. Plasmid transfer rates were observed to be greater in soils with high clay content (i.e. finer textured soils) [124], suggesting that HGT rates may be similarly enhanced in freshwater lakes with finer suspended matter (e.g. lakes receiving extremely fine glacial powder from runoff). Concurrent sorption of HGT participants and chemical agents capable of influencing gene transfer may further modulate the effect of suspended particulate matter on freshwater HGT. In freshwater lakes, suspended floc, which are aggregates of cells, organic matter and minerals, may be highly enriched in trace elements relative to bottom sediments and the bulk water phase [125, 126], potentially resulting in localized stress-driven HGT changes. At the lower end of the particle size scale, nanomaterials in water are capable of influencing HGT. Nanoalumina has been demonstrated to increase in vitro conjugative plasmid transfer across bacterial genera via increased stress and alteration of gene expression [127].

Concentrations of suspended particulate matter may vary substantially over time and space within freshwater lakes. Climatic variability (e.g. storms), anthropogenic point and non-point sources (e.g. combined sewer outfalls, agricultural runoff), phytoplankton blooms, as well as seasonal events including the turnover of stratified systems and springtime melting of snow and ice in temperate climates, may temporarily and locally or generally stimulate HGT by increasing concentrations of suspended particulate matter in freshwater systems via the resuspension of settled particles or increased particulate matter loads in inflowing waters. Concentrations of suspended particulate matter, and rates of HGT, may be relatively high in the metalimnion of thermally stratified lakes due to the accumulation of particulates from the epilimnion at the thermocline where maximal thermally related density differences inhibit mixing of water layers. Microbial communities are commonly concentrated in this region, taking advantage of the local build up of organic carbon and nutrients as well as the unique biogeochemical and energetic conditions that are often present. Similarly, suspended particulate matter is typically enriched at the air-water [128] and sediment-water interfaces of freshwater systems, and may also be generally elevated in the hypolimnion of stratified freshwater lakes due to internal seiches or large water circulation movements driving the resuspension of settled particles [129].

3.6. pH

No in situ studies examining the influence of pH on HGT in freshwater systems have been published to date. Rates of in vitro transformation and conjugation involving bacteria from freshwater river water or epilithic biofilms as donor cells were not significantly affected by pH in the range (6 - 8) found in most freshwater lakes [130-133]. In situ bacterial production was similar across acidified and limed humic lakes (pH range 5 - 7), suggesting that pH does not appreciably influence HGT via a general effect on prokaryotic metabolic activity [134]. However, freshwater pH may influence HGT by controlling the net surface charge present on prokaryotic cells, phage particles, and extracellular DNA, as well as the mineral and organic surfaces they adsorb to, altering surface-facilitated interactions between mediators of HGT and recipient cells and protecting HGT participants from degradation. Romanowski et al. [51] observed that the adsorption of plasmid DNA to sand and groundwater aquifer material in microcosms was inversely dependent on the pH of the solution (pH range 5 - 9), reflecting the increasing negative charge on mineral particle surfaces typically observed with increasing pH, which repulses the negatively charged DNA molecules [51]. This finding suggests that less DNA may be available for transformation in freshwater lacustrine systems and system compartments with higher pH due to the lack of protection from degradation conferred by adsorption to mineral surfaces. The bioavailability of trace elements, which dictates their potential for influencing HGT rates in freshwater prokaryotic communities, is also pH-dependent. The pH of a freshwater lake reflects both its chemical and biological characteristics, and is variable over depth dependent spatial scales as well as diel, seasonal and annual timeframes. This variability may be reflected in changing freshwater HGT rates. For example, diel variation in HGT may occur in lacustrine zones of high photosynthetic activity, as fluctuations in pH may be substantial (i.e. 1 - 2 pH unit shifts) over this timescale, with pH increasing occurring during lighted hours and decreasing at night when respiration dominates [135].

3.7. Ionic Composition

As with pH, the total dissolved ion concentration (salinity) of a freshwater environment can affect the adsorption of prokaryotic cells, phage particles, and extracellular DNA to mineral and organic surfaces, and in turn, the potential for HGT [36,51,136,137]. Freshwater systems, in contrast to marine systems, are highly dilute with the highest concentrations of the most abundant typical ion, Ca^{2+}, in the mM range (in contrast to water itself at 25 M). The adsorption of DNA to sand and groundwater aquifer material in microcosms was enhanced in the presence of divalent cations, with concentrations for half-maximum DNA adsorption of 0.5 mM for Mg^{2+} and 0.4 mM for Ca^{2+}[51,112,120]. In vitro transformation of the soil bacterium Azotobacter vinelandii by DNA adsorbed to natural organic matter significantly increased in the presence of 1 mM Ca^{2+} [122]. Similarly, in vitro transformation of Bacillus subtilis was optimal with media containing 20 mM Mg^{2+} [138]. Together, these findings suggest that transformation rates in freshwater prokaryotic communities may be influenced by the concentrations of divalent cations in their local environment. Metal cations, particularly Ca^{2+} and Mg^{2+}, are involved in competence development, and

Ca^{2+} is necessary for transformation to occur in bacteria [139-141]. Trombe et al. [142] determined that 1 mM Ca^{2+} was sufficient to induce competence in vitro using Streptococcus pneumoniae, while Page and von Tigerstrom [139] found that 0.5 to 1 mM Ca^{2+} resulted in optimal competence induction in vitro using Azotobacter vinelandii. Transformation of Escherichia coli with plasmid DNA was demonstrated in vitro using fresh river water containing concentrations of Ca^{2+} as low as 1.4 mM, with rates of transformation generally being directly related to Ca^{2+} concentration [140]. The authors noted that rivers sampled in a calcareous area had Ca^{2+} concentrations of 2 - 2.5 mM, while those from a granitic area (minerals with low [Ca^{2+}]) had concentrations of 0.3 - 1.5 mM, suggesting that transformation rates may be higher in lacustrine systems with drainage basins rich in calcium carbonate rock [140]. Concentrations of Ca^{2+} can also vary appreciably with depth in stratified hardwater (i.e. Ca^{2+} rich) lakes, with photosynthetically driven calcium carbonate ($CaCO_3$) precipitation (associated with increasing pH) decreasing concentrations in the euphotic regions of the epilimnion and metalimnion during lighted hours, and decomposition (commonly decreases pH through CO_2 generation which increases $CaCO_3$ solubility) increasing concentrations in the hypolimnion. In highly productive hardwater lakes, Ca^{2+} concentrations can substantially decrease during periods of intense biological activity, which can strip out the Ca^{2+} in $CaCO_3$precipitates, potentially inhibiting transformation.

3.8. Prokaryotic Cell Density and Mode of Growth

Higher rates of HGT are typically found in more dense populations of prokaryotic cells, primarily due to the increased probability of contact between mediators of HGT and recipient cells [143]. Using a continuous culture model under conditions relevant to those found in freshwater environments, Replicon et al. [144] found that transduction rates were directly related to the recipient cell concentration. Prokaryotic cells are typically unevenly distributed throughout freshwater lacustrine systems. In the bulk water phase, prokaryotic cell concentrations can vary over space and time [145] but are typically relatively low, resulting in substantial spatial isolation that reduces the probability of HGT. Local regions of relatively high cell density include physical interfaces within the water column (e.g. boundaries of the metalimnion), the air-water and sediment-water interfaces, and biofilms both suspended within the water column (e.g. lake snow) and present at abiotic and biotic solid surfaces. Bacterial and viral abundance and production have been found to vary significantly among strata in stratified water columns, but consistently appear to be elevated in the metalimnion [146-148], indicating this region is a hotspot for HGT and that rates of gene transfer likely differ among epilimnetic, metalimnetic and hypolimnetic regions. A greater number of transconjugants was detected at the air-water interface relative to the bulk underlying water in a model freshwater system, suggesting that the surface microlayer represents a distinct microzone for HGT in freshwater lakes [29].

In aquatic environments, prokaryotes overwhelmingly tend to form and be found in biofilms at surfaces [149]. As these structures maintain a high density of metabolically active cells surrounded by a matrix containing potentially transformable extracellular DNA and capable of facilitating extracellular communication, providing protection from predation, and trapping nutrients and other chemicals, they are potential hotspots for HGT [150]. Underscoring the prevalence of HGT in biofilm settings, conjugation machinery [151], conjugative plasmids [152], and phage particles [153] have important roles in biofilm formation. Increased rates of HGT have been observed in biofilms relative to planktonic cells [154-157]. In a microcosm study designed to mimic a marine environment, conjugation frequencies were typically higher in biofilms compared to the bulk water phase, particularly at the biofilm surface in contact with the surrounding environment [158]. Biofilm stability and the stage of biofilm development may influence HGT. Regular disturbance of the spatial organization of prokaryotic cells in an in vitro biofilm strongly improved plasmid invasiveness, suggesting that the periodic breakup of biofilms (e.g. due to wind/solar driven turbulence) may promote HGT [157]. In cultured Escherichia coli and Pseudomonas aeruginosa mixed-biofilm communities, phage particles were observed by transmission electron microscopy to be closely associated with the biofilm matrix [159]. Following their release from lysogenic Staphylococcus aureus cells, phage particles remained detectable for longer periods in biofilm cultures than in planktonic cultures due to protection from degradation [160], suggesting that transduction may be more prevalent in freshwater biofilms relative to the surrounding bulk water phase.

Within biofilms, microscale gradients in oxygen, pH, and concentrations of nutrients, electron donors and acceptors, signalling molecules, trace elements, and antibiotics can produce microenvironments with distinct biogeochemistry [161-164] that may differentially influence local rates of HGT. Sites of biofilm formation and potential HGT hotspots in freshwater lakes include suspended particulate matter, submerged bulk abiotic surfaces (e.g. rocks) and biotic surfaces (e.g. epilithon of aquatic plants), and the internal structures of aquatic plants and animals (e.g. digestive tract of filter feeding bivalves) [23,165] (**Figure 1**(b)). As noted by Hermansson and Linberg [166] and Hill and Top [124], rates of HGT likely differ substantially among the numerous prokaryotic biofilms present in a lacustrine system, reflecting in part the highly variable environmental conditions that are present.

3.9. Nutrient Availability

The availability of nutrients for prokaryotic cell growth is potentially an important determinant of HGT rates in freshwater lakes. In vitro conjugation rates in pure cultures of heterotrophic bacteria have been positively correlated with the availability of rich nutrient sources containing phosphorus, nitrogen, and organic carbon [30,67], as well as with organic carbon alone [157,167]. Additionally, conjugation rates between introduced heterotrophic bacteria were found to be significantly higher in sterile nutrient-rich sewage water compared to sterile lake water [168]. Interestingly, the positive relationship between nutrient availability and conjugative plasmid transfer appears to be attenuated when nutrient concentrations rise above a particular threshold value in vitro [169,170], suggesting that a similar threshold may exist in natural environments. The latency period between phage infection and cell lysis, as well as the resulting number of released phage particles, are reduced in starved bacterial cells compared to actively metabolizing cells, indicating that transduction rates are reduced under conditions of nutrient limitation

[171]. In addition to serving as a vector for transformation, extracellular DNA is a source of exogenous nucleotides, labile carbon, nitrogen, and phosphorus for microbial communities [119]. Reduced availability of these nutrients may increase the utilization of extracellular DNA by prokaryotes as a nutrient source may increase, potentially reducing the amount of transformable DNA available for HGT [172]. Collectively, these findings suggest that HGT rates in freshwater lacustrine systems are directly related to the availability of nutrients for prokaryotic growth and that HGT rates in heterotrophic prokaryotes are likely to be higher in nutrient/organic carbon rich systems. In freshwater lakes, bacterial production and growth rates can be constrained by the availability of nitrogen, phosphorus, and/or organic carbon [173], with phosphorus most commonly being the limiting nutrient [174]. It is important to note that most investigations of the effects of nutrient availability on HGT employed a single chemoorganoheterotrophic species of bacteria to assess the effects of organic carbon or general nutrient limitation, which precluded the investigation of other heterotrophic and autotrophic prokaryotes as well as any specific effects of phosphorus and nitrogen limitation on HGT in freshwater prokaryotic communities.

Nutrient limitation can also stimulate HGT, blurring the relationship between nutrient availability and HGT that has been established in vitro and exemplifying the likely complexity of HGT regulation within natural environments. As with DNA damage resulting from exposure to genotoxic chemicals or ultraviolet radiation, starvation can lead to the activation of bacterial stress responses that in turn can promote HGT. Amino acid and thymine deprivation can induce prophage and thereby promote transduction [175]. The development of competence in vitro has been observed in cells in association with a reduction in metabolic activity, which can result from nutrient limitation [50,176]. Conjugation has been demonstrated in laboratory seawater microcosms even after starvation for up to 15 days [177], indicating that nutrient limitation does not preclude the potential for conjugation in a freshwater prokaryotic community. Hausner and Wuertz [154] found that conjugation rates were not appreciably diminished by nutrient limitation under laboratory conditions in bacterial biofilms, and suggested that although conjugation requires energy, starved cells that constitutively express their conjugation apparatus are still able to participate in HGT.

Viral infection rates and bacterial abundance and metabolic activity are generally higher in nutrient-rich freshwater systems [36,178], indicating greater rates of HGT. Within lacustrine systems, HGT rates may reflect spatial and temporal variations in nutrient availability. Epilimnetic waters of highly productive stratified lakes are typically depleted of nutrients during summer stratification as organic matter settles out of this compartment into the underlying metalimnion and hypolimnion [179]. The availability of phosphorus, nitrogen, and organic carbon is generally increased during turnover events, storm events, and the spring melting of snow and ice in temperate climates, as nutrient-rich material is introduced or redistributed within a system. Nutrient inputs may be particularly high in urban and agriculturally impacted areas, and point sources within lakes (e.g. wastewater treatment plant effluent) may create localized HGT hotspots. Concentrations of organic and inorganic nutrients tend to be greater in interfacial regions including the metalimnion, air-water interface [128,180], and sedimentwater interface [181] compared to the bulk water column. Further, biofilms, commonly observed at such interfaces, can sorb and accumulate nutrients from the surrounding environment and the extracellular polymeric substances of which they are composed can serve as a nutrient source (e.g. extracellular DNA) [47].

3.10. Ecological Interactions

In freshwater lacustrine environments, interactions within prokaryotic communities and among prokaryotes and plants, algae, and protozoa can potentially influence rates of prokaryotic HGT. Bacterial quorum sensing (intercellular communication) systems can regulate conjugative transfer [182,183], prophage induction [184], and transformation [185]. Further, addition of Escherichia coli cell-free supernatant was observed to increase the frequency of transformation in Bacillus subtilis cells nearly ten-fold, suggesting that organic factors produced by one species in a freshwater prokaryotic community can stimulate transformation in a distantly related species [138]. Investigations of HGT to date have largely utilized single bacterial species, such that the influence of interacttions between different prokaryotic taxa on gene transfer remains poorly defined.

The rhizosphere of terrestrial plants has been identified as a hotspot for HGT, in part due to the availability of nutrient-rich plant root exudates [124,186]. Similarly, in lacustrine systems, aquatic plants and algae may enhance HGT in nearby prokaryotic communities. Conjugative plasmid transfer between bacteria in a microcosm was enhanced in the presence of the aquatic marsh plant Echinochlora crusgalli [187]. The presence of algae or algal exudates can enhance in vitro rates of conjugation [188] and transformation [189] in prokaryotes, suggesting that algal blooms may be hotspots for HGT. Under laboratory conditions, cocultivation of Bacillus subtilis with the ciliate protozoa Tetrahymena thermophila or metabolites from co-cultures decreased transformation rates, with the inhibitory effect of the metabolites being attenuated when the photosynthetic alga Euglena gracilis was present in culture [190]. In contrast, the presence of ciliates, which are ubiquitous in freshwater lakes, increased conjugation rates in cultures of Escherichia coli by facilitating contact between donor and recipient strains through co-accumulation in its vesicles [191]. Intercellular communication between bacteria is enhanced in biofilms [192] and freshwater biofilm communities typically include prokaryotes, algae, and protozoa, suggesting that they are important sites for ecological regulation of HGT in freshwater lacustrine systems.

More generally, low molecular weight organic acids (e.g. tartrate, citrate) derived from organic matter and associated with regions of high biological activity can compete with DNA for adsorption to cells, potentially inhibiting transformation [193]. Chitin, a biopolymer found in fungi, insects, molluscs, and crustaceans in aquatic environments, has been shown to induce competence in bacteria capable of colonizing and degrading it in vitro [194]. This suggests that transformation rates may be influenced by the distribution and abundance of chitin producing organisms within freshwater lakes.

3.11. Prokaryotic Community Composition

Generally speaking, HGT in the majority of freshwater prokaryotes remains uncharacterized and taxon-specific variables that may influence HGT remain to be determined. However, several lines of evidence suggest that the composition of freshwater prokaryotic communities influences the character of HGT taking place within them, although much more research is necessary to firmly establish such a relationship. First, metabolisms limited to particular taxa, as discussed previously, can differentially alter local biogeochemical conditions, in turn potentially influencing HGT among impacted prokaryotes. Second, the formation of consortia dependent on the presence of particular community members may impact HGT rates through changes in donor and recipient cell numbers, metabolic activities, and proximity [124]. For example, chemoorganoheterotrophic bacteria tend to accumulate around cyanobacterial heterocysts, taking advantage of secreted organic carbon and nutrients [195]. Third, plasmids and phages can have specific host ranges [196] and transforming DNA can be tagged such that it will be preferentially taken up by specific taxa [197], such that the extent to which they mediate HGT within a prokaryotic community is dependent on the presence and relative abundance of particular taxa. High prokaryotic diversity can potentially inhibit transduction rates via cells not susceptible to infection by a particular phage interfering with initial virus-host contact [37]. Fourth, nuclease activity and DNA release may differ depending on prokaryotic community composition, influencing transformation rates via the availability of transformable DNA [176]. Finally, as discussed previously, communication within prokaryotic communities via extracellular signalling molecules with taxa-specific production and responses can differentially regulate HGT.

Prokaryotic community composition can differ substantially between freshwater lakes [198,199] and within systems can vary with depth and site location, as well as over seasonal and annual timescales [200-203]. Further, community composition can vary between epilimnetic and hypolimnetic regions within summer stratified lakes [204-206] and planktonic bacterial communities can differ from those associated with suspended particulate matter [207,208]. Finally, regular and episodic system perturbations (e.g.

seasonal lake mixing events, addition of prokaryotes and nutrients via wastewater input) may appreciably alter community composition [206,209]. These spatial and temporal variations in lacustrine prokaryotic community composition may be associated with different HGT rates and participants.

4. CONCLUDING REMARKS

Although the three principal mechanisms of HGT have been demonstrated in freshwater environments, field studies linking HGT to specific biogeochemical parameters are relatively scarce and experimental studies assessing natural prokaryotic communities are equally rare. These linkages, although established in many cases using laboratory-based approaches with chemoorganoheterotrophic organisms, have yet to be confirmed in freshwater lacustrine systems. Consequently, how and to what extent individual biogeochemical factors influence HGT in freshwater lakes remains poorly understood. An appreciation for the substantial spatial and temporal variability in factors capable of influencing HGT within freshwater lakes is necessary to fully characterize environmental gene transfer. In particular, interfacial regions, whether between thermally stratified zones of water, between bottom sediments and overlying water, or between the water surface and the atmosphere, are areas that should be targeted for further investigation of freshwater lacustrine HGT. Further, greater exploration of HGT within autotrophic guilds, a key component of environmental prokaryotic communities, is required to fully constrain these processes within freshwater systems. Culture-independent molecular techniques (e.g. [27,210,211]) provide a means of in situ assessment of HGT in freshwater prokaryotic communities. It is anticipated that the coupling of these techniques with biogeochemical characterization will further understanding of HGT occurrence, controls and impacts within freshwater lakes.

REFERENCES

1. E. V. Koonin and Y. I. Wolf, "Genomics of Bacteria and Archaea: The Emerging Dynamic View of the Prokaryotic World," Nucleic Acids Research, Vol. 36, No. 21, 2008, pp. 6688-6719.

2. O. Zhaxybayeva and W. F. Doolittle, "Lateral Gene Transfer," Current Biology, Vol. 21, No. 7, 2011, pp. R242-R246.

3. O. Summers, "Genetic Linkage and Horizontal Gene Transfer, the Roots of the Antibiotic Multi-Resistance Problem," Animal Biotechnology, Vol. 17, No. 2, 2006, pp. 125-135.

4. D. Springael and E. M. Top, "Horizontal Gene Transfer and Microbial Adaptation to Xenobiotics: New Types of Mobile Genetic Elements and Lessons from Ecological Studies," Trends in Microbiology, Vol. 12, No. 2, 2004, pp. 53-58.

5. B. Liang, G. Wang, Y. Zhao, K. Chen, S. Li and J. Jiang, "Facilitation of Bacterial Adaptation to ChlorothalonilContaminated Sites by Horizontal Transfer of the Chlorothalonil Hydrolytic Dehalogenase Gene," Applied and Environmental Microbiology, Vol. 77, No. 12, 2011, pp. 4268-4272.

6. J. M. Monier, D. Bernillon, E. Kay, A. Faugier, O. Rybalka, Y. Dessaux, P. Simonet and T. M. Vogel, "Detection of Potential Transgenic Plant DNA Recipients among Soil Bacteria," Environmental Biosafety Research, Vol. 6, No. 1-2, 2007, pp. 71-83.

7. F. Donnarumma, D. Paffetti, G. Stotzky, R. Giannini and C. Vettori, "Potential Gene Exchange between Bacillus thuringiensis subsp. kurstaki and Bacillus spp. in Soil in Situ," Soil Biology and Biochemistry, Vol. 42, No. 8, 2010, pp. 1329-1337.

8. J. B. Cotner and B. A. Biddanda, "Small Players, Large Role: Microbial Influence on Biogeochemical Processes in Pelagic Aquatic Ecosystems," Ecosystems, Vol. 5, No. 2, 2002, pp. 105-121.

9. US Environmental Protection Agency, "Great Lakes," 2011. http://www.epa.gov/greatlakes/statsrefs.html

10. Pruden, R. Pei, H. Storteboom and K. H. Carlson, "Antibiotic Resistance Genes as Emerging Contaminants: Studies in Northern Colorado,"

Environmental Science and Technology, Vol. 40, No. 23, 2006, pp. 7445-7450.

11. J. L. Martinez, "Environmental Pollution by Antibiotics and by Antibiotic Resistance Determinants," Environmental Pollution, Vol. 157, No. 11, 2009, pp. 2893-2902.

12. H. K. Allen, J. Donato, H. H. Wang, K. A. Cloud-Hansen, J. Davies and J. Handelsman, "Call of the Wild: Antibiotic Resistance Genes in Natural Environments," Nature Reviews Microbiology, Vol. 8, No. 4, 2010, pp. 251-259.

13. N. G. Taylor, D. W. Verner-Jeffreys and C. Baker-Austin, "Aquatic Systems: Maintaining, Mixing and Mobilising Antimicrobial Resistance?" Trends in Ecology and Evolution, Vol. 26, No. 6, 2011, pp. 278-284.

14. Lupo, S. Coyne and T. U. Berendonk, "Origin and Evolution of Antibiotic Resistance: The Common Mechanisms of Emergence and Spread in Water Bodies," Frontiers in Microbiology, Vol. 3, 2012, p. 18.

15. L. S. Frost and G. Koraimann, "Regulation of Bacterial Conjugation: Balancing Opportunity with Adversity," Future Microbiology, Vol. 5, No. 7, 2010, pp. 1057-1071.

16. O. Johnsborg, V. Eldholm and L. S. Håvarstein, "Natural Genetic Transformation: Prevalence, Mechanisms and Function," Research in Microbiology, Vol. 158, No. 10, 2007, pp. 767-778.

17. C. Canchaya, G. Fournous, S. Chibani-Chennoufi, M. L. Dillmann and H. Brüssow, "Phage as Agents of Lateral Gene Transfer," Current Opinion in Microbiology, Vol. 6, No. 4, 2003, pp. 417-424.

18. M. Martinez-Garcia, B. K. Swan, N. J. Poulton, M. L. Gomez, D. Masland, M. E. Sieracki and R. Stepanauskas, "High-Throughput Single-Cell Sequencing Identifies Photoheterotrophs and Chemoautotrophs in Freshwater Bacterioplankton," The ISME Journal, Vol. 6, No. 1, 2012, pp. 113-123.

19. D. J. Saye, O. Ogunseitan, G. S. Sayler and R. V. Miller, "Potential for Transduction of Plasmids in a Natural Freshwater Environment: Effect of Plasmid Donor Concentration and a Natural Microbial Community on

Transduction in Pseudomonas aeruginosa," Applied and Environmental Microbiology, Vol. 53, No. 5, 1987, pp. 987- 995.

20. S. B. O'Morchoe, O. Ogunseitan, G. S. Sayler and R. V. Miller, "Conjugal Transfer of R68.45 and FP5 between Pseudomonas aeruginosa Strains in a Freshwater Environment," Applied and Environmental Microbiology, Vol. 54, No. 8, 1988, pp. 1923-1929.

21. H. G. Williams, M. J. Day, J. C. Fry and G. J. Stewart, "Natural Transformation in River Epilithon," Applied and Environmental Microbiology, Vol. 62, No. 8, 1996, pp. 2994-2998.

22. A.Caro-Quintero, J. Deng, J. Auchtung, I. Brettar, M. G. Höfle, J. Klappenbach and K. T. Konstantinidis, "Unprecedented Levels of Horizontal Gene Transfer among Spatially Co-Occurring Shewanella Bacteria from the Baltic Sea," The ISME Journal, Vol. 5, No. 1, 2011, pp. 131-140.

23. K. E. Ashelford, J. C. Fry, M. J. Day, K. E. Hill, M. A. Learner, J. R. Marchesi, C. D. Perkins and A. J. Weightman, "Using Microcosms to Study Gene Transfer in Aquatic Habitats," FEMS Microbiology Ecology, Vol. 23, No. 2, 1997, pp. 81-94.

24. J. D. van Elsas, J. Fry, P. Hirsch and S. Molin, "Ecology of Plasmid Transfer and Spread," In: C. M. Thomas, Ed., The Horizontal Gene Pool, Harwood, Amsterdam, 2000, pp. 175-206.

25. M. Shintani, N. Fukushima, M. Tezuka, H. Yamane and H. Nojiri, "Conjugative Transfer of the IncP-7 Carbazole Degradative Plasmid, pCAR1, in River Water Samples," Biotechnology Letters, Vol. 30, No. 1, 2008, pp. 117-122.

26. T. Kenzaka, K. Tani and M. Nasu, "High-Frequency Phage-Mediated Gene Transfer in Freshwater Environments Determined at Single-Cell Level," The ISME Journal, Vol. 4, No. 5, 2010, pp. 648-659.

27. F. Maruyama, K. Tani, T. Kenzaka, N. Yamaguchi and M. Nasu, "Quantitative Determination of Free-DNA Uptake in River Bacteria at the Single-Cell Level by in Situ Rolling-Circle Amplification," Applied and Environmental Microbiology, Vol. 72, No. 9, 2006, pp. 6248-6256.

28. F. Maruyama, K. Tani, T. Kenzaka, N. Yamaguchi and M. Nasu, "Application of Real-Time Long and Short Polymerase Chain Reaction

for Sensitive Monitoring of the Fate of Extracellular Plasmid DNA Introduced into River Waters," Microbes and Environments, Vol. 23, No. 3, 2008, pp. 229-236.

29. G. W. Jones, L. Baines and F. J. Genthner, "Heterotrophic Bacteria of the Freshwater Neuston and Their Ability to Act as Plasmid Recipients under Nutrient Deprived Conditions," Microbial Ecology, Vol. 22, No. 1, 1991, pp. 15- 25.

30. M. J. Bale, J. C. Fry and M. J. Day, "Plasmid Transfer between Strains of Pseudomonas aeruginosa on Membrane Filters Attached to River Stones," Journal of General Microbiology, Vol. 133, No. 11, 1987, pp. 3099- 3107.

31. M. J. Bale, M. J. Day and J. C. Fry, "Novel Method for Studying Plasmid Transfer in Undisturbed River Epilithon," Applied and Environmental Microbiology, Vol. 54, No. 11, 1988, pp. 2756-2758.

32. M. J. Bale, J. C. Fry and M. J. Day, "Transfer and Occurrence of Large Mercury Resistance Plasmids in River Epilithon," Applied and Environmental Microbiology, Vol. 54, No. 4, 1988, pp. 972-978.

33. M. G. Jobling, S. E. Peters and D. A. Ritchie, "Plasmid-Borne Mercury Resistance in Aquatic Bacteria," FEMS Microbiology Letters, Vol. 49, No. 1, 1988, pp. 31-37.

34. J. C. Fry and M. J. Day, "Plasmid Transfer in the Epilithon," In: J. C. Fry and M. J. Day, Eds., Bacterial Genetics in Natural Environments, Chapman and Hall, London, 1990, pp. 172-181.

35. M. G. Weinbauer, "Ecology of Prokaryotic Viruses," FEMS Microbiology Reviews, Vol. 28, No. 2, 2004, pp. 127-181.

36. K. E. Wommack and R. R. Colwell, "Virioplankton: Viruses in Aquatic Ecosystems," Microbiology and Molecular Biology Reviews, Vol. 64, No. 1, 2000, pp. 69-114.

37. R. V. Miller, "Environmental Bacteriophage-Host Interactions: Factors Contribution to Natural Transduction," Antonie Van Leeuwenhoek, Vol. 79, No. 2, 2001, pp. 141- 147.

38. W. D. Morrison, R. V. Miller and G. S. Sayler, "Frequency of F116-Mediated Transduction of Pseudomonas aeruginosa in a Freshwater

Environment," Applied and Environmental Microbiology, Vol. 36, No. 5, 1978, pp. 724-730.

39. D. J. Saye, O. Ogunseitan, G. S. Sayler and R. V. Miller, "Transduction of Linked Chromosomal Genes between Pseudomonas aeruginosa Strains during Incubation in Situ in a Freshwater Habitat," Applied and Environmental Microbiology, Vol. 56, No. 1, 1990, pp. 140-145.

40. S. Ripp and R. V. Miller, "Effects of Suspended Particulates on the Frequency of Transduction among Pseudomonas aeruginosa in a Freshwater Environment," Applied and Environmental Microbiology, Vol. 61, No. 4, 1995, pp. 1214-1219.

41. M. K. Amin and M. J. Day, "Donor and Recipient Effects on Transduction Frequency in Situ," REGEM 1 Programme for the 1st International Conference on the Release of Genetically Engineered Microorganisms, Cardiff, 5-8 April 1988, p. 11.

42. K. M. Nielsen, P. J. Johnsen, D. Bensasson and D. Daffonchio, "Release and Persistence of Extracellular DNA in the Environment," Environmental Biosafety Research, Vol. 6, No. 1-2, 2007, pp. 37-53.

43. D. M. Karl and M. D. Bailiff, "The Measurement and Distribution of Dissolved Nucleic Acids in Aquatic Environments," Limnology and Oceanography, Vol. 34, No. 3, 1989, pp. 543-558.

44. J. H. Paul, S. C. Jiang and J. B. Rose, "Concentration of Viruses and Dissolved DNA from Aquatic Environments by Vortex Flow Filtration," Applied and Environmental Microbiology, Vol. 57, No. 8, 1991, pp. 2197-2204.

45. W. Siuda, R. J. Chróst and H. Güde, "Distribution and Origin of Dissolved DNA in Lakes of Different Trophic States," Aquatic Microbial Ecology, Vol. 15, No. 1, 1998, pp. 89-96.

46. S. C. Jiang and J. H. Paul, "Viral Contribution to Dissolved DNA in the Marine Kingdom as Determined by Differential Centrifugation and Kingdom Probing," Applied and Environmental Microbiology, Vol. 61, No. 1, 1995, pp. 317-325.

47. H. C. Flemming and J. Wingender, "The Biofilm Matrix," Nature Reviews Microbiology, Vol. 8, No. 9, 2010, pp. 623-633.

48. T. Hara and S. Ueda, "A Study on the Mechanism of DNA Excretion from P. aeruginosa KYU-1. Effect of Mitomycin C on Extracellular DNA Production," Agricultural Chemistry and Biotechnology, Vol. 45, No. 11, 1981, pp. 2457-2461.

49. M. G. Lorenz, D. Gerjets and W. Wackernagel, "Release of Transforming Plasmid and Chromosomal DNA from Two Cultured Soil Bacteria," Archives of Microbiology, Vol. 156, No. 4, 1991, pp. 319-326.

50. G. J. Stewart, "Transformation in Natural Environments," In: E. M. H. Wellington and J. D. van Elsas, Eds., Genetic Interactions among Microorganisms in the Natural Environment, Pergamon Press, Oxford, 1992, pp. 216-234.

51. G. Romanowski, M. G. Lorenz and W. Wackernagel, "Adsorption of Plasmid DNA to Mineral Surfaces and Protection against DNase I," Applied and Environmental Microbiology, Vol. 57, No. 4, 1991, pp. 1057-1061.

52. V. Torsvik, L. Øvreås and T. F. Thingstad, "Prokaryotic Diversity—Magnitude, Dynamics, and Controlling Factors," Science, Vol. 296, No. 5570, 2002, pp. 1064-1066.

53. E. J. Biers, S. Sun and E. C. Howard, "Prokaryotic Genomes and Diversity in Surface Ocean Waters: Interrogating the Global Ocean Sampling Metagenome," Applied and Environmental Microbiology, Vol. 75, No. 7, 2009, pp. 2221-2229.

54. R. J. Newton, S. E. Jones, A. Eiler, K. D. McMahon and S. Bertilsson, "A Guide to the Natural History of Freshwater Lake Bacteria," Microbiology and Molecular Biology Reviews, Vol. 75, No. 1, 2011, pp. 14-49.

55. L. A. Warren and E. A. Haack, "Biogeochemical Controls on Metal Behaviour in Freshwater Environments," EarthScience Reviews, Vol. 54, No. 4, 2001, pp. 261-320.

56. Bissett, A. Reimer, D. de Beer, F. Shiraishi and G. Arp, "Metabolic Microenvironmental Control by Photosynthetic Biofilms under Changing Macroenvironmental Temperature and pH Conditions," Applied and Environmental Microbiology, Vol. 74, No. 20, 2008, pp. 6306-6312.

57. D. A. Fike, C. L. Gammon, W. Ziebis and V. J. Orphan, "Micron-Scale Mapping of Sulfur Cycling across the Oxycline of a Cyanobacterial Mat: A Paired nanoSIMS and CARD-FISH Approach," The ISME Journal, Vol. 2, No. 7, 2008, pp. 749-759.

58. T. Borch, R. Kretzschmar, A. Kappler, P. Van Cappellen, M. Ginder-Vogel, A. Voegelin and K. Campbell, "Biogeochemical Redox Processes and Their Impact on Contaminant Dynamics," Environmental Science and Technology, Vol. 44, No. 1, 2010, pp. 15-23.

59. T. Lawley, B. M. Wilkins and L. S. Frost, "Bacterial Conjugation in Gram-Negative Bacteria," In: B. E. Funnell and G. J. Philips, Eds., Plasmid Biology, ASM Press, Washington DC, 2004, pp. 203-226.

60. G. Alonso, K. Baptista, T. Ngo and D. E. Taylor, "Transcriptional Organization of the Temperature-Sensitive Transfer System from the IncHI1 Plasmid R27," Microbiology, Vol. 151, No. 11, 2005, pp. 3563-3573.

61. F. W. Shiah and H. W. Ducklow, "Temperature Regulation of Heterotrophic Bacterioplankton Abundance, Production, and Specific Growth Rate in Chesapeake Bay," Limnology and Oceanography, Vol. 39, No. 6, 1994, pp. 1243-1258.

62. S. P. Ram, D. Boucher, T. Sime-Ngando, D. Debroas and J. C. Romagoux, "Phage Bacteriolysis, Protistan Bacterivory Potential, and Bacterial Production in a Freshwater Reservoir: Coupling with Temperature," Microbial Ecology, Vol. 50, No. 1, 2005, pp. 64-72.

63. W. F. Yue, M. Du and M. J. Zhu, "High Temperature in Combination with UV Irradiation Enhances Horizontal Transfer of stx2 Gene from E. coli O157:H7 to NonPathogenic E. coli," PLoS ONE, Vol. 7, No. 2, 2012, Article ID: e31308.

64. Ahrenholtz, M. G. Lorenz and W. Wackernagel, "The Extracellular Nuclease of Serratia marcescens: Studies on the Activity in Vitro and Effect on Transforming DNA in a Groundwater Aquifer Microcosm," Archives of Microbiology, Vol. 161, No. 2, 1994, pp. 176-183.

65. Matsui, M. Honjo and Z. Kawabata, "Estimation of the Fate of Dissolved DNA in Thermally Stratified Lake Water from the Stability of Exogenous

Plasmid DNA," Aquatic Microbial Ecology, Vol. 26, No. 1, 2001, pp. 95-102.

66. Auzat, S. Chapuy-Regaud, G. Le Bras, D. D. Santos, A. D. Ogunniyi, I. L. Thomas, J. R. Garel, J. C. Paton and M. C. Trombe, "The NADH Oxidase of Streptococcus pneumoniae: Its Involvement in Competence and Virulence," Molecular Microbiology, Vol. 34, No. 5, 1999, pp. 1018-1028.

67. D. R. Stallions and R. Curtiss, "Bacterial Conjugation under Anaerobic Conditions," Journal of Bacteriology, Vol. 111, No. 1, 1972, pp. 294-295.

68. E. Król, H. D. Nguyen, L. M. Rogers, H. Beyenal, S. M. Krone and E. M. Top, "Increased Transfer of a Multidrug Resistance Plasmid in Escherichia coli Biofilms at the Air-Liquid Interface," Applied and Environmental Microbiology, Vol. 77, No. 15, 2011, pp. 5079-5088.

69. S. Frost and J. Simon, "Studies on the Pili of the Promiscuous Plasmid RP4," In: C. I. Kado and J. H. Crosa, Eds., Molecular Mechanisms of Bacterial Virulence, Kluwer Academic Publishers, Dordrecht, 1994, pp. 47-65.

70. P. Viljanen and J. Boratynski, "The Susceptibility of Conjugative Resistance Transfer in Gram-Negative Bacteria to Physicochemical and Biochemical Agents," FEMS Microbiology Letters, Vol. 88, No. 1, 1991, pp. 43-54.

71. C. A. Pinedo and B. F. Smets, "Conjugal TOL Transfer from Pseudomonas putida to Pseudomonas aeruginosa: Effects of Restriction Proficiency, Toxicant Exposure, Cell Density Ratios, and Conjugation Detection Method on Observed Transfer Efficiencies," Applied and Environmental Microbiology, Vol. 71, No. 1, 2005, pp. 51-57.

72. Schäfer, J. Kalinowski and A. Pühler, "Increased Fertility of Corynebacterium glutamicum Recipients in Intergeneric Matings with Escherichia coli after Stress Exposure," Applied and Environmental Microbiology, Vol. 60, No. 2, 1994, pp. 756-759.

73. R. A. Edwards, R. A. Helm and S. R. Maloy, "Increasing DNA Transfer Efficiency by Temporary Inactivation of Host Restriction," Biotechniques, Vol. 26, No. 5, 1999, pp. 892-900.

74. P. Claverys, M. Prudhomme and B. Martin, "Induction of Competence Regulons as a General Response to Stress in Gram-Positive Bacteria," Annual Review of Microbiology, Vol. 60, 2006, pp. 451-475.

75. Prudhomme, L. Attaiech, G. Sanchez, B. Martin and J. P. Claverys, "Antibiotic Stress Induces Genetic Transformability in the Human Pathogen Streptococcus pneumonia," Science, Vol. 313, No. 5783, 2006, pp. 89-92.

76. C. Úbeda, E. Maiques, E. Knecht, Í. Lasa, R. P. Novick and J. R. Penadés, "Antibiotic-Induced SOS Response Promotes Horizontal Dissemination of Pathogenicity Island-Encoded Virulence Factors in Staphylococci," Molecular Microbiology, Vol. 56, No. 3, 2005, pp. 836-844.

77. E. Maiques, C. Úbeda, S. Campoy, N. Salvador, Í. Lasa, R. P. Novick, J. Barbé and J. R. Penadés, "β-Lactam Antibiotics Induce the SOS Response and Horizontal Transfer of Virulence Factors in Staphylococcus aureus," Journal of Bacteriology, Vol. 188, No. 7, 2006, pp. 2726- 2729.

78. J. W. Beaber, B. Hochhut and M. K. Waldor, "SOS Response Promotes Horizontal Dissemination of Antibiotic Resistance Genes," Nature, Vol. 427, No. 6969, 2004, pp. 72-74.

79. T. G. Rossman, M. Molina and L. W. Meyer, "The Genetic Toxicology of Metal Compounds: I. Induction of λ Prophage in E. coli WP2S (λ)," Environmental Mutagenesis, Vol. 6, No. 1, 1984, pp. 59-69.

80. T. G. Rossman, M. Molina, L. Meyer, P. Boone, C. B. Klein, Z. Wang, F. Li, W. C. Lin and P. L. Kinney, "Performance of 133 Compounds in the Lambda Prophage Induction Endpoint of the Microscreen Assay and a Comparison with S. typhimurium Mutagenicity and Rodent Carcinogenicity Assays," Mutation Research, Vol. 260, No. 4, 1991, pp. 349-367.

81. D. M. DeMarini and H. G. Brooks, "Induction of Prophage Lambda by Chlorinated Organics: Detection of Some Single-Species/Single-Site Carcinogens," Environmental and Molecular Mutagenesis, Vol. 19, No. 2, 1992, pp. 98-111.

82. Soberón, R. Martín and J. E. Suárez, "New Method for Evaluation of Genotoxicity, Based on the Use of RealTime PCR and Lysogenic Gram-

Positive and GramNegative Bacteria," Applied and Environmental Microbiology, Vol. 73, No. 9, 2007, pp. 2815-2819.

83. R. Mesak, V. Miao and J. Davies, "Effects of Subinhibitory Concentrations of Antibiotics on SOS and DNA Repair Gene Expression in Staphylococcus aureus," Antimicrobial Agents and Chemotherapy, Vol. 52, No. 9, 2008, pp. 3394-3397.

84. Z. Baharoglu and D. Mazel, "Vibrio cholerae Triggers SOS and Mutagenesis in Response to a Wide Range of Antibiotics: A Route towards Multiresistance," Antimicrobial Agents and Chemotherapy, Vol. 55, No. 5, 2011, pp. 2438-2441.

85. V. M. F. Vargas, S. B. Migliavacca, A. C. de Melo, R. C. Horn, R. R. Guidobono, I. C. F. de Sá Ferreira and M. H. D. Pestana, "Genotoxicity Assessment in Aquatic Environments under the Influence of Heavy Metals and Organic Contaminants," Mutation Research, Vol. 490, No. 2, 2001, pp. 141-158.

86. E. Kristiansson, J. Fick, A. Janzon, R. Grabic, C. Rutgersson, B. Weijdegård, H. Söderström and D. G. J. Larsson, "Pyrosequencing of Antibiotic-Contaminated River Sediments Reveals High Levels of Resistance and Gene Transfer Elements," PLoS ONE, Vol. 6, No. 2, 2011, e17038.

87. K. Smalla, A. S. Haines, K. Jones, E. Krögerrecklenfort, H. Heuer, M. Schloter and C. M. Thomas, "Increased Abundance of IncP-1β Plasmids and Mercury Resistance Genes in Mercury-Polluted River Sediments: First Discovery of IncP-1β Plasmids with a Complex mer Transposon as the Sole Accessory Element," Applied and En vironmental Microbiology, Vol. 72, No. 11, 2006, pp. 7253-7259.

88. S. Wright, C. Baker-Austin, A. H. Lindell, R. Stepanauskas, H. W. Stokes and J. V. McArthur, "Influence of Industrial Contamination on Mobile Genetic Elements: Class 1 Integron Abundance and Gene Cassette Structure in Aquatic Bacterial Communities," The ISME Journal, Vol. 2, No. 4, 2008, pp. 417-428.

89. C. P. Rosewarne, V. Pettigrove, H. W. Stokes and Y. M. Parsons, "Class 1 Integrons in Benthic Bacterial Communities: Abundance, Association with Tn402-Like Transposition Modules and Evidence for Coselection

with Heavy-Metal Resistance," FEMS Microbiology Ecology, Vol. 72, No. 1, 2010, pp. 35-46.

90. F. Burton, M. J. Day and A. T. Bull, "Distribution of Bacterial Plasmids in Clean and Polluted Sites in a South Wales River," Applied and Environmental Microbiology, Vol. 44, No. 5, 1982, pp. 1026-1029.

91. R. P. Schwarzenbach, B. I. Escher, K. Fenner, T. B. Hofstetter, C. A. Johnson, U. von Gunten and B. Wehrli, "The Challenge of Micropollutants in Aquatic Systems," Science, Vol. 313, No. 5790, 2006, pp. 1072-1077.

92. K. Kümmerer, "Antibiotics in the Aquatic Environment—A Review—Part I," Chemosphere, Vol. 75, No. 4, 2009, pp. 417-434.

93. B. L. Skjelkvåle, T. Andersen, E. Fjeld, J. Mannio, A. Wilander, K. Johansson, J. P. Jensen and T. Moiseènko, "Heavy Metal Surveys in Nordic Lakes; Concentrations, Geographic Patterns and Relation to Critical Limits," Ambio, Vol. 30, No. 1, 2001, pp. 2-10.

94. C. Alder, A. Bruchet, M. Carballa, M. Clara, A. Joss, D. Löffler, C. S. McArdell, K. Miksch, F. Omil, T. Tuhkanen and T. A. Ternes, "Consumption and Occurrence," In: T. A. Ternes and A. Joss, Eds., Human Pharmaceuticals, Hormones and Fragrances: The Challenge of Micropollutants in Urban Water Management, IWA Publishing, London, 2009, pp. 15-54.

95. D. E. Armstrong and A. W. Elzerman, "Trace Metal Accumulation in Surface Microlayers," Journal of Great Lakes Research, Vol. 8, No. 2, 1982, pp. 282-287.

96. J. Eggleton and K. V. Thomas, "A Review of Factors Affecting the Release and Bioavailability of Contaminants during Sediment Disturbance Events," Environment International, Vol. 30, No. 7, 2004, pp. 973-980.

97. Wurl and J. P. Obbard, "A Review of Pollutants in the Sea-Surface Microlayer (SML): A Unique Habitat for Marine Organisms," Marine Pollution Bulletin, Vol. 48, No. 11-12, 2004, pp. 1016-1030.

98. E. Haack and L. A. Warren, "Biofilm Hydrous Manganese Oxyhydroxides and Metal Dynamics in Acid Rock Drainage," Environmental Science and Technology, Vol. 37, No. 18, 2003, pp. 4138-4147.

99. D. B. Wunder, V. A. Bosscher, R. C. Cok and R. M. Hozalski, "Sorption of Antibiotics to Biofilm," Water Research, Vol. 45, No. 6, 2011, pp. 2270-2280.

100. J. M. Harrington, S. E. Fendorf and R. F. Rosenzweig, "Biotic Generation of Arsenic(III) in Metal(loid)-Contaminated Freshwater Lake Sediments," Environmental Science and Technology, Vol. 32, No. 16, 1998, pp. 2425- 2430.

101. R. P. Rastogi, Richa, A. Kumar, M. B. Tyagi and R. P. Sinha, "Molecular Mechanisms of Ultraviolet RadiationInduced DNA Damage and Repair," Journal of Nucleic Acids, 2010, Article ID: 592980.

102. R. Maranger, P. A. del Giorgio and D. F. Bird, "Accumulation of Damaged Bacteria and Viruses in Lake Water Exposed to Solar Radiation," Aquatic Microbial Ecology, Vol. 28, No. 3, 2002, pp. 213-227.

103. K. E. Wommack, R. T. Hill, T. A. Muller and R. R. Colwell, "Effects of Sunlight on Bacteriophage Viability and Structure," Applied and Environmental Microbiology, Vol. 62, No. 4, 1996, pp. 1336-1341.

104. S. W. Wilhelm, M. G. Weinbauer, C. A. Suttle and W. H. Jeffrey, "The Role of Sunlight in the Removal and Repair of Viruses in the Sea," Limnology and Oceanography, Vol. 43, No. 4, 1998, pp. 586-592.

105. J. L. Ravanat, T. Douki and J. Cadet, "Direct and Indirect Effects of UV Radiation on DNA and Its Components," Journal of Photochemistry and Photobiology B: Biology, Vol. 63, No. 1-3, 2001, pp. 88-102.

106. Tapper and R. E. Hicks, "Temperate Viruses and Lysogeny in Lake Superior Bacterioplankton," Limnology and Oceanography, Vol. 43, No. 1, 1998, pp. 95-103.

107. S. W. Wilhelm and R. E. H. Smith, "Bacterial Carbon Production in Lake Erie is Influenced by Viruses and Solar Radiation," Canadian Journal of Fisheries and Aquatic Sciences, Vol. 57, No. 2, 2000, pp. 317-326.

108. D. P. Häder, H. D. Kumar, R. C. Smith and R. C. Worrest, "Effects of Solar UV Radiation on Aquatic Ecosystems and Interactions with Climate Change," Photochemical and Photobiological Sciences, Vol. 6, No. 3, 2007, pp. 267-285.

109. D. P. Morris, H. Zagarese, C. E. Williamson, E. G. Balseiro, B. R. Hargreaves, B. Modenutti, R. Moeller and C. Queimalinos, "The Attenuation of Solar UV Radiation in Lakes and the Role of Dissolved Organic Carbon," Limnology and Oceanography, Vol. 40, No. 8, 1995, pp. 1381-1391.

110. M. Scully and D. R. S. Lean, "The Attenuation of Ultraviolet Radiation in Temperate Lakes," Archiv für Hydrobiologie, Vol. 43, 1994, pp. 135-144.

111. M. G. Lorenz, B. W. Aardema and W. Wackernagel, "Highly Efficient Genetic Transformation of Bacillus subtilis Attached to Sand Grains," Journal of General Microbiology, Vol. 134, No. 1, 1988, pp. 107-112.

112. B. Chamier, M. G. Lorenz and W. Wackernagel, "Natural Transformation of Acinetobacter calcoaceticus by Plasmid DNA Adsorbed on Sand and Groundwater Aquifer Material," Applied and Environmental Microbiology, Vol. 59, No. 5, 1993, pp. 1662-1667.

113. C. Vettori, G. Stotzky, M. Yoder and E. Gallori, "Interaction between Bacteriophage PBS1 and Clay Minerals and Transduction of Bacillus subtilis by Clay-Phage Complexes," Environmental Microbiology, Vol. 1, No. 4, 1999, pp. 347-355.

114. C. Vettori, E. Gallori and G. Stotzky, "Clay Minerals Protect Bacteriophage PBS1 of Bacillus subtilis Against Inactivation and Loss of Transducing Ability by UV Radiation," Canadian Journal of Microbiology, Vol. 46, No. 8, 2000, pp. 770-773.

115. D. Kirchman and R. Mitchell, "Contribution of Particle-Bound Bacteria to Total Microheterotrophic Activity in Five Ponds and Two Marshes," Applied and Environmental Microbiology, Vol. 43, No. 1, 1982, pp. 200-209.

116. B. C. Crump, J. A. Baross and C. A. Simenstad, "Dominance of Particle-Attached Bacteria in the Columbia River Estuary, USA," Aquatic Microbial Ecology, Vol. 14, No. 1, 1998, pp. 7-18.

117. N. D. Seeley and S. B. Primrose, "A Review: The Isolation of Bacteriophages from the Environment," Journal of Applied Bacteriology, Vol. 53, No. 1, 1982, pp. 1-17.

118. L. Kernegger, I. Zweimüller and P. Peduzzi, "Effects of Suspended Matter Quality and Virus Abundance on Microbial Parameters: Experimental Evidence from a Large European River," Aquatic Microbial Ecology, Vol. 57, No. 2, 2009, pp. 161-173.

119. G. Pietramellara, J. Ascher, F. Borgogni, M. T. Ceccherini, G. Guerri and P. Nannipieri, "Extracellular DNA in Soil and Sediment: Fate and Ecological Relevance," Biology and Fertility of Soils, Vol. 45, No. 3, 2009, pp. 219-235.

120. G. Romanowski, M. G. Lorenz and W. Wackernagel, "Plasmid DNA in a Groundwater Aquifer Microcosm— Adsorption, DNAase Resistance and Natural Genetic Transformation of Bacillus subtilis," Molecular Ecology, Vol. 2, No. 3, 1993, pp. 171-181.

121. C. Crecchio and G. Stotzky, "Binding of DNA on Humic Acids: Effect on Transformation of Bacillus subtilis and Resistance to DNase," Soil Biology and Biochemistry, Vol. 30, No. 8-9, 1998, pp. 1061-1067.

122. N. Lu, J. L. Zilles and T. H. Nguyen, "Adsorption of Extracellular Chromosomal DNA and Its Effects on Natural Transformation of Azotobacter vinelandii," Applied and Environmental Microbiology, Vol. 76, No. 13, 2010, pp. 4179-4184.

123. Cai, Q. Huang, W. Chen, D. Zhang, K. Wang, D. Jiang and W. Liang, "Soil Colloids-Bound Plasmid DNA: Effect on Transformation of E. coli and Resistance to DNase I Degradation," Soil Biology and Biochemistry, Vol. 39, No. 5, 2007, pp. 1007-1013.

124. K. E. Hill and E. M. Top, "Gene Transfer in Soil Systems Using Microcosms," FEMS Microbiology Ecology, Vol. 25, No. 4, 1998, pp. 319-329.

125. J. M. Plach, A. V. C. Elliott, I. G. Droppo and L. A. Warren, "Physical and Ecological Controls on Freshwater Floc Trace Metal Dynamics," Environmental Science and Technology, Vol. 45, No. 6, 2011, pp. 2157-2164.

126. V. C. Elliott, J. M. Plach, I. G. Droppo and L. A. Warren, "Comparative Floc-Bed Sediment Trace Element Partitioning across Variably Contaminated Aquatic Ecosystems," Environmental Science and Technology, Vol. 46, No. 1, 2012, pp. 209-216.

127. Z. Qiu, Y. Yu, Z. Chen, M. Jin, D. Yang, Z. Zhao, J. Wang, Z. Shen, X. Wang, D. Qian, A. Huang, B. Zhang and J. W. Li, "Nanoalumina Promotes the Horizontal Transfer of Multiresistance Genes Mediated by Plasmids across Genera," Proceedings of the National Academy of Sciences of the United States of America, Vol. 109, No. 13, 2012, pp. 4944-4949.

128. Södergren, "Role of Aquatic Surface Microlayer in the Dynamics of Nutrients and Organic Compounds in Lakes, with Implications for the Ecotones," Hydrobiologia, Vol. 251, No. 1-3, 1993, pp. 217-225.

129. G. A. Weyhenmeyer, L. Hakanson and M. Meili, "A Validated Model for Daily Variations in the Flux, Origin, and Distribution of Settling Particles Within Lakes," Limnology and Oceanography, Vol. 42, No. 7, 1997, pp. 1517-1529.

130. M. K. Amin and M. J. Day, "Influence of pH Value on Viability and Transduction Frequency of Pseudomonas aeruginosa Phage F116," Letters in Applied Microbiology, Vol. 6, No. 4, 1988b, pp. 93-96.

131. Rochelle, M. J. Day and J. C. Fry, "Occurrence, Transfer and Mobilization in Epilithic Strains of Acinetobacter of Mercury-Resistance Plasmids Capable of Transformation," Journal of General Microbiology, Vol. 134, No. 11, 1988, pp. 2933-2941.

132. Rochelle, J. C. Fry and M. J. Day, "Factors Affecting Conjugal Transfer of Plasmids Encoding Mercury Resistance from Pure Cultures and Mixed Natural Suspensions of Epilithic Bacteria," Journal of General Microbiology, Vol. 135, No. 2, 1989, pp. 409-424.

133. Fernandez-Astorga, A. Muela, R. Cisterna, J. Iriberri and I. Barcina, "Biotic and Abiotic Factors Affecting Plasmid Transfer in Escherichia coli Strains," Applied and Environmental Microbiology, Vol. 58, No. 1, 1992, pp. 392-398.

134. L. J. Tranvik, W. Graneli and G. Gahnstroem, "Microbial Activity in Acidified and Limed Humic Lakes," Canadian Journal of Fisheries and Aquatic Sciences, Vol. 51, No. 11, 1994, pp. 2529-2536.

135. Maberly, "Diel, Episodic and Seasonal Changes in pH and Concentrations of Inorganic Carbon in a Productive Lake," Freshwater Biology, Vol. 35, No. 3, 1996, pp. 579-598.

136. L. Mills, J. S. Herman, G. M. Hornberger and T. H. DeJesús, "Effect of Solution Ionic Strength and Iron Coatings on Mineral Grains on the Sorption of Bacterial Cells to Quartz Sand," Applied and Environmental Microbiology, Vol. 60, No. 9, 1994, pp. 3300-3306.

137. H. Nguyen and K. L. Chen, "Role of Divalent Cations in Plasmid DNA Adsorption to Natural Organic Matter-Coated Silica Surface," Environmental Science and Technology, Vol. 41, No. 15, 2007, pp. 5370-5375.

138. X. Wang, M. Li, Q. Yan, X. Chen, J. Geng, Z. Xie and P. Shen, "Across Genus Plasmid Transformation between Bacillus subtilis and Escherichia coli and the Effect of Escherichia coli on the Transforming Ability of Free Plasmid DNA," Current Microbiology, Vol. 54, No. 6, 2007, pp. 450-456.

139. W. J. Page and M. von Tigerstrom, "Optimal Conditions for Transformation of Azotobacter vinelandii," Journal of Bacteriology, Vol. 139, No. 3, 1979, pp. 1058-1061.

140. Baur, K. Hanselmann, W. Schlimme and B. Jenni, "Genetic Transformation in Freshwater: Escherichia coli Is Able to Develop Natural Competence," Applied and Environmental Microbiology, Vol. 62, No. 10, 1996, pp. 3673-3678.

141. J. M. Solomon and A. D. Grossman, "Who's Competent and When: Regulation of Natural Genetic Competence in Bacteria," Trends in Genetics, Vol. 12, No. 4, 1996, pp. 150-155.

142. M. C. Trombe, C. Clavé and J. M. Manias, "Calcium Regulation of Growth and Differentiation in Streptococcus pneumonia," Journal of General Microbiology, Vol. 138, No. 1, 1992, pp. 77-84.

143. M. Thomas and K. M. Nielsen, "Mechanisms of, and Barriers to, Horizontal Gene Transfer between Bacteria," Nature Reviews Microbiology, Vol. 3, No. 9, 2005, pp. 711-721.

144. J. Replicon, A. Frankfater and R. V. Miller, "A Continuous Culture Model to Examine Factors that Affect Transduction among Pseudomonas aeruginosa Strains in Freshwater Environments," Applied and Environmental Microbiology, Vol. 61, No. 9, 1995, pp. 3359-3366.

145. Tuomi, T. Torsvik, M. Heldal and G. Bratbak, "Bacterial Population Dynamics in a Meromictic Lake," Applied and Environmental Microbiology, Vol. 63, No. 6, 1997, pp. 2181-2188.

146. M. G. Weinbauer and M. G. Höfle, "Significance of Viral Lysis and Flagellate Grazing as Factors Controlling Bacterioplankton Production in a Eutrophic Lake," Applied and Environmental Microbiology, Vol. 64, No. 2, 1998, pp. 431-438.

147. M. G. Weinbauer and M. G. Höfle, "Size-Specific Mortality of Lake Bacterioplankton by Natural Virus Communities," Aquatic Microbial Ecology, Vol. 15, No. 2, 1998, pp. 103-113.

148. Y. Bettarel, T. Sime-Ngando, C. Amblard and J. Dolan, "Viral Activity in Two Contrasting Lake Ecosystems," Applied and Environmental Microbiology, Vol. 70, No. 5, 2004, pp. 2941-2951.

149. J. W. Costerton, Z. Lewandowski, D. E. Caldwell, D. R. Korber and H. M. Lappin-Scott, "Microbial Biofilms," Annual Review of Microbiology, Vol. 49, 1995, pp. 711- 745.

150. J. Sørensen, M. Bailey, L. H. Hansen, N. Kroer and S. Wuertz, "Studying Plasmid Horizontal Transfer in Situ: A Critical Review," Nature Reviews Microbiology, Vol. 3, No. 9, 2005, pp. 700-710.

151. K. B. Barken, S. J. Pamp, L. Yang, M. Gjermansen, J. J. Bertrand, M. Klausen, M. Givskov, C. B. Whitchurch, J. N. Engel and T. Tolker-Nielsen, "Roles of Type IV Pili, Flagellum-Mediated Motility and Extracellular DNA in the Formation of Mature Multicellular Structures in Pseudomonas aeruginosa Biofilms," Environmental Microbiology, Vol. 10, No. 9, 2008, pp. 2331-2343.

152. J. M. Ghigo, "Natural Conjugative Plasmids Induce Bacterial Biofilm Development," Nature, Vol. 412, No. 6845, 2001, pp. 442-445.

153. J. Gödeke, K. Paul, J. Lassak and K. M. Thormann, "Phage-Induced Lysis Enhances Biofilm Formation in Shewanella oneidensis MR-1," The ISME Journal, Vol. 5, No. 4, 2011, pp. 613-626.

154. M. Hausner and S. Wuertz, "High Rates of Conjugation in Bacterial Biofilms as Determined by Quantitative in Situ Analysis," Applied and Environmental Microbiology, Vol. 65, No. 8, 1999, pp. 3710-3713.

155. Molin and T. Tolker-Nielsen, "Gene Transfer Occurs with Enhanced Efficiency in Biofilms and Induces Enhanced Stabilisation of the Biofilm Structure," Current Opinion in Biotechnology, Vol. 14, No. 3, 2003, pp. 255- 261.

156. Maeda, M. Ito, T. Ando, Y. Ishimoto, Y. Fujisawa, H. Takahashi, A. Matsuda, A. Sawamura and S. Kato, "Horizontal Transfer of Nonconjugative Plasmids in a Colony Biofilm of Escherichia coli," FEMS Microbiology Letters, Vol. 255, No. 1, 2006, pp. 115-120.

157. R. E. Fox, X. Zhong, S. M. Krone and E. M. Top, "Spatial Structure and Nutrients Promote Invasion of IncP-1 Plasmids in Bacterial Populations," The ISME Journal, Vol. 2, No. 10, 2008, pp. 1024-1039.

158. M. L. Angles, K. C. Marshall and A. E. Goodman, "Plasmid Transfer between Marine Bacteria in the Aqueous Phase and Biofilms in Reactor Microcosms," Applied and Environmental Microbiology, Vol. 59, No. 3, 1993, pp. 843-850.

159. M. K. Kay, T. C. Erwin, R. J. C. McLean and G. M. Aron, "Bacteriophage Ecology in Escherichia coli and Pseudomonas aeruginosa Mixed-Biofilm Communities," Applied and Environmental Microbiology, Vol. 77, No. 3, 2011, pp. 821-829.

160. Resch, B. Fehrenbacher, K. Eisele, M. Schaller and F. Götz, "Phage Release from Biofilm and Planktonic Staphylococcus aureus Cells," FEMS Microbiology Letters, Vol. 252, No. 1, 2005, pp. 89-96.

161. P. S. Stewart and M. J. Franklin, "Physiological Heterogeneity in Biofilms," Nature Reviews Microbiology, Vol. 6, No. 3, 2008, pp. 199-210.

162. G. Hidalgo, A. Burns, E. Herz, A. G. Hay, P. L. Houston, U. Wiesner and L. W. Lion, "Functional Tomographic Fluorescence Imaging of pH Microenvironments in Microbial Biofilms by Use of Silica Nanoparticle Sensors," Applied and Environmental Microbiology, Vol. 75, No. 23, 2009, pp. 7426-7435.

163. P. Hitchcock, J. J. Dynes, J. R. Lawrence, M. Obst, G. D. W. Swerhone, D. R. Korber and G. G. Leppard, "Soft X-Ray Spectromicroscopy of Nickel Sorption in a Natural River Biofilm," Geobiology, Vol. 7, No. 4, 2009, 432-453.

164. G. Chen, X. Chen, Y. Yang, A. G. Hay, X. Yu and Y. Chen, "Sorption and Distribution of Copper in Unsaturated Pseudomonas putida CZ1 Biofilms as Determined by X-ray Fluorescence Microscopy," Applied and Environmental Microbiology, Vol. 77, No. 14, 2011, pp. 4719-4727.

165. J. D. van Elsas and M. J. Bailey, "The Ecology of Transfer of Mobile Genetic Elements," FEMS Microbiology Ecology, Vol. 42, No. 2, 2002, pp. 187-197.

166. M. Hermansson and C. Linberg, "Gene Transfer in the Marine Environment," FEMS Microbiology Ecology, Vol. 15, No. 1-2, 1994, pp. 47-54.

167. F. Smets, B. E. Rittmann and D. A. Stahl, "Quantification of the Effect of Substrate Concentration on the Conjugal Transfer Rate of the TOL Plasmid in Short-Term Batch Mating Experiments," Letters in Applied Microbiology, Vol. 21, No. 3, 1995, pp. 167-172.

168. M. R. Shakibaie, K. A. Jalilzadeh and S. M. Yamakanamardi, "Horizontal Transfer of Antibiotic Resistance Genes among Gram Negative Bacteria in Sewage and Lake Water and Influence of Some Physico-Chemical Parameters of Water on Conjugation Process," Journal of Environmental Biology, Vol. 30, No. 1, 2009, pp. 45-49.

169. Normander, B. B. Christensen, S. Molin and N. Kroer, "Effect of Bacterial Distribution and Activity on Conjugal Gene Transfer on the Phylloplane of the Bush Bean (Phaseolus vulgaris)," Applied and Environmental Microbiology, Vol. 64, No. 5, 1998, pp. 1902-1909.

170. R. Johnsen and N. Kroer, "Effects of Stress and Other Environmental Factors on Horizontal Plasmid Transfer Assessed by Direct Quantification of Discrete Transfer Events," FEMS Microbiology Ecology, Vol. 59, No. 3, 2007, pp. 718-728.

171. Kokjohn and G. S. Sayler, "Attachment and Replication of Pseudomonas aeruginosa Bacteriophages under Conditions Simulating Aquatic Environments," Journal of General Microbiology, Vol. 137, No. 3, 1991, pp. 661- 666.

172. Finkel and R. Kolter, "DNA as a Nutrient: Novel Role for Bacterial Competence Gene Homologs," Journal of Bacteriology, Vol. 183, No. 21, 2001, pp. 6288-6293.

173. J. J. Elser, L. B. Stabler and R. P. Hassett, "Nutrient Limitation of Bacterial Growth and Rates of Bacterivory in Lakes and Oceans: A Comparative Study," Aquatic Microbial Ecology, Vol. 9, No. 2, 1995, pp. 105-110.

174. M. Smith and Y. T. Prairie, "Bacterial Metabolism and Growth Efficiency in Lakes: The Importance of Phosphorus Availability," Limnology and Oceanography, Vol. 49, No. 1, 2004, pp. 137-147.

175. N. E. Melechen and G. Go, "Induction of Lambdoid Prophages by Amino Acid Deprivation: Differential Inducibility; Role of recA," Molecular and General Genetics, Vol. 180, No. 1, 1980, pp. 147-155.

176. J. H. Paul, M. E. Frischer and J. M. Thurmond, "Gene Transfer in Marine Water Column and Sediment Microcosms by Natural Plasmid Transformation," Applied and Environmental Microbiology, Vol. 57, No. 5, 1991a, pp. 1509-1515.

177. E. Goodman, K. C. Marshall and M. Hermansson, "Gene Transfer among Bacteria under Conditions of Nutrient Depletion in Simulated and Natural Aquatic Environments," FEMS Microbiology Ecology, Vol. 15, No. 1- 2, 1994, pp. 55-60.

178. J. J. Cole, M. L. Pace, N. F. Caraco and G. S. Steinhart, "Bacterial Biomass and Cell Size Distributions in Lakes: More and Larger Cells in Anoxic Waters," Limnology and Oceanography, Vol. 38, No. 8, 1993, pp. 1627-1632.

179. P. Carlsson and D. A. Caron, "Seasonal Variation of Phosphorus Limitation of Bacterial Growth in a Small Lake," Limnology and Oceanography, Vol. 46, No. 1, 2001, pp. 108-120.

180. Münster, E. Heikkinen and J. Knulst, "Nutrient Composition, Microbial Biomass and Activity at the Air-Water Interface of Small Boreal Forest Lakes," Hydrobiologia, Vol. 363, No. 1-3, 1997, pp. 261-270.

181. Boström, J. M. Andersen, S. Fleischer and M. Jansson, "Exchange of Phosphorus across the Sediment-Water Interface," Hydrobiologia, Vol. 170, No. 1, 1988, pp. 229- 244.

182. J. E. González and M. M. Marketon, "Quorum Sensing in Nitrogen-Fixing Rhizobia," Microbiology and Molecular Biology Reviews, Vol. 67, No. 4, 2003, pp. 574-592.

183. J. P. Ramsay, J. T. Sullivan, N. Jambari, C. A. Ortori, S. Heeb, P. Williams, D. A. Barrett, I. L. Lamont and C. W. Ronson, "A LuxRI-Family Regulatory System Controls Excision and Transfer of the Mesorhizobium loti Strain R7A Symbiosis Island by Activating Expression of Two Conserved Hypothetical Genes," Molecular Microbiology, Vol. 73, No. 6, 2009, pp. 1141-1155.

184. Ghosh, K. Roy, K. E. Williamson, S. Srinivasiah, K. E. Wommack and M. Radosevich, "Acyl-Homoserine Lactones Can Induce Virus Production in Lysogenic Bacteria: An Alternative Paradigm for Prophage Induction," Applied and Environmental Microbiology, Vol. 75, No. 22, 2009, pp. 7142-7152.

185. S. Antonova and B. K. Hammer, "Quorum-Sensing Autoinducer Molecules Produced by Members of a Multispecies Biofilm Promote Horizontal Gene Transfer to Vibrio cholerae," FEMS Microbiology Letters, Vol. 322, No. 1, 2011, pp. 68-76.

186. L. Mølbak, S. Molin and N. Kroer, "Root Growth and Exudate Production Define the Frequency of Horizontal Plasmid Transfer in the Rhizosphere," FEMS Microbiology Ecology, Vol. 59, No. 1, 2007, pp. 167-176.

187. N. Kroer, T. Barkay, S. Sørensen and D. Weber, "Effect of Root Exudates and Bacterial Metabolic Activity on Conjugal Gene Transfer in the Rhizosphere of a Marsh Plant," FEMS Microbiology Ecology, Vol. 25, No. 4, 1998, pp. 375-384.

188. M. Ueki, K. Matsui, K. Choi and Z. Kawabata, "The Enhancement of Conjugal Plasmid pBHR1 Transfer between Bacteria in the Presence of Extracellular Metabolic Products Produced by Microcystis aeruginosa," FEMS Microbiology Ecology, Vol. 51, No. 1, 2004, pp. 1-8.

189. K. Matsui, N. Ishii and Z. Kawabata, "Release of Extracellular Transformable Plasmid DNA from Escherichia coli Cocultivated with Algae," Applied and Environmental Microbiology, Vol. 69, No. 4, 2003, pp. 2399- 2404.

190. K. Matsui, N. Ishii and Z. Kawabata, "Microbial Interactions Affecting the Natural Transformation of Bacillus subtilis in a Model Aquatic Ecosystem," FEMS Microbiology Ecology, Vol. 45, No. 3, 2003, pp. 211-218.

191. J. Matsuo, S. Oguri, S. Nakamura, T. Hanawa, T. Fukumoto, Y. Hayashi, K. Kawaguchi, Y. Mizutani, T. Yao, K. Akizawa, H. Suzuki, C. Simizu, K. Matsuno, S. Kamiya and H. Yamaguchi, "Ciliates Rapidly Enhance the Frequency of Conjugation between Escherichia coli Strains through Bacterial Accumulation in Vesicles," Research in Microbiology, Vol. 161, No. 8, 2010, pp. 711-719.

192. T. R. de Kievit, "Quorum Sensing in Pseudomonas aeruginosa Biofilms," Environmental Microbiology, Vol. 11, No. 2, 2009, pp. 279-288.

193. P. Cai, Q. Huang, J. Zhu, D. Jiang, X. Zhou, X. Rong and W. Liang, "Effects of Low-Molecular-Weight Organic Ligands and Phosphate on DNA Adsorption by Soil Colloids and Minerals," Colloids and Surfaces B: Biointerfaces, Vol. 54, No. 1, 2007, pp. 53-59.

194. K. L. Meibom, M. Blokesch, N. A. Dolganov, C. Y. Wu and G. K. Schoolnik, "Chitin Induces Natural Competence in Vibrio cholerae," Science, Vol. 310, No. 5755, 2005, pp. 1824-1827.

195. H. W. Paerl and J. L. Pinckney, "A Mini-Review of Microbial Consortia: Their Roles in Aquatic Production and Biogeochemical Cycling," Microbial Ecology, Vol. 31, No. 3, 1996, pp. 225-247.

196. L. de Gelder, F. P. Vandecasteele, C. J. Brown, L. J. Forney and E. M. Top, "Plasmid Donor Affects Host Range of Promiscuous IncP-1β Plasmid pB10 in an ActivatedSludge Microbial Community," Applied and Environmental Microbiology, Vol. 71, No. 9, 2005, pp. 5309- 5317.

197. Chen and D. Dubnau, "DNA Uptake during Bacterial Transformation," Nature Reviews Microbiology, Vol. 2, No. 3, 2004, pp. 241-249.

198. C. Yannarell and E. W. Triplett, "Withinand Between-Lake Variability in the Composition of Bacterioplankton Communities: Investigations Using Multiple Spatial Scales," Applied and Environmental Microbiology, Vol. 70, No. 1, 2004, pp. 214-223.

199. R. C. Crump, H. E. Adams, J. E. Hobbie and G. W. Kling, "Biogeography of Bacterioplankton in Lakes and Streams of an Arctic Tundra Catchment," Ecology, Vol. 88, No. 6, 2007, pp. 1365-1378.

200. Dominik and M. G. Höfle, "Changes in Bacterioplankton Community Structure and Activity with Depth in a Eutrophic Lake as Revealed by 5S rRNA Analysis," Applied and Environmental Microbiology, Vol. 68, No. 7, 2002, pp. 3606-3613.

201. C. Yannarell, A. D. Kent, G. H. Lauster, T. K. Kratz and E. W. Triplett, "Temporal Patterns in Bacterial Communities in Three Temperate Lakes of Different Trophic Status," Microbial Ecology, Vol. 46, No. 4, 2003, pp. 391-405.

202. Zwisler, N. Selje and M. Simon, "Seasonal Patterns of the Bacterioplankton Community Composition in a Large Mesotrophic Lake," Aquatic Microbial Ecology, Vol. 31, No. 3, 2003, pp. 211-225.

203. S. R. Mueller-Spitz, G. W. Goetz and S. L. McLellan, "Temporal and Spatial Variability in Nearshore Bacterioplankton Communities of Lake Michigan," FEMS Microbiology Ecology, Vol. 67, No. 3, 2009, pp. 511-522.

204. Konopka, T. Bercot and C. Nakatsu, "Bacterioplankton Community Diversity in a Series of Thermally Stratified Lakes," Microbial Ecology, Vol. 38, No. 2, 1999, pp. 126-135.

205. Boucher, L. Jardillier and D. Debroas, "Succession of Bacterial Community Composition over Two Consecutive Years in Two Aquatic Systems: A Natural Lake and a Lake-Reservoir," FEMS Microbiology Ecology, Vol. 55, No. 1, 2006, pp. 79-97.

206. Shade, S. E. Jones and K. D. McMahon, "The Influence of Habitat Heterogeneity on Freshwater Bacterial Community Composition and Dynamics," Environmental Microbiology, Vol. 10, No. 4, 2008, pp. 1057-1067.

207. F. DeLong, D. G. Franks and A. L. Alldredge, "Phylogenetic Diversity of Aggregate-Attached vs. Free-Living Marine Bacterial Assemblages," Limnology and Oceanography, Vol. 38, No. 5, 1993, pp. 924-934.

208. C. Crump, E. V. Armbrust and J. A. Baross, "Phylogenetic Analysis of Particle-Attached and Free-Living Bacterial Communities in the Columbia River, Its Estuary, and the Adjacent Coastal Ocean," Applied and Environmental Microbiology, Vol. 65, No. 7, 1999, pp. 3192- 3204.

209. A. Pearce, C. J. van der Gast, K. Woodward and K. K. Newsham, "Significant Changes in the Bacterioplankton Community Structure of a Maritime Antarctic Freshwater Lake Following Nutrient Enrichment," Microbiology, Vol. 151, No. 10, 2005, pp. 3237-3248.

210. J. Haagensen, S. K. Hansen, T. Johansen and S. Molin, "In Situ Detection of Horizontal Transfer of Mobile Genetic Elements," FEMS Microbiology Ecology, Vol. 42, No. 2, 2002, pp. 261-268.

211. Haug, S. A. Tanner, C. Lacroix, L. Meile and M. J. A. Stevens, "Construction and Characterization of Enterococcus faecalis CG110/gfp/pRE25*, a Tool for Monitoring Horizontal Gene Transfer in Complex Microbial Ecosystems," FEMS Microbiology Letters, Vol. 313, No. 2, pp. 111-119.

Chapter 8

Biofloc Technology (BFT): A Review for Aquaculture Application and Animal Food Industry

Maurício Emerenciano[1, 2], Gabriela Gaxiola[3] and Gerard Cuzon[4]

[1] Posgrado en Ciencias del Mar y Limnología, Universidad Nacional Autónoma de México (UNAM), Unidad Multidisciplinaria de Docencia e Investigación (UMDI), UNAM, Sisal, Yucatán, Mexico

[2] Santa Catarina State University, Centro de Educação Superior da Região Sul (CERES), Laguna, Santa Catarina, Brazil

[3] UMDI, Facultad de Ciencias, UNAM, Sisal, Yucatán, México

[4] Ifremer (Institut Français de Recherche pour l'Exploitation de la Mer) Taravao, Tahiti, French Polynesia

1. INTRODUCTION

The aquaculture industry is growing fast at a rate of ~9% per year since the 1970s [1]. However, this industry has come under scrutiny for contribution to environmental degradation and pollution. As a result, requirement for more ecologically sound management and culture practices remains fully necessary. Moreover, the expansion of aquaculture is also restricted due to land costs and by its strong dependence on fishmeal and fish oil [2,3]. Such ingredients are one of the prime constituents of feed for commercial aquaculture [4]. Feed costs represent at least 50% of the total aquaculture production costs, which is predominantly due to the cost of protein component in commercial diets [5].

Interest in closed aquaculture systems is increasing, mostly due to biosecurity, environmental and marketing advantages over conventional extensive and semi-intensive systems [6]. When water is reused, some risks such as pathogen introduction, escapement of exotic species and discharging of waste water (pollution) are reduced and even eliminated. Furthermore, because of high productivity and reduced water use, marine species can be raised at inland locations [6]. A classic example is the currently expansion of marine shrimp farms at inland location in USA, which allows local farmers market fresh never frozen shrimp in metropolitan locations with good profitability.

The environmental friendly aquaculture system called "Biofloc Technology (BFT)" is considered as an efficient alternative system since nutrients could be continuously recycled and reused. The sustainable approach of such system is based on growth of microorganism in the culture medium, benefited by the minimum or zero water exchange. These microorganisms (biofloc) has two major roles: (i) maintenance of water quality, by the uptake of nitrogen compounds generating "*in situ*" microbial protein; and (ii) nutrition, increasing culture feasibility by reducing feed conversion ratio and a decrease of feed costs.

As a closed system, BFT has primordial advantage of minimizing the release of water into rivers, lakes and estuaries containing escaped animals, nutrients, organic matter and pathogens. Also, surrounding areas are benefitted by the "vertically growth" in terms of productivity, preventing coastal or inland area destruction, induced eutrophication and natural resources losses. Drained water from ponds and tanks often contains relatively high concentrations of nitrogen and phosphorous, limiting nutrients that induce algae growth, which may cause severe eutrophication and further anaerobic conditions in natural water bodies. In BFT, minimum water discharge and reuse of water prevent environment degradation and convert such system in a real "environmentally friendly system" with a "green" approach. Minimum water exchange maintain the heat and fluctuation of temperature is prevented [7], allowing growth of tropical species in cold areas.

Currently, BFT has received alternate appellation such as ZEAH or Zero Exchange Autotrophic Heterotrophic System [8-10], active-sludge or suspended bacterial-based system [11], single-cell protein production system [12], suspended-growth systems [13] or microbial floc systems [14,15]. However, researches are trying to keep the term "BFT or Biofloc Technology" in order to establish a key reference, mainly after the book release *"Biofloc Technology – A Practical Guide Book"* in 2009 [16]. Moreover, BFT has been focus of intensive research in nutrition field as a protein source in compounded feeds. Such source is produced in a form of "biofloc meal", mainly in bioreactors [17]. In addition, the fast spread and the large number of BFT farms worldwide induced significant research effort of processes involved in BFT production systems [14].

The objective of this chapter is to review the application of Biofloc Technology (BFT) in aquaculture; and describes the utilization of biofloc biomass (also described in this chapter as "biofloc meal") as an ingredient for compounded feeds. An addition goal is to help students, researchers and industry to clarify the basic aspects of such technology, aiming to encourage further research.

2. HISTORY OF BFT

According to [18], BFT was first developed in early 1970s at Ifremer-COP (French Research Institute for Exploitation of the Sea, Oceanic Center of Pacific) with different penaeid species including*Penaeus monodon*, *Fenneropenaeus merguiensis*, *Litopenaeus vannamei* and *L. stylirostris* [19,20]. Such culture system was compared with an "external rumen", but now applied for shrimp [21]. At the same period, Ralston Purina developed a system based on nitrifying bacteria while keeping shrimp in total darkness. In connection with Aquacop, such system was applied to *L. stylirostris* and *L. vannamei* both in Crystal River (USA) and Tahiti, leding considerations on benefits of biofloc for shrimp culture [22]. In 1980, a French scientific program 'Ecotron' was initiated by Ifremer to better understand such system. Several studies enabled a comprehensive approach of BFT and explained interrelationships between different compartments such as water and bacteria, as well as shrimp nutritional physiology. Also in 1980s and beginning of 1990s, Israel and USA (Waddell Mariculture Center) started R&D in BFT with tilapia and white shrimp *L. vannamei*, respectively, in which water limitation, environmental concerns and land costs were the main causative agents that promoted such research (Fig. 1).

Figure 1. Biofloc technology at Ifremer, Tahiti (A), Sopomer farm, Tahiti (B), Waddell Mariculture Center (C) and Israel (D) (Photos A and B: Gerard Cuzon; C: courtesy of Wilson Wasielesky; and D: courtesy of Yoram Avnimelech)

Regarding to commercial application of BFT, in 1988 Sopomer farm in Tahiti (French Polynesia) using $1000m^2$ concrete tanks and limited water exchange achieved a world record in production (20–25 ton/ha/year with two crops) [22, 23]. On the other hand, Belize Aquaculture farm or "BAL" (located at Belize, Central America), probably the most famous case of BFT commercial application in the world, produced around 11-26 ton/ha/cycle using 1.6 ha lined grow-out ponds. Much of know-how of running worldwide commercial scale BFT shrimp ponds is derived from BAL experience. In small-scale BFT greenhouse-based farms, Marvesta farm (located at Maryland, USA), probably is the well-known successful indoor BFT shrimp farm in USA, can produce around 45 ton of fresh never frozen shrimp per year using ~570 m^3 indoor race-ways [24]. Nowadays, BFT have being successfully expanded in large-scale shrimp farming in Asia, Latin and Central America, as well as in small-scale greenhouses in USA, South Korea, Brazil, Italy, China and others (Fig 2). In addition, many research centers and universities are intensifying R&D in BFT, mostly applied to key fields such as grow-out management, nutrition, BFT applied to reproduction, microbial ecology, biotechnology and economics.

3. THE ROLE OF MICROORGANISMS

The particulate organic matter and other organisms in the microbial food web have been proposed as potential food sources for aquatic animals [25]. In BFT, microorganisms present a key role in nutrition of cultured animals. The macroaggregates (biofloc) is a rich protein-lipid natural source available *"in situ"* 24 hours per day [14]. In the water column occurs a complex interaction between organic matter, physical substrate and large range of microorganisms such as phytoplankton, free and attached bacteria, aggregates of particulate organic matter and grazers, such as rotifers, ciliates and flagellates protozoa and copepods [26] (Fig 3). This natural productivity play an important role recycling nutrients and maintaining the water quality [27,28].

Figure 2. Biofloc technology commercial-scale at BAL (A) and Malaysia (B), and pilot-scale in Mexico (C and D) (Photos A, B and D: Maurício Emerenciano; and C: courtesy of Manuel Valenzuela)

The consumption of biofloc by shrimp or fish has demonstrated innumerous benefits such as improvement of growth rate [10], decrease of FCR and associated costs in feed [9]. Growth enhancement has been attributed to both bacterial and algae nutritional components, which up to 30% of conventional feeding ration can be lowered due to biofloc consumption in shrimp [29]. In reference [9] was reported that more than 29% of daily food consumed for *L. vannamei* could be biofloc. In tilapia, in [30] was estimated that feed utilization is higher in BFT at a rate of 20% less than conventional water-exchange systems. Also, consumption of macroaggregates can increase nitrogen retention from added feed by 7-13% [31, 32]. In this context, BFT has driven opportunities to use alternative diets. Low protein feeds and feeds with alternative protein sources different than marine-based products (i.e. fishmeal, squid meal, etc) have been successfully applied in BFT [28, 33-35], leading "green" market opportunities.

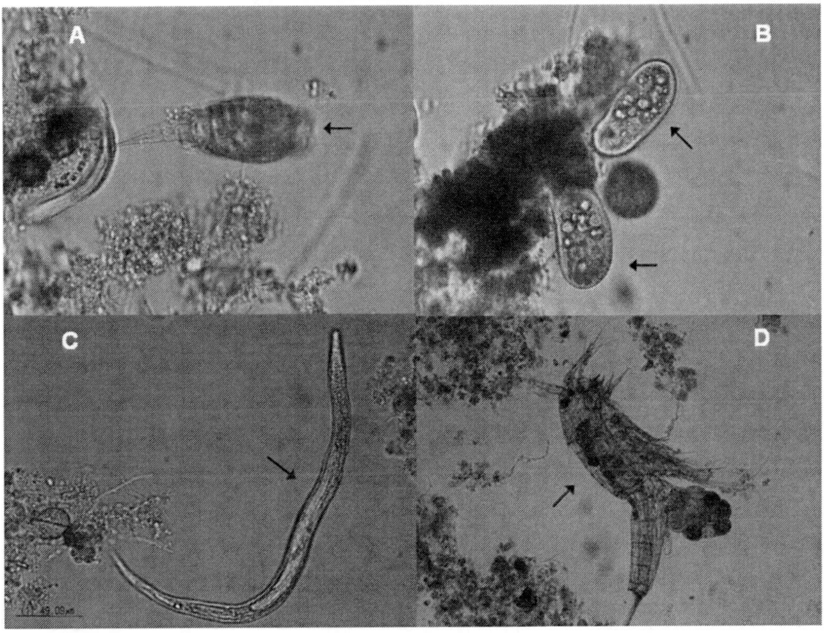

Figure 3. Grazers often observed in BFT such as flagellates protozoa (A), ciliates protozoa (B), nematodes (C) and copepods (D) (10x magnification) (Source: Maurício Emerenciano)

Regarding to maintenance of water quality, control of bacterial community over autotrophic microorganisms is achieved using a high carbon to nitrogen ratio (C:N) [30], which nitrogenous by-products can be easily taken up by heterotrophic bacteria [36]. High carbon to nitrogen ratio is required to guarantee optimum heterotrophic bacteria growth [14, 37], using this energy for maintenance (respiration, feeding, movement, digestion, etc), but also for growth and to produce new cells. High carbon concentration in water could supersede the carbon assimilatory capacity of algae, contributing to bacteria growth. Aerobic microorganisms are efficient in converting feed to new cell material (40-60% of conversion efficiency), rather than higher organisms that spend about 10-15% to rise in weight [16]. Bacteria and other microorganisms act as very efficient "biochemical systems" to degrade and metabolize organic residues [36]. In other words, they recycle very efficiently nutrients in a form of organic and inorganic matter (un-consumed and non-digested feed, metabolic residues and carbon sources applied as fertilizers) into new microbial cells.

The carbon sources applied in BFT are often by-products derived from human and/or animal food industry, preferentially local available. Cheap sources of carbohydrates such as molasses, glycerol and plant meals (i.e. wheat, corn, rice, tapioca, etc) will be applied before fry/post-larvae stocking and during grow-out phase, aiming to maintain a high C:N ratio (~15-20:1) and to control N compounds peaks. Also, a mix of plant meals can be pelletized ("green-pellet") and applied into ponds [38]; or low protein diets containing high C:N ratio can also be carried out [16,33]. The carbon source serves as a substrate for operating BFT systems and production of microbial protein cells [36]. There are many considerations for its selection such as costs, local availability, biodegradability and efficiency of bacteria assimilation. In Table 1 is summarized some studies with different species and carbon source applied in BFT system.

Table 1. Different carbon sources applied on BFT system (Source: adapted from [**36**])

Carbon source	Culture specie	Reference
Acetate	*Macrobrachium rosenbergii*	[39]
Cassava meal	*Penaeus monodon*	[40]
Cellulose	Tilapia	[12]
Corn flour	Hybrid bass and hybrid tilapia	[41, 42]
Dextrose	*Litopenaeus vannamei*	[43]
Glycerol and Glycerol+*Bacillus*	*M. rosenbergii*	[39]
Glucose	*M. rosenbergii*	[39]
Molasses	*L. vannamei and P. monodon*	[9, 29, 44]
Sorghum meal	Tilapia	[12]
Tapioca	*L. vannamei* and *M. rosenbergii*	[31, 45]
Wheat flour	Tilapia (*O. niloticus*)	[33]
Wheat bran + molasses	*Farfantepenaeus brasiensis, F. paulensis and F. duorarum*	[37, 46, 47]
Starch	Tilapia *O. niloticus x O. aureus* and tilapia (Mozambique)	[7, 14]

Not all species are candidates to BFT. Some characteristics seems to be necessary to achieve a better growth performance such as resistance to high density, tolerance to intermediate levels of dissolved oxygen (~3-6 mg/L), settling solids in water (~10 with a maximum of 15 mL/L of "biofloc volume", measured in Imhoff cones) [38] and N-compounds, presence of filtering apparatus (i.e. tilapia), omnivorous habits and/or digestive system adaptable to better assimilate the microbial particles.

4. APPLICATIONS IN AQUACULTURE

4.1. Nursery snd Grow-Out

Nursery phase is defined as an intermediate step between hatchery-reared early postlarvae and grow-out phase [48]. Such phase presents several benefits such as optimization of farm land, increase in survival and enhanced growth performance in grow-out ponds [49-51]. BFT has been applied successfully in nursery phase in different shrimp species such as *L. vannamei* [44, 48], *P. monodon* [51],*F. paulensis* [15, 46], *F. brasiliensis* [37, 52] and *F. setiferus* [34]. The primary advantage observed is related to a better nutrition by continuous consumption of biofloc, which might positively influence grow-out performance *a posteriori* [53], but was not always the case [54]. In addition, optimization of farm facilities provided by the high stocking densities in BFT nursery phase seems to be an important advantage to achieve profitability in small farms, mainly in cold regions or when farmers are operating indoor facilities.

In [46] was observed that presence of bioflocs resulted in increases of 50% in weight and almost 80% in final biomass in *F. paulensis* early postlarval stage when compared to conventional clear-water system. This trend was observed even when postlarvae were not fed with a commercial feed (biofloc without commercial feed). In *L. vannamei* nursery in BFT conditions, references [48] and [55] reported survival rates ranging from 55.9% to 100% and 97% and 100%, respectively. In [51] was demonstrated that the addition of substrates in BFT systems increased growth and further enhanced production, while also contributing to more favorable water quality conditions. According to the same study, growth and survival was not affected by stocking density (2500 *vs* 5000 PL/m^2), therefore greater production outputs were achieved at the higher density. Furthermore, in [37] was found that *F. brasiliensis*postlarvae grow similarly with or without pelletized feed in biofloc conditions during 30-d of nursery phase, which was 40% more than conventional clear-water continuous exchange system.

In grow-out, BFT has been also shown nutritional and zootechnical benefits. In [9] was estimated that more than 29% of the daily food intake of *L. vannamei* consisted of microbial flocs, decreasing FCR and reducing costs in feed. The reference [10] showed that juveniles of *L. vannamei* fed with 35% CP pelletized feed grew significantly better in biofloc conditions as compared to clear-water conditions. In [28] was showed that controlling the concentration of particles in super-intensive shrimp culture systems can significantly improve shrimp production and water quality. Also, the same authors demonstrated that environmentally friendly plant-based diet can produce results comparable to a fish-based feed in BFT conditions. In [56] was evaluated the stocking density in a 120d of *L. vannamei*BFT culture, reporting consistent survival of 92, 81 and 75% with 150, 300 and 450 shrimp/m^2, respectively. Moreover, the study [57] performed in a heterotrophic-based condition detected no significant difference in FCR when feeding *L. vannamei* 30% and 45% CP diets and 39% and 43% CP diets, respectively. With these results in mind, floc biomass might provide a complete source of cellular nutrition as well as various bioactive compounds even at high density. It is not known exactly how microbial flocs enhance growth. Growth might be enhanced by continuous consumption of "native protein", protein source without previous treatment [18], which could possess a "growth factor" similar to the one investigated in squid [58]. Is well known that protein, peptides and aminoacids participate fully in synthesis of new membranes, somatic growth and immune function and biofloc can potentially provide such ingredients.

For fish and other species, BFT also has been demonstrated encouraged results. Intensive BFT*Oreochromis niloticus* tilapia culture could produce an equivalent of 155 ton/ha/crop [11]. Besides high yields, decrease of FCR and decreased of protein content in diets have also been observed. In [30] was estimated that feed utilization by tilapia is higher in BFT with a ration

20% less than conventional water exchange system. Studying the effect of BFT in juveniles tilapia, the reference [33] showed no difference in fish growth/production between 35% and 24% CP fed tanks under BFT, but both were higher than clear-water control without biofloc with 35% CP. Moreover, in [7] was investigated the effectiveness of BFT for maintaining good water quality in over-wintering ponds for tilapia. The authors concluded that BFT emerge as an alternative to overcome over-wintering problems, particularly mass mortality of fish due to low temperatures. In the study [14] was observed that biofloc consumed by fish (tilapia) may represent a very significant feed source, constituting about 50% of the regular feed ration of fish (assuming daily feeding of 2% body weight).

In *M. rosenbergii* larviculture was evaluated the effect of different carbon sources in a BFT culture conditions [39]. The authors found that using glucose or a combination of glycerol plus *Bacillus* as a carbon source in bioreactors led to higher biofloc protein content, higher n-6 fatty acids, which resulted in improved survival rates. In a study with a Brazilian endemic tropical fish species tambaqui (*Colossoma macropomum*) was observed that BFT did not improve fish growth/production as compared to clear-water conditions [59], although some water quality problems in such study remained unsolved (i.e. turbidity and nitrite). The authors showed no differences in 44% CP fed tanks under BFT and clear-water conditions, as well as 28% CP BFT. Certainly further research is needed to clarify the effect of BFT in *Colossoma macropomum*. On the other hand, *Piaractus brachypomus* or pirapitinga seems to be a candidate species to BFT [60].

4.2. Breeding

The BFT has been successfully applied for grow-out, but little is known about biofloc benefits on breeding. For example, in the shrimp industry with the global spread of viruses, the use of closed-life cycle broodstock appeared as a priority to guarantee biosecurity, avoiding vertical transmissions. Moreover, such industry places a considerable interest on penaeid breeding program, often performed in closed facilities, controlling the production plan through successive generations. These programs were frequently associated with large animals, disease resistance as well as the enhancement of reproductive performance. However, nutritional problems remain unresolved [61] and alternatives should be evaluated.

As an alternative for continuous *in situ* nutrition during the whole life-cycle, breeders raised in BFT limited or zero water exchange system are nutritional benefited by the natural productivity (biofloc) available 24 hours per day. Biofloc in a form of rich-lipid-protein source could be utilized for first stages of broodstock's gonads formation and ovary development. Furthermore, production of broodstock in BFT could be located in small areas close to hatchery facilities, preventing spread of diseases caused by shrimp transportation.

In conventional systems breeders used to be produced in large ponds at low density. However, risks associate with accumulation of organic matter, cyanobacteria blooms and fluctuations of some water quality parameters (such as temperature, DO, pH and N-compounds) remains high and could affect the shrimp health in outdoor facilities. Once the system is stable (sufficient particulate microbiota biomass measured in Imhoff cones), BFT provides stabilized parameters of water quality when performed in indoor facilities such as greenhouses, guaranteeing shrimp health.

According to studies performed with the blue shrimp *L. stylirostris* [18] and the pink shrimp *F. duorarum* broodstock [62], BFT could enhance spawning performance as compared to the conventional pond and tank-reared system, respectively (i.e. high number of eggs per spawn and high spawning activity; Fig 4). Such superior performance might be caused by better control of water quality parameters and continuous availability of food (biofloc) in a form of fatty acids protected against oxidation, vitamins, phospholipids and highly diverse "native protein", rather than conventional systems which "young" breeders are often limited to pelletized feed. These nutrients are required to early gonad formation in young breeders and subsequent ovary development. The continuous availability of nutrients could promote high nutrient storage in hepatopancreas, transferred to hemolymph and directed to ovary, resulting in a better sexual tissue formation and reproduction activity [18].

Regarding to shrimp broodstock management, one of the most important management procedures is related to control of solids and stocking density. High levels of solids negatively affect shrimp health, particularly with shrimp weight higher than 15g [47]. Settling solids or "biofloc volume" should be managed below than 15mL/L (measured in Imhoff cones) [38, 47]. Excess of particulate organic matter covered breeder's gills and could limit oxygen exchange, might resulting in mortalities.

Stocking density has to be carefully managed, mainly in sub-adult/adult phase (i.e. >15g). High density or high biomass will lead to an increase in organic matter, TSS levels and N-compounds in tanks or in ponds [63]. Moreover, physical body damages are prevented at low density, improving breeder's health. For review, a suggested stocking density is well described in [64].

For fish, no literature is available regarding BFT and application in breeders. The same trend observed in penaeid shrimp might be observed in fish. The continuous consumption of diverse microbiota (biofloc) should improve nutrients transfer, gonad formation and reproduction performance in fish. Lipid is a well-known nutrient that plays a key role in reproduction of aquatic species. In tilapia, breeders fed with crude palm oil based-feed (n-6 fatty acid rich source) presented high concentration of acid arachidonic or "ARA" (C20:4 n-6) in gonads, eggs and larvae of tilapia as compared to fish oil or linseed oil-based feeds [65]. As a result, better reproductive performance was observed in terms of higher total number of eggs per fish, larger gonad sizes, shorter latency period, inter spawning interval and higher spawning frequency. ARA is an essential fatty acid crucial in reproduction, acting as hormone precursor [66]. In the study [33] was found high ARA content in biofloc harvested in tilapia culture freshwater tanks. Bioflocs in freshwater bioreactors contained high ARA content using glucose and glycerol as a carbon source [67]. These findings suggested that biofloc (according its nutritional profile, for review see section 5.0) might positively influence the reproductive performance in fish, supplying nutrients for gonad development, possibly also enhancing larval quality *a posteriori*. BFT in tilapia broodstock could be an effective method to increase tilapia fry production and further research is need in this field.

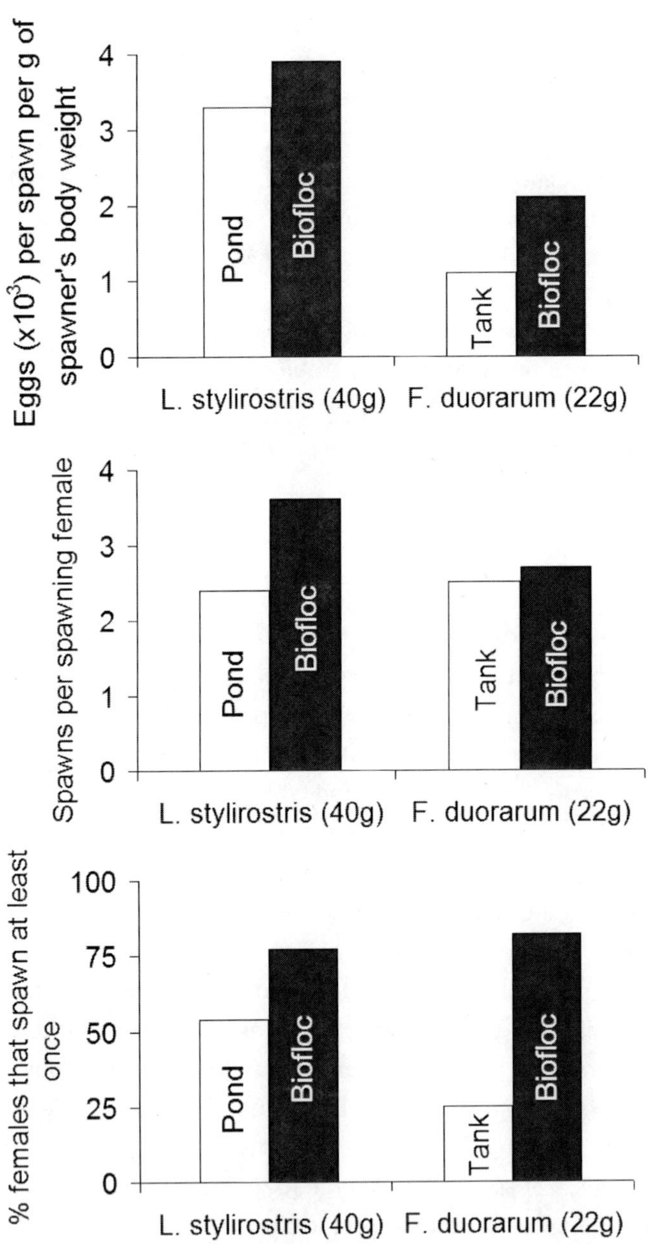

Figure 4. Spawning performance of *F. duorarum* (tank-reared vs biofloc) and *L. stylirostris* (pond-reared vs biofloc) performed in 45 and 30 days after ablation, respectively. Mean weights in parenthesis (more details in [**18**] and [**62**]).

4.3. The "Natural Probiotic" Effect Of Biofloc

Biofloc can be a novel strategy for disease management in contrast to conventional approaches such as antibiotic, antifungal, probiotic and prebiotic application. The "natural probiotic" effect in BFT could act internally and/or externally against, i.e., to *Vibrio sp.* and ectoparasites, respectively. This effect is promoted by large groups of microorganisms, but mainly bacteria that is considered the first trophic level in the system.

Internally, bacteria and its synthesized compounds could act similar to organic acids and might be effective bio-control agents, also given beneficial host's microbial balance in the gut [68]. The regular addition of carbon in the water is known to select for polyhydroxyalkanoates (PHA) accumulating bacteria and other groups of bacteria that synthesize PHA granules. The microbial storage product poly-ß-hydroxybutyrate (PHB), a biodegradable polymer belonging to the polyesters class, is only one compound of a whole family of polyhydroxyalkanoates. PHB is produced by a widely variety of microorganisms such as *Bacillus sp., Alcaligenes sp., Pseudomonas sp.* from soluble organic carbon and is also involved in bacterial carbon metabolism and energy storage [68]. This polymer could comprise ~80% of the bacteria's cell dry matter and up to 16% on biofloc dry weight [69]. Different carbon sources or structures of carbon substrate will result in varying types of PHA [69].

Such granules are synthesized under conditions of physiological and nutrient stress, i.e., when an essential nutrient like nitrogen is limited in the presence of an excess carbon source [68]. When these polymers are degraded in the gut, they could have antibacterial activity similar to short chain fatty acids (SCFAs) or organic acids. The breakdown of PHA inside the gastrointestinal tract can be carried out via chemical and enzymatic hydrolysis [70].

Chemical hydrolysis can be carried out by treating the polymers with, i.e., NaOH, in which could significantly accelerate its digestibility [70]. On the other hand, enzyme hydrolysis is generally carried out by extracellular depolymerases activities which are widely distributed among bacteria and fungi, acting as a preventive or curative protector against *Vibrio sp.* infections and stimulate growth and survival of shrimp and fish larvae [69].

The working mechanism of PHAs with respect to their antibacterial activity is not well understood [68]. As they could act similarly to SCFA, some studies speculated the working mechanism by (i) reduction of pH, in which antibacterial activity increases with decreasing pH value [71]; (ii) inhibiting the growth of pathogenic bacteria by interference on cell membrane structure and membrane permeability, as well as instability of internal protons balance, lowering ATP and depletion of cellular energy [72]; and (iii) down-regulate virulence factor expression and positively influence the gut health of animals [73]. Further research is need to maximizing PHA content in bioflocs applied, i.e., for fish/shrimp feed, characterizing and analyzing their bio-control efficacy in different host-microbe systems [68].

Externally, the working mechanism of biofloc microorganisms against pathogens seems to be by competition of space, substrate and nutrients. Some essentials nutrients such as nitrogen are required by both groups (i.e. heterotrophic bacteria *vs Vibrio sp.*) limiting their growth. Inhibiting compounds excreted by BFT microorganisms, light intensity and type of carbon source also could reduce pathogens growth. Unfortunately, limited information is available on this field. In a study with fish fingerlings [74] was reported that tilapia (initial weight 0.98 ± 0.1g) reared under BFT limited water-exchange condition (FLOC) presented less ectoparasites in gills and ectoderm's mucous as compared to conventional water-exchange system (CW) after 60 days (Fig 5).

4.4. Aquaponics

Aquaponics is a sustainable food production system that combines a traditional aquaculture with hydroponics in a symbiotic environment. The water is efficiently recirculated and reused for maximum benefits through natural biological filtration and recirculation. The waste that is excreted by aquatic species or uneaten feed is naturally converted into nitrate and other beneficial nutrients in the water. Those nutrients are then absorbed by the vegetables and fruits in a "natural fertilization way".

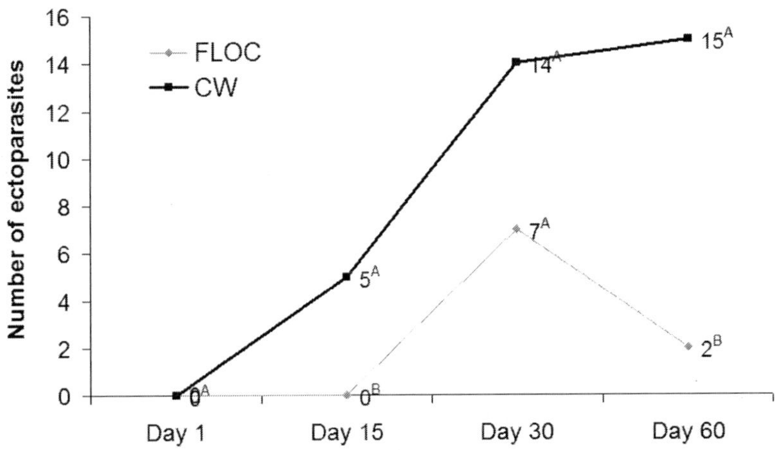

Figure 5. Number of total ectoparasites in gills and ectoderm's mucous of fry tilapia reared under BFT limited water-exchange condition (FLOC) and conventional water-exchange system (CW) after 60 days (more details in [**74**])

Aquaculture species including fish, crayfish, freshwater prawns or shrimp are usually reared in tanks and the water directed into separated race-ways of hydroponics vegetables. A worldwide well-known aquaponics system was successfully developed by University of Virgin Islands (Fig 6). Typical plants raised in aquaponics include lettuce, chard, tomato, fruits such as passion fruit, strawberry, water melon, etc.; and a large variety of spices. Size of aquaculture tanks varies according aquatic species/vegetables demand and usual shapes includes round, square or rectangular tanks.

Nowadays, BFT have been successfully applied in aquaponics. The presence of rich-biota (microorganisms of biofloc) and a variety of nutrients such as micro and macronutrients originated from un-eaten or non-digested feed seems to contribute in plant nutrition. A well-known example of biofloc and aquaponics interaction was also developed by UVI. However, the application of BFT in aquaponics needs particular attention, mainly on management of solid levels in water (for review, see [28]). High concentration of solids may cause excessive adhesion of microorganism on plants roots (biofilm), causing its damage, lowering oxygenation and poor growth. Filtering and settling devices are often needed (Fig 7).

Figure 6. Aquaponics system at University of Virgin Islands

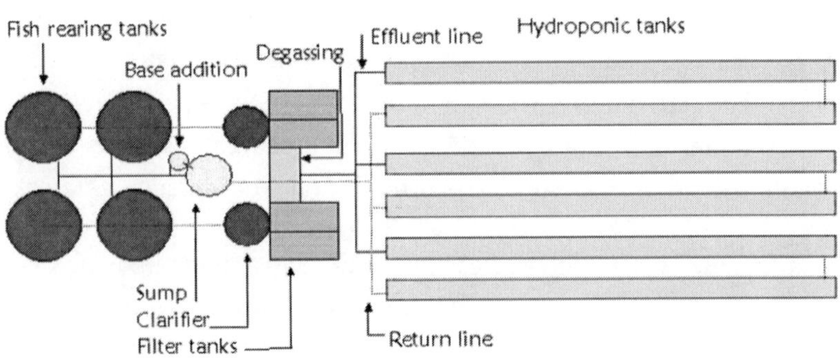

Figure 7. Scheme of worldwide well-known UVI Aquaponics System

5. MICROBIAL BIOMASS APPLICATION IN ANIMAL FOOD INDUSTRY

The cost of diets in several animal cultures is predominantly due to the cost of protein component [75]. In the case of aquaculture, its massive expansion in the last decades has begun to face some important limitations like increasing prices of fishmeal, a raw material prime component of aquaculture diets. However, pressure caused in natural stocks (over-fishing) has depleted fishmeal production and, as a consequence, continuous increase in prices has been observed [76]. Moreover, growth of aquafeed industry (driven by an increase in fish/shrimp demand as the global population continues to growth), the competition with other animal cultures (such as swine and poultry) and differences in fishmeal quality also collaborated with increase in prices of fishmeal. The quality attributed to fishmeal includes high palatability, high content of digestible protein, highly unsaturated fatty acids (HUFA) and minerals.

In this context, alternatives should be evaluated opposing this non-optimism scenario. Aquaculture industry needs to investigate alternative source of proteins to replace less sustainable ones. Candidates of protein sources might have good digestibility, palatability, energy content, low ash content and present a well-balanced essential amino acids profile (EAA) [77].

In the past years, BFT has been emerged not only as promising alternative to grow-out system, but also as a method to obtain protein for compounds diets originated from its diverse microbiota. Collected in tanks/ponds [46, 62] or produced in bioreactors [17, 39, 67] biofloc (Fig 8) is a raw material to produce "biofloc meal". In bioreactors, biofloc production can clean up effluent waters from aquaculture facilities, converting dissolved nutrients into single-cell protein [78]. Usually, two types of bioreactors have been employed: sequencing batch reactors (SBRs) and membrane batch reactors (MBRs), both controlling ammonia, nitrite and suspended solids with great efficacy (for review of bioreactors and its employ, see Kuhn et al 2012). Moreover, excess of solids removed from culture tanks or ponds and/or concentrated into solid removal devices [28] could also be a recyclable source of biofloc for biofloc meal production. This sustainable approach of protein source is getting more attention in the aquaculture industry. The microbial particles can provide important nutrients such as protein [33, 46], lipids [10, 37], aminoacids [80] and fatty acids [33, 67, 81].

Biofloc meal (also called "single-celled" protein), added to compounded feed is currently focus of intensive research in nutrition fields [17, 78]. However, to produce this protein ingredient some processes are required such as drying, milling and storage. In this context, nutritional characteristics could be affected (by i.e. temperature during drying), which the "native" properties could be altered.

Nutritional composition of biofloc differs according to environmental condition, carbon source applied, TSS level, salinity, stocking density, light intensity, phytoplankton and bacteria communities and ratio, etc. Regarding to age of bioflocs, in "young" biofloc heterotrophic bacteria is mainly presented as compared to "old" biofloc dominated by fungi [79]. In biofloc particles, protein, lipid and ash content could vary substantially (12 to 49, 0.5 to 12.5 and 13 to 46%, respectively; Table 2). The same trend occurs with fatty acids (FA) profile. Essential FA such as linoleic acid (C18:2 n-6 or LA), linolenic acid (C18:3 n-3 or ALA), arachidonic acid (C20:4 n-6 or ARA), eicosapentanoic acid (C20:5 n-3 or EPA) and docosahexaenoic acid (C22:6 n-3 or DHA), as well as sum of n-3 and sum of n-6 differ considerably between 1.5 to 28.2, 0.04 to 3.3, 0.06 to 3.55, 0.05 to 0.5, 0.05 to 0.77, 0.4 to 4.4 and 2.0 to 27.0% of total FA. Type of carbon source, freshwater or marine water and production of biofloc biomass (in bioreactors or culture tanks) definitely influence the FA profile (Table 3 and 4). Vitamin and amino acids profile from biofloc produced in large-scale commercial bioreactors [82] in given inTable 5.

Information is still scarce about how microorganisms profile and its nutritional composition could impact animal growth. However, is already known that microorganisms in biofloc might partially replace protein content in shrimp diets, although were not always the case [10, 88]. Recent studies determined how reducing the protein content of diet would affect growth performance of shrimp reared in biofloc conditions. In the study [15] was found that at least 10% of protein content in pelletized feed can be reduced when *F. paulensis* postlarvae are raised in BFT conditions. In

[89] was observed that shrimp fed with less than 25% crude protein under biofloc conditions performed similarly to shrimp raised under regular clear-water intensive culture with a 37%-protein diet. The biofloc system also delivered more consistent survival rates, especially at higher density. A low-protein biofloc meal-based pellet (25% CP) was evaluated as a replacement of conventional high-protein fishmeal diet (40% CP) for *L. vannamei* in a relatively low temperature (25°C) under biofloc conditions [35]. The results showed that is possible to replace 1/3 part of a conventional diet by alternative low-protein biofloc meal pellet without interfering survival and shrimp performance.

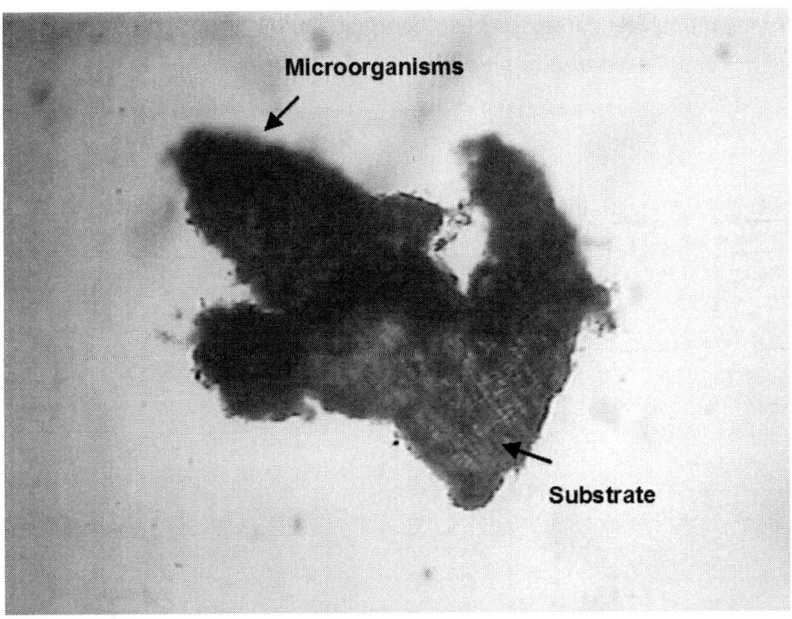

Figure 8. Biofloc particle (10x magnification) (Source: [**54**])

Also, recent studies have been demonstrated that fishmeal in shrimp diets can be partially replaced by other protein sources under biofloc conditions or by biofloc meal. In [90] was evaluated two fishmeal replacement levels (40 and 100% of replacement) by other ingredients (soyabean meal and viscera meals) in diets for *Litopenaeus vannamei* reared in a biofloc system. The authors observed that fishmeal can be replaced in a level of 40.0% without interfering on growth performance and water quality. On the other hand, incorporating treated solids (microbial flocs) generated from tilapia effluent into shrimp feed, [91] demonstrated that shrimp performance was significantly increased as compared to untreated solids (settling basins of tilapia culture units). In [92] a trial performed in clear-water conditions detected that fishmeal can be completely replaced with soy protein concentrate and biofloc meal (obtained from super-intensive shrimp farm effluent) in 38% CP diets without adverse effects on*L. vannamei* performance. Moreover, [17] observed that biofloc produced in SBRs bioreactors using tilapia effluent and sugar as a growth media could offer an alternative protein source to shrimp feeds. Microbial floc-based diets significantly outperformed control fishmeal-based diets in terms of weight gain per week with no differences in survival.

Table 2. Proximate analysis of biofloc particles in different studies.

Crude protein (%)	Carbohydrates (%)	Lipids (%)	Crude fiber (%)	Ash (%)	Reference
43.0	-	12.5	-	26.5	[27]
31.2	-	2.6	-	28.2	[83]
12.0 - 42.0	-	2.0 - 8.0	-	22.0 - 46.0	[84]
31.1	23.6	0.5	-	44.8	[10]
26.0 - 41.9	-	1.2 - 2.3	-	18.3 - 40.7	[80]
30.4	-	1.9	12.4*	38.9	[85]
49.0	36.4	1.13	12.6	13.4	[17]
38.8	25.3	<0.1	16.2	24.7	[78]
28.8 - 43.1	-	2.1 - 3.6	8.7 - 10.4	22.1 - 42.9	[86]
30.4	29.1	0.5	0.8	39.2	[37]
18.2-29.3	22.8-29.9	0.4-0.7	1.5-3.5	43.7-51.8	[47]
18.4-26.3	20.2-35.7	0.3-0.7	2.1-3.4	34.5-41.5	[87]
28.0-30.4	18.1-22.7	0.5-0.6	3.1-3.2	35.8-39.6	[62]

Regarding to biofloc meal production, one bottleneck seems to be the large amount of wet biofloc biomass required to produce 1kg of dry biofloc meal.

Estimative indicates that biofloc plug in 1L settling cones contained only 1.4% of dry matter [14]. The reference [17] indicated that 1 kg of microbial floc could be produced per 1.49 kg of sucrose in bioreactors. Certainly more research is needed on this field. On the other hand, other applications of biofloc meal in animal industry should be evaluated, mainly considering its nutritional profile and relatively low costs as compared to other protein sources (i.e. fishmeal) [17]. In aquaculture, biofloc meal could be included into broodstock pelletized feed, prior or after eyestalk ablation. Further research is encouraged in this field.

Table 3. Fatty acid profile of biofloc (produced in experimental bioreactors) using different carbon source in marine water and freshwater

Fatty Acid	% of total fatty acid							
C14:0	0.10	0.60	0.80	0.45	1.43	0.69	0.61	0.43
C15:0	0.15	0.25	0.25	0.30	0.31	0.31	0.17	0.26
C16:0	2.2	17.0	26.0	15.0	6.06	8.01	6.34	8.86
C16:1	4.0	3.7	3.0	5.0	6.61	2.61	1.61	1.54
C17:0	0.05	0.4	0.5	0.2	0.20	0.23	0.14	0.68
C18:0	0.5	4.0	7.1	6.0	2.37	4.82	3.94	6.27
C18:1 n-7	1.5	3.0	1.9	2.7	3.96	1.72	2.71	4.19
C18:1 n-9	1.8	19.0	30.0	18.0	3.34	7.26	8.12	12.05
C18:2 n-6 (LA)	5.0	19.0	28.2	11.0	1.91	17.24	11.95	21.87
C18:3 n-3 (ALA)	0.04	0.5	0.45	2.0	0.23	0.99	0.20	0.21
C20:0	-	0.10	0.20	0.20	0.06	0.34	0.33	0.49
C20:1 n-9	0.05	0.10	0.15	0.10	0.25	0.20	0.06	0.02
C20:3 n-6	0.15	0.10	0.06	0.07	0.55	0.36	0.15	0.04
C20:4 n-6 (ARA)	0.7	0.3	0.15	0.20	0.77	0.87	0.17	0.06
C20:5 n-3 (EPA)	0.10	0.11	0.05	0.25	0.15	0.15	0.19	0.12
C22:6 n-3 (DHA)	0.05	-	0.07	0.05	0.18	0.06	0.18	0.10
∑ Saturated	22.08	22.99	35.35	22.45	10.76	14.85	11.53	16.99
∑ Monounsaturated	8.16	26.22	35.45	27.15	16.51	14.21	12.5	17.8
∑ n-3	0.4	0.6	0.7	0.65	1.04	2.02	0.60	0.43
∑ n-6	7.0	20.0	27.0	12.0	4.03	19.03	12.27	21.97
Type of water	freshwater	freshwater	freshwater	freshwater	freshwater	freshwater	marine	marine
Carbon source	Acetate	Glycerol	(Glycerol+ *Bacillus*)	Glucose	Glucose	Glycerol	Glucose	Glycerol
Collection	bioreactors	bioreactors	bioreactors	bioreactors	bioreactors	bioreactors	bioreactors	bioreactors
Reference	[39]				[67]			

Table 4. Fatty acid profile of biofloc (collected in tanks) using different carbon source in marine water and freshwater

Fatty Acid	% of total fatty acid		
C14:0	2.02-2.48	13.8-16.1	5.4-6.2
C15:0	0.70-0.77	1.1-1.5	1.1-1.3
C16:0	17.88-19.10	45.4-53.5	48.7-49.3
C16:1	7.15-7.74	9.9-15.3	16.5-21.6
C17:0	-	0.7	0.9-1.0
C18:0	6.24-7.27	3.4-3.5	3.7-4.5
C18:1 n-7	11.05-11.28	-	-
C18:1 n-9	8.51-10.08	8.8-9.2	7.7-10.8
C18:2 n-6 9 (LA)	15.38-16.68	1.5-2.5	2.2-2.6
C18:3 n-3 (ALA)	0.65-0.73	2.0-2.3	2.2-3.3
C20:0	0.87-1.44	0.2-0.4	0.4
C20:1 n-9	0.74-0.80	0.3-0.4	0.5
C20:3 n-6	0.40-0.46	0.2	0.2
C20:4 n-6 (ARA)	3.11-3.55	0.3-0.4	0.3-0.4
C20:5 n-3 (EPA)	0.39-0.46	0.3-0.5	0.5
C22:6 n-3 (DHA)	0.74-0.77	0.2-0.4	0.3-0.4
Σ Saturated	30.2-34.92	67.6-73.0	61.5-61.9
Σ Monounsaturated	28.10-29-38	19.7-25.0	28.3-30.5
Σ n-3	1.38-1.91	2.8-3.4	3.2-4.4
Σ n-6	23.5-25.81	2.0-3.0	2.7-3.1
Type of water	freshwater	marine	marine
Carbon source	Wheat flour	molasses	molasses
Collection	Tilapia tanks	shrimp tanks	shrimp tanks
Reference	[33]	[87]	[62]

Table 5. Example of vitamin and amino acids profile from biofloc produced in large-scale commercial bioreactors [**82**].

Amino Acids	As Fed (%)
Alanine	3.82
Arginine	3.60
Aspartic acid	6.36
Glutamic acid	8.04
Glycine	2.81
Histidine	1.46
Isoleucine	3.38
Leucine	5.06
Lysine	4.34
Methionine	1.41
Cysteine	0.55
Phenylalanine	3.29
Proline	2.77
Serine	2.82
Taurine	0.25
Threonine	3.11
Tryptophan	0.98
Tyrosine	2.83
Valine	3.52
Total	60.4
Vitamins	
Niacin	83.3 mg/kg
Thiamine B1	7.7 mg/kg
Riboflavin	39.0 mg/kg
Vitamin B12	12.0 mg/kg
Vitamin E	29.8 IU/kg

6. CONCLUSIONS AND PERSPECTIVES OF BFT

Biosecurity is a priority in aquaculture industry. For example, in shrimp farming, considerable impact of disease outbreaks during the past two decades greatly affected the operational management of shrimp farms worldwide [10]. Infected PLs and incoming water seem to be the main pathway for pathogen introduction. This scenario forced farmers to look for more biosecure culture practices to minimize the risk associated with exposure to pathogens [2]. Biofloc technology brings an obvious advantage of minimizing consumption and release of water, recycling *in situ* nutrients and organic matter. Furthermore, pathogens introduction is reduced, improving the farm biosecurity.

Biofloc technology will enable aquaculture grow towards an environmental friendly approach. Consumption of microorganisms in BFT reduces FCR and consequently costs in feed. Also, microbial community is able to rapidly utilize dissolved nitrogen leached from shrimp faeces and uneaten food and convert it into microbial protein. These qualities make minimal-exchange BFT system an alternative to extensive aquaculture. Microorganisms in biofloc might partially replace protein content in diets or decrease its dependence of fishmeal.

Related to biofloc meal and its perspectives, the study [17] detected initial estimates of cost for producing a metric ton of biofloc meal is approximately $400 to $1000. The same authors cited that global soymeal market varied approximately from $375 to $550/metric ton from January 2008 through May 2009. During the same time period, fishmeal varied approximately from $1000 to $1225, suggesting feasibility on replacement of either soybean and/or fish meal by biofloc meal. Moreover, generated from a process that cleans aquaculture effluents [17, 39] biofloc meal production avoids discharge of waste water and excessive damage to natural habitats [4]. This ingredient seems to be free of deleterious levels of mycotoxins, antinutritional factors and other constituents that limit its use in aquafeeds [79]. Large-scale production of biofloc meal for use in aquaculture could result in environmental benefits to marine and coastal ecosystems, as the need for wild fish as an aquafeed ingredient is reduced [79, 92].

Sensorial quality of BFT products is also an important issue. BFT may bring higher profit if fresh non-frozen shrimp/fish is sold to near-by market, mainly at inland locations. These advantages certainly should be more explored and niche markets achieved, contributing to social sustainability.

ACKNOWLEDGEMENT

The authors would like to thank CONCYTEY (Consejo de Ciencia y Tecnología del Estado de Yucatán), Coordenação de Aperfeiçoamento de Pessoal de Nível Superior-CAPES, Brazilian Ministry of Education (PhD grant number 4814061 provided to the primary author) and Consejo Nacional de Ciencia y Tecnología-CONACyT, México (grant 60824) for research support. The authors also would like to thank Wilson Wasielesky, Yoram Avnimelech and Manuel Valenzuela for photos courtesy and Miguel Arévalo, Maite Mascaró, Elsa Noreña, Santiago Capella, Adriana Paredes, Gabriela Palomino, Korynthia Aguiar, Moisés Cab, Nancy Aranda Cirerol, Concepción Burgos, Manuel Valenzuela and all staff of Programa Camarón-UMDI for their contribution towards researches performed at UMDI-UNAM cited in this chapter.

REFERENCES

1. Food and Agriculture Organization of the United Nations FAO (2008) Cultured aquaculture species information programmePenaeus vannamei (Boone,1931Food and Agriculture Organization of the United Nations. Available at http://www.fao.org

2. C. L Browdy, D Bratvold, A. D Stokes, R. P Mcintosh, 2001Perspectives on the application of closed shrimp culture systems. In: Jory ED, Browdy CL, editors. The new Wave, Proceedings of the Special Session on Sustainable Shrimp Culture, The World Aquaculture Society, Baton Rouge, LA, USA. 2034

3. P De Schryver, R Crab, T Defoirdt, N Boon, W Verstraete, 2008The basics of bio-flocs technology: the added value for aquaculture. Aquaculture 277125137

4. R. L Naylor, R. J Goldburg, J. H Primavera, N Kautsk, Beveridge MCM, Clay J, Folke C, Lubchencoi J, Mooney H, Troell M (2000Effect of aquaculture on world fish supplies. Nature 40510171024

5. J Bender, R Lee, M Sheppard, K Brinkley, P Philips, Y Yeboah, R. C Wah, 2004A waste effluent treatment system based on microbial mats for black sea bass Centropristis striata recycled water maricultureAquac Eng 317382

6. A Ray, 2012Biofloc technology for super-intensive shrimp culture. In: Avnimelech Y, editor. Biofloc Technology- a practical guide book, 2nd ed., The World Aquaculture Society, Baton Rouge, Louisiana, USA. 167188

7. R Crab, M Kochva, W Verstraete, Y Avnimelech, 2009Bio-flocs technology application in over-wintering of tilapiaAquac Eng 40105112

8. M. A Burford, P. J Thompson, R. P Mcintosh, R. H Bauman, D. C Pearson, 2003Nutrient and microbial dynamics in high-intensity, zero-exchange shrimp ponds in BelizeAquaculture 219393411

9. M. A Burford, P. J Thompson, R. P Mcintosh, R. H Bauman, D. C Pearson, 2004The contribution of flocculated material to shrimp (Litopenaeus vannamei) nutrition in a high-intensity, zero-exchange systemAquaculture 232525537

10. W Wasielesky, Jr, Atwood H, Stokes A, Browdy CL (2006Effect of natural production in a zero exchange suspended microbial floc based super-intensive culture system for white shrimp Litopenaeus vannamei Aquaculture 258396403

11. J. E Rakocy, D. S Bailey, E. S Thoman, R. C Shultz, 2004Intensive tank culture of tilapia with a suspended, bacterial based treatment process: new dimensions in farmed tilapia. In: Bolivar R, Mair G, Fitzsimmons K, editors. Proceedings of the Sixth International Symposium on Tilapia in Aquaculture. 584596

12. Y Avnimelech, S Mokady, G. L Schoroder, 1989Circulated ponds as efficient bioreactors for single cell protein productionBamdigeh. 415866

13. J. A Hargreaves, 2006Photosynthetic suspended-growth systems in aquacultureAquac Eng 34344363

14. Y Avnimelech, 2007Feeding with microbial flocs by tilapia in minimal discharge bioflocs technology ponds. Aquaculture 264140147

15. Ballester ELCAbreu PC, Cavalli RO, Emerenciano M, Abreu L, Wasielesky W (2010Effect of practical diets with different protein levels on the performance of Farfantepenaeus paulensis juveniles nursed in a zero exchange suspended microbial flocs intensive systemAquac Nut 16163172

16. Yoram Avnimelech (2012Biofloc Technology- A Practical Guide Booknd ed. The World Aquaculture Society, Baton Rouge, Louisiana, EUA. 272p.

17. D. D Kuhn, G. D Boardman, A. L Lawrence, L Marsh, G. J Flick, 2009Microbial flocs generated in bioreactors is a superior replacement ingredient for fishmeal or soybean meal in shrimp feed. Aquaculture 2965157

18. M Emerenciano, G Cuzon, J Goguenheim, G Gaxiola, Aquacop (2011Floc contribution on spawning performance of blue shrimp Litopenaeus stylirostrisAquac Res (published online first DOI:j.13652109x)

19. Aquacop (1975Maturation and spawning in captivity of penaeid shrimp: Penaeus merguiensis de ManPenaeus japonicus Bate, Penaeus aztecus Ives, Metapenaeus ensis de Haan and Penaeus semisulcatus de Haan. In: Avault W, Miller R, editors. Proceedings of the Sixth Annual Meeting World Mariculture Society, Lousiana State University, Baton Rouge, 123129

20. L Sohier, 1986Microbiologie appliquée à l'aquaculture marine intensive. 119Thèse Doctorat d'Etat, Université Aix-Marseille II Marseille, France.

21. G Cuzon, A Lawrence, G Gaxiola, C Rosas, J Guillaume, 2004Nutrition of Litopenaeus vannamei reared in tanks or in pondsAquaculture 235513551

22. B Rosenberry, 2010Controlling pH in biofloc ponds. The shrimp news international.http://www.shrimpnews.com/FreeReportsFolder/phContronBioflocPonds.html.Acessed in 11/02/2012.

23. P Garen, Aquacop (1993Nuevos resultados en la cría intensiva de camarón Penaeus vannamei y P. stylirostris. In: Calderón JV, Sandoval VC, editors. Memorias del I Congresso Ecuatoriano de Acuicultura, Guayaquil, 18-23 Octubre 1992. Escuela Superior Politécnica del Litoral, Guayaquil, 137145

24. B Rosenberry, 2007Marvesta Shrimp Farms Delivers "Fresh" Shrimp to Restaurants. The shrimp news international.http://www. shrimpnews. com/FreeNewsFolder/FreeNewsBackIssues/2007BackIssues/FreeNewsApril200720.html.Acessed in 12/02/2012.

25. Moriarty DJW (1997The role of microorganisms in aquaculture pondsAquaculture151333349

26. J Ray, G Seaborn, J. W Leffler, S. B Wilde, A Lawson, C. L Browdy, 2010Characterization of microbial communities in minimal-exchange, intensive aquaculture systems and the effects of suspended solids management2010310130138

27. D Mcintosh, T. M Samocha, E. R Jones, A. L Lawrence, D. A Mckee, S Horowitz, A Horowitz, 2000The effect of a bacterial supplement on the high-density culturing of Litopenaeus vannamei with low-protein diet in outdoor tank system and no water exchange. Aquac Eng 21215227

28. J Ray, B. L Lewis, C. L Browdy, J. W Leffler, 2010Suspended solids removal to improve shrimp (Litopenaeus vannamei) production and an evaluation of a plant-based feed in minimal-exchange, superintensive culture systemsAquaculture2998998

29. P Panjaitan, 2004Field and laboratory study of Penaeus monodon culture with zero water exchange and limited water exchange model using molasses as a carbon source. Ph.D. Thesis, Charles Darwin Univ., Darwin, NT, Australia.

30. Y Avnimelech, M Kochva, S Diab, 1994Development of controlled intensive aquaculture systems with a limited water exchange and adjusted carbon to nitrogen ratioBamidgeh 46119131

31. B Hari, B. M Kurup, J. T Varghese, J. W Schrama, Verdegem MCJ (2004Effects of carbohydrate addition on production in extensive shrimp culture systemsAquaculture 241179194

32. Schneider, V Sereti, E. H Eding, Verreth JAJ (2005Analysis of nutrient flows in integrated intensive aquaculture systemsAquac Eng 32379401

33. M. E Azim, D. C Little, 2008The biofloc technology (BFT) in indoor tanks: Water quality, biofloc composition, and growth and welfare of Nile tilapia (Oreochromis niloticus)Aquaculture2832935

34. M Emerenciano, L Vinatea, A. G Gálvez, A Shuler, A Stokes, J Venero, J Haveman, J Richardson, B Thomas, J Leffler, 2009Effect of two different diets fish meal based and "organic" plant based diets in Litopenaeus setiferus earlier post-larvae culture under biofloc, green-water and clear-water conditions. CD of abstracts of World Aquaculture Society Meeting 2009, Veracruz, México.

35. M Emerenciano, G Cuzon, K López-aguiar, E Norena-barroso, M Máscaro, G Gaxiola, 2011Biofloc meal pellet and plant-based diet as an alternative nutrition for shrimp under limited water exchange systems. CD of abstracts of World Aquaculture Society Meeting 2011, Natal, RN, Brazil.

36. Y Avnimelech, 1999Carbon and nitrogen ratio as a control element in aquaculture systems. Aquaculture 176227235

37. M Emerenciano, Ballester ELC, Cavalli RO, Wasielesky W (2012Biofloc technology application as a food source in a limited water exchange nursery system for pink shrimp Farfantepenaeus brasiliensis (Latreille, 1817)Aquac Res 43447457

38. N Taw, 2010Biofloc technology expanding at white shrimp farms. Global Advocate may/june, 24-26 (available inhttp://www.gaalliance.org/ mag/May_June2010.pdf)

39. R Crab, B Chielens, M Wille, P Bossier, W Verstraete, 2010The effect of different carbon sources on the nutritional value of bioflocs, a feed for Macrobrachium rosenbergii postlarvaeAquac Res 41559567

40. Y Avnimelech, S Mokady, 1988Protein biosynthesis in circulated ponds. In: Pullin RSV, Bhukaswan T, Tonguthai K, Maclean JL, editors. Proceedings of Second International Symposium on Tilapia in

Aquaculture, Departament of Fisheries of Thailand and ICLARM, Manila, Philippines, 301309

41. A Milstein, Y Anvimelech, M Zoran, D Joseph, 2001Growth performance of hybrid bass and hybrid tilapia in conventional and active suspension intensive pondsBamidgeh 53147157

42. M Asaduzzaman, M. M Rahman, M. E Azim, M. A Islam, M. A Wahab, Verdegem MCJ, Verreth JAJ (2010Effects of C/N ratio and substrate addition on natural food communities in freshwater prawn monoculture pondsAquaculture306127136

43. S. M Suita, 2009O uso da Dextrose como fonte de carbono no desenvolvimento de bio-flocos e desempenho do camarão-branco (Litopenaeus vannamei) cultivado em sistema sem renovação de água. Tese de mestrado. Universidade Federal do Rio Grande, Rio Grande do Sul, Brasil.

44. T. M Samocha, S Patnaik, M Speed, A. M Ali, J. M Burger, R. V Almeida, Z Ayub, M Harisanto, A Horowitz, D. L Brock, 2007Use of molasses as carbon source in limited discharge nursery and grow-out systems for Litopenaeus vannamei. Aquac Eng 36184191

45. M Asaduzzaman, M. A Wahab, Verdegem MCJ, Huque S, Salam MA, Azim ME (2008C/N ratio control and substrate addition for periphyton development jointly enhance freshwater prawn Macrobrachium rosenbergii production in pondsAquaculture280117123

46. M Emerenciano, Ballester ELC, Cavalli RO, Wasielesky W (2011bEffect of biofloc technology (BFT) on the early postlarval stage of pink shrimp Farfantepenaeus paulensis: growth performance, floc composition and salinity stress toleranceAquac Int 19891901

47. M Emerenciano, G Cuzon, A Paredes, G Gaxiola, 2012Biofloc technology applied to intensive broodstock farming of pink shrimp Farfantepenaeus duorarum (Part I): grow-out, water quality, microorganisms profile and proximate analysis of biofloc. Aquac Res (submitted)

48. J. K Mishra, T. M Samocha, S Patnaik, M Speed, R. L Gandy, A Ali, 2008Performance of an intensive nursery system for the Pacific white

shrimp, Litopenaeus vannamei, under limited discharge conditionAquac Eng 38215

49. F. D Apud, J. H Primavera, P. L Torres, 1983Farming of Prawns and ShrimpsExtension Manual (5), SEAFDEC Aquaculture Department, Iloilo, Philippines. 67pp.

50. P. A Sandifer, A. D Stokes, J. S Hopkins, 1991Further intensification of pond shrimp culture in South Carolina. In: Sandifer PA, editor. Shrimp Culture in North America and the Caribbean, World Aquaculture Society, Baton Rouge, LA, USA. 8495

51. S. J Arnold, F. E Coman, C. J Jackson, S. A Groves, 2009High-intensity, zero water-exchange production of juvenile tiger shrimp, Penaeus monodon: An evaluation of artificial substrates and stocking density. Aquaculture 2934248

52. D. M Souza, S. M Suita, Leite FPL, Romano LA, Wasielesky W, Ballester ELC (2011The use of probiotics during the nursery rearing of the pink shrimp Farfantepenaeus brasiliensis (Latreille, 1817) in a zero exchange systemAquac Res (published online first doi:10.1111/j.13652109x)

53. AQUACOPLe Moullac G, Damez D (1991Modélisation de la résistance au chocs de salinité des postlarves de Penaeus vannamei. Aquat Living Resour 4169173

54. Emerenciano MGCWasielesky W, Soares RB, Ballester EC, Cavalli RO, Izeppi EM (2007Crescimento e sobrevivêcia do camarão-rosa Farfantepenaeus paulensis na fase de berçário em meio heterotrófico. Acta Scientiarum Biological Sciences 2917

55. J Cohen, T. M Samocha, J. M Fox, R. L Gandy, A. L Lawrence, 2005Characterization of water quality factors during intensive raceway production of juvenile L. vannamei using limited discharge and biosecure management tools. Aquac Eng 32425442

56. D Krummenauer, S Peixoto, R. O Cavalli, L. H Poersch, W Wasielesky, 2011Superintensive culture of white shrimp, Litopenaeus vannamei, in a biofloc technology system in southern brazil at different stocking densitiesJ. World Aquacult. Soc. 42726733

57. T. M Samocha, S Patnaik, R. L Gandy, 2004Heterotrophic intensification of pond shrimp production. Book of abstract of Fifth International

Conference on Recirculating Aquaculture, 2225July 2004, Roanoke, Virginia, USA.

58. L. E Cruz-ricque, J Guillaume, G Cuzon, 1987Squid protein effect on growth of four penaeid shrimp. J. World Aquacult. Soc. 18209217

59. L Itani, Neto ETA, Silva SL, Araújo ML, Lima AF, Barbosa JM (2010Efeito do sistema heterotrófico no crescimento do Tambaqui (Colossoma macropomum). X JORNADA DE ENSINO, PESQUISA E EXTENSÃO- JEPEX 2010UFRPE: Recife.

60. G Poleo, J. V Aranbarrio, L Mendoza, O Romero, 2011Cultivo de cachama blanca en altas densidades y en dos sistemas cerradosPesq. Agropec. Bras. 46429437

61. R Wouters, P Lavens, J Nieto, P Sorgeloos, 2001aPenaeid shrimp broodstock nutrition: an updated review on research and development.Aquaculture 202121

62. M Emerenciano, G Cuzon, M Arévalo, G Gaxiola, 2012Biofloc technology applied to intensive broodstock farming of pink shrimp Farfantepenaeus duorarum (Part II): spawning performance, biochemical composition and fatty acid profile. Aquac Res (submitted)

63. L Vinatea, A. O Gálvez, C. L Browdy, A Stokes, J Venero, J Haveman, B. L Lewis, A Lawson, A Shuler, J. W Leffler, 2010Photosynthesis, water respiration and growth performance of Litopenaeus vannamei in a super-intensive raceway culture with zero water exchange: Interaction of water quality variablesAquac Eng 421724

64. M Emerenciano, G Gaxiola, G Cuzon, 2012Biofloc Technology applied to shrimp broodstock. In: Avnimelech Y, editor. Biofloc Technology-apractical guide book, 2nd ed., The World Aquaculture Society, Baton Rouge, Louisiana, USA. 217230

65. W Ng, Y Wang, 2011Inclusion of crude palm oil in the broodstock diets of female Nile tilapia, Oreochromis niloticus, resulted in enhanced reproductive performance compared to broodfish fed diets with added fish oil or linseed oilAquaculture314122131

66. B. D Glencross, 2009Exploring the nutritional demand for essential fatty acids by aquaculture species. Rev Aquac 171124

67. J Ekasari, R Crab, W Verstraete, 2010Primary nutritional content of bio-flocs cultured with different organic carbon sources and salinityHAYATI Journal of Biosciences17125130

68. K Sinha, K Baruah, P Bossier, 2008Horizon Scanning: the potential use of biofloc as an anti-infective strategy in aquaculture- an overview. Aquac Health Int 13810

69. P De Schryver, N Boon, W Verstraete, P Bossier, 2012The biology and biotechnology behind bioflocs. In: Avnimelech Y, editor. Biofloc Technology- a practical guide book, 2nd ed., The World Aquaculture Society, Baton Rouge, Louisiana, USA. 217230

70. J Yu, D Plackett, Chen LXL (2005Kinetics and mechanism of the monomeric products from abiotic hydrolysis of poly [(R)-3-hydroxybutyrate] under acidic and alkaline conditions. Polym Degrad Stabil 89289299

71. S. C Ricke, 2003Perspectives on the use of organic acids and shortchain fatty acids as antimicrobials. Poult Sci 82632639

72. J. B Russel, 1992Another explanation for the toxicity of fermentationacids at low pH: anion accumulation versus uncoupling. J Appl Bacteriol 73363370

73. J. E Teitelbaum, W. A Walker, 2002Nutritional impact of pre and pro-biotics as protective gastrointestinal organisms. Annu Rev Nutr 22107138

74. M Emerenciano, Y Avnimelech, R Gonzalez, Leon ATD, Cuzon G, Gaxiola G (2009Effect of bio-floc technology (BFT) in ectoparasite control in Nile tilapia Oreochromis niloticus culture. CD of Abstracts of World Aquaculture Society Meeting 2009, Veracruz, Veracruz, Mexico.

75. Tacon AGJ (1990Standard methods for the nutrition and feeding of farmed fish and shrimpWashington DC, Argent Laboratories Press, 454 pp.

76. Tacon AGJ (1997Feeding tomorrow's fish: keys for sustainabilityIn: Tacon A, Basurco B, editors. Feeding tomorrow's fish. Proceedings of the workshop of the CIHEAM Network on Technology of Aquaculture in the Mediterranean (TECAM), jointly organized by CIHEAM, FAO and IEO Mazarron, Spain, 24-26 June 1996, CIHEAM, Apodo, Spain, 1133

77. R. W Hardy, 2010Utilization of plant proteins in fish diets effects of global demand and supplies of fishmeal.Aquac Res 41770776

78. D. D Kuhn, A. L Lawrence, G. D Boardman, S Patnaik, L Marsh, G. J Flick, 2010Evaluation of two types of bioflocs derived from biological treatment of fish effluent as feed ingredients for Pacific white shrimp, Litopenaeus vannameiAquaculture3032833

79. D. D Kuhn, A Lawrence, 2012Ex-situ biofloc technology. In: Avnimelech Y, editor. Biofloc Technology- a practical guide book, 2nd ed., The World Aquaculture Society, Baton Rouge, Louisiana, USA. 217230

80. Z. Y Ju, I Forster, L Conquest, W Dominy, W. C Kuo, F. D Horgen, 2008Determination of microbial community structures of shrimp floc cultures by biomarkers and analysis of floc amino acid profilesAquac Res 39118133

81. M Izquierdo, I Forster, S Divakaran, L Conquest, O Decamp, A Tacon, 2006Effect of green and clear water and lipid source on survival, growth and biochemical composition of Pacific white shrimp Litopenaeus vannameiAquac Nut 12192202

82. J Logan, A Lawrence, W Dominy, Tacon AGJ 2010Single-cell proteins from food by-products provide protein in aquafeed. Global Advocate 135657

83. Tacon AGJCody JJ, Conquest LD, Divakaran S, Forster IP, Decamp OE (2002Effect of culture system on the nutrition and growth performance of Pacific white shrimp Litopenaeus vannamei (Boone) fed different dietsAquac Nut 8121137

84. R Soares, C Jackson, F Coman, N Preston, 2004Nutritional composition of flocculated material in experimental zero-exchange system for Penaeus monodon. In: Proceedings of Australian Aquaculture, 2004, WAS, Sydney 89

85. Z. Y Ju, I Forster, L Conquest, W Dominy, 2008Enhanced growth effects on shrimp (Litopenaeus vannamei) from inclusion of whole shrimp floc or floc fractions to a formulated dietAquac Nut 14533543

86. P. F Maicá, M. R Borba, W Wasielesky, 2012Effect of low salinity on microbial floc composition and performance of Litopenaeus vannamei

(Boone) juveniles reared in a zero-water-exchange super-intensive systemAquac Res 43361370

87. M Emerenciano, G Cuzon, M Arévalo, M. M Miquelajauregui, G Gaxiola, 2012Effect of short-term fresh food supplementation on reproductive performance, biochemical composition and fatty acid profile of Litopenaeus vannamei (Boone) reared under biofloc conditions.Aquac Int (submitted).

88. D Mcintosh, T. M Samocha, E. R Jones, A. L Lawrence, S Horowitz, A Horowitz, 2001Effects of two commercially available low protein diets (21% and 31%) on water sediment quality, and on the production of Litopenaeus vannamei in an outdoor tank system with limited water discharge. Aquac Eng 256982

89. Nunes AJPCastro LF, Sabry-Neto H (2010Microbial flocs spare protein in white shrimp diets. Global Advocate 102830

90. R. B Scopel, R Schveitzer, W. Q Seiffert, V Pierri, R. F Arantes, F. N Vieira, L. A Vinatea, 2011Substituição da farinha de peixe em dietas para camarões marinhos cultivados em sistema bioflocos. Pesq Agropec Bras 46928934

91. D. D Kuhn, G. D Boardman, S. R Craig, G. J Flick, E Mclean, 2008Use of microbial flocs generated from tilapia effluent as a nutritional supplement for shrimp, Litopenaeus vannamei, in recirculating aquaculture systems. J. World Aquacult. Soc. 397282

92. W Bauer, C Prentice-hernandez, M. B Tesser, W Wasielesky, Poersch LHS (2012Substitution of fishmeal with microbial floc meal and soy protein concentrate in diets for the pacific white shrimp Litopenaeus vannamei. Aquaculture 342343

Chapter 9

Causalities between Price, Pond Area and Employment in Aquaculture Production

Nik Hashim Nik Mustapha[1], Azlina Abd Aziz[1], Nik Mohd Hazrul Hashim[2]

[1]Department of Economics, Faculty of Management and Economics, University of Malaysia Terengganu, Kuala Terengganu, Malaysia

[2]Graduate School of Business, National University of Malaysia, Bandar Baru Bangi, Malaysia.

ABSTRACT

The role of aquaculture industry is becoming more prominent in order to supplement marine capture in meeting the food need for the growing Malaysian population. In an attempt to minimize depletion of marine fisheries, only traditional vessels are allowed to fish along the coastal area while bigger vessels are relegated to deep-sea fishing. During the 9th Malaysian Plan (2006-2010) aquaculture has been recognized as the engine of growth in the national food sector's development strategy. Future fisheries policy is expected to focus more on aquaculture production, marketing and technological improvement as an alternative to marine capture. This paper investigates the causalities between the selected freshwater fish prices, aquaculture area and production. The study aspires to establish whether or not market price is a key contributor to a rise in the aquaculture area and production. Aquaculture firms comprising the individual culturists are generally motivated by the economic potential of the industry which is reflected in excess of price over cost of production. Our hypothesis is that government policy and initiation rather than prices had give rise to greater participation of culturists and hence augmented the level of employment. However, production increase has a negative implication on environment degradation. Thus there is a conflicting view as regards to the employment opportunity generated by aquaculture undertakings and the need for sustainable development arising from this growing industry. Multivariate time series analysis was used in this investigation.

KEYWORDS

Aquaculture; Unit Root; Cointegration; Vector Error Correction Model; Granger's Causality

1. INTRODUCTION

The popularity of aquaculture practice in Malaysia comes with its commercialization during the last few decades. As such aquaculture is dominated by several fishery species associated with the market potential which is primarily reflected in high prices offered to the breeders. These fisheries are in fact experiencing shortage in supply. Species such as giant prawn (Macrobrachium rosenbergii) and tiger prawn (Penaeus monodon) are delicacies of hotels and restaurants due to the high demand while the supply is limited. Another underlying factor for their high demand is that special fishery species like the catfish (Clarias gariepinus) is believed among the majority of its medicinal healing capability. However, the success of culturing these aquatic resources either freshwater or brackish-water fisheries is always constrained by the economic factors such as high capital outlay to start the business and the expertise knowledge required in their breeding and maintenance. Availability of fingerlings, cost of feeding ingredients, suitability of land area in terms of location and lastly problems related to marketing and distributional system could be the salient factors that need to be considered in aquaculture practices.

In macro perspective aquaculture fishery is an essential component of the economy food security and sustainability. In 2009 the aquaculture statistics on area, production and the number of culturist were 4916.2 km^2, 21,051 metric tons and 70361.5 persons respectively. The retail value accounted for about RM476, 549.45 for 2009 [1]. These statistics show continuous growth trends over the years indicating that the aquaculture industry is escalating perhaps due to the fact investors are attracted to the profitability of the prospective species. With an increase in demand for food owing to ever-growing population and heavy reliance of the present fishery policy on marine capture, the crucial function of aquaculture as supplementary to marine fisheries becomes obvious. By the end of the 9th Malaysia Plan 2010 Malaysia targeted to produce some 507,550 metric tons of aquaculture and aquatic products under the National Agricultural Policy [2]. Aquaculture is seen as an alternative that could reduce the pressure of over fishing on marine fisheries resource. The role of aquaculture in food economy is stipulated to become more important in future in view of depleting marine fisheries. In the light of this changing scenario which the government has agreed to adopt this as a new fisheries policy it deems necessary to investigate the extent to which interactions between pricing, area and employment are intertwined in order to provide economic rationale.

The objective of this study is to investigate the interdependencies of economic policy variables namely prices of freshwater fisheries, area invested with aquaculture ponds and the number of culturists. The general hypothesis is that although aquaculture provides the best alternative to marine capture its growth depends on the economic interplays of fish prices and aquaculture area. In big cities land competes for alternative investments pushing land price extremely high. Since the investment in aquaculture activity generally requires high initial capital, technical knowhow on fisheries rearing, difficulty of getting seedlings, high percentage of fish mortality due to river pollution, investing in aquaculture is almost impossible. Culturally majority of Malaysian favor marine fisheries it is therefore, hypothesized that the probability of a significant increase in aquaculture operators and thus employment is minimal. The results of this investigation are useful for decision makers in formulating strategic fisheries policies. If price is highly significant and correlated with employment and area cultured or otherwise, then strategic policy can be under-taken to further support such relationships.

2. FRESHWATER CATCH AND PRODUCTION

Statistics provided by AQUASTAT [3] show that freshwater catch in 1990 was estimated at 12,995 metric tons about 10 percent of the total catch of 3,783,743 metric tons in Asia (excluding Middle-East). Freshwater catch for 2000 figures were 22,636 metric tons and 5,959,055 metric tons for Malaysia and Asia respectively. The catch difference is about 12 percent but there is an improvement of about 2 percent per annum over the period 1990-2000. The total share of aquaculture for Malaysia in terms of Asian aquaculture production in 1997 was relatively small which accounted for only 5 percent.

One of the more recent developments in aquaculture species of market potential is the red hybrid tilapia (Oreochromis sp.), a genetic improvement with attractive red coloration. In 2007 Malaysia produced 32,258 metric tons (80% red tilapia and 20% black tilapia) with the wholesale value of MYR170.6 million [4]. According to the same source fresh water red tilapia production constitutes about 46 percent of total aquaculture production and 12 percent of this total is from brackish water. In value term the production of tilapia is currently leading all other aquaculture species constituting 49.4 percent, followed by catfish 37 percent and carps 10 percent. The major drawbacks in aquaculture, in particular, the red hybrid tilapia production are soaring feeds cost, shortage of trained workers, diseases resistance fries and the supply of fingerlings has to be bought from commercial firms. The Institute of Tropical Aquaculture of the University

Malaysia Terengganu (UMT) is doing research to produce diseases resistant seeds of tilapia.

The major drawback during the earlier part of freshwater development in this country originated from the domination of the marine capture over freshwater species and the consumers' preference for the former. Over the last decade marine capture had experienced drastic changes due to depletion of the resource partly caused by the increased demand as population and income grew. On the supply side depletion of the resource is attributable to several factors, namely the continuous application of efficient fishing methods such as trawlers, the encroachment of efficient vessels from outsiders within the Malaysian 200 nautical miles Exclusive Economic Zone (EEZ), ineffectiveness policing of the fishing grounds owing to insufficient enforcement officers to guard the assigned EEZ. Aquaculture investment had been in Malaysia for a long period of time but had been tried on a relatively small scale because of the high risk and large capital needed for the industry. Hence, with existing requirement for additional supply of fish in addition to marine capture this industry started to flourish partly due to the aquaculture potential for sea food industry.

Experience shows that the success of aquaculture investment is likely to benefit the large scale operators more relative to small operators pertaining to the technical knowhow, culturing and breeding practices. In delivering their products contracts with the local hotels are made and for exporting fish products to destinations outside the country approval and procedures need to be adhered. The small operators with limited working capital, technical knowhow and access to the target market would be greatly at the disadvantage. Alternatively small operators could make this aquaculture undertaking a success through fishermen Association or Cooperatives. Otherwise the government policy that attempts to motivate culturists on an individual basis would be constrained by those farming practices, technicalities of making contracts and the delivery system requirements.

3. MATERIALS AND METHODS

The theoretical discussion of the econometric multivariate model is discussed in Enders [5]. For a general presentation we begin with Pesaran and Pesaran [6] augmented vector autoregressive model of order ρ that is AVAR (ρ)

$$y_t = \alpha_0 + \alpha_{1t} + \sum_{i=1}^{\rho} \omega_i y_{t-1} + \tau x_t + \mu_t \tag{1}$$

where y_t is an n × 1 vector of endogenous (jointly determined) variables, a_0 is a drift, t is a linear time trend, x_t is an m × 1 vector of exogenous variables, and μ_t is m × 1 vector of unobserved disturbance terms. The assumptions for the disturbance terms are the mean is zero, $E(\mu_t) = 0$, not serially correlated $E(\mu_t\mu_s) = 0$, the variance matrix is time-invariant that is it does not vary with time. Since not all time series data representing these variables are stationary they have to be tested using Dickey-Fuller (DF) and augmented Dickey-Fuller (ADF) unit root statistics which is necessary in identifying their long-run cointegration [7]. Moreover, equation one contains a constant drift a_0 and the linear time trend which has to be de-trended and to achieve the necessary condition for cointegration the non-stationary variables have to be differenced. The difference multivariate econometric equations exemplifying the general vector error correction model (VECM) [5] is shown in Equation (2) below

$$\Delta y_t = \kappa_0 + \kappa_1 t + \sum_{i=1}^{\rho-1} \psi_i \Delta z_{t-1} + \gamma w_0 + \mu_t \tag{2}$$

where Δy_t is an n × 1 vector of endogenous (jointly determined) I(1) variables.

$\Delta z_{t-i} = \begin{pmatrix} y_{t-i} \\ x_{t-i} \end{pmatrix}$ are n_y × 1 vector of endogenous I(1) and n_x × 1 exogenous I(1) variables respectively.

w_t is m × 1 vector of exogenous I(0) variables.

κ_0 & κ_1 are intercepts and deterministic linear trends.

μ_t is n × 1 vector of unobserved disturbance terms.

T = 1, 2, ..., T represents time periods.

Analysis of aquaculture development uses time series data 1976-2007 obtained from the Department of Fisheries annual statistics, the Ministry of Agriculture Malaysia. The objective of this investigation is to identify causalities between the freshwater fish prices namely that of giant prawn, catfish, river carps (Cyprinus carpio, Pangasius sp.) and goby (Gobiidae) with the total area (in cubic meter) of ponds devoted to rearing of these fresh0 water fisheries and the total number of culturists involved in the production activities. The multivariate equations for the model for the long run variables are presented in Equation (3).

An increase in the area of pond will attract more production of aquaculture into the market that acts to bring down the price of river carps and vice versa. Changes in the price of catfish will have both direct and indirect effect on the price of river carps depending on whether catfish compete or complement the river carps. While a rise in the number of culturists would increase aquaculture production that will ultimately reduce the price of river carps. Similar rationales can be used to explain the economic interrelationships for the rest of the equations in (3).

A direct impact on number of culturist a proxy for employment is anticipated from an increase in the area of pond. On the other hand, changes in either price of catfish or river carps will have a meaningful impact in terms of policy implication on the employment attraction to the aquaculture industry if the prices were found to be significantly associated with the number of culturists and vice versa. In the final analysis only the price of catfish (LPCF) was retained while the price of river carps (LPRC) was omitted from the endogenous system of equation for the long run cointegration analysis.

4. RESULTS AND DISCUSSIONS

4.1. Stationary Tests

Table 1 shows results of unit root tests for the selection of variables that will be used in the multivariate equation model both with trend and without trend. At form level I(0) the number of culturists, the price of river carps and price of catfish were found to be stationary without trend. However, when the linear time trend is considered, with the exception of river carps price, all investigated variables were stationary at level I(0). The results at first difference revealed that all selected variables fulfilled the requirement for the stationary test since the estimated values were found to be greater than the Dickey-Fuller or the augmented Dickey-Fuller critical values as shown in **Table 1** (see the Equation (3) below).

where, $LARP_t$ is annual area of freshwater pond in cubic meters at time t.

$LPRC_t$ is annual average price of river carps in MYR

$$\begin{bmatrix} \Delta LARP_t \\ \Delta LPRC_t \\ \Delta LPCF_t \\ \Delta LCUL_t \\ \Delta LPGP_t \\ \Delta LARP_t \end{bmatrix} = \begin{bmatrix} \alpha_1 \\ \alpha_2 \\ \alpha_3 \\ \alpha_4 \\ \alpha_5 \\ \alpha_6 \end{bmatrix} + \begin{bmatrix} \beta_{11}\beta_{12}\beta_{13}\beta_{14}\beta_{15}\beta_{16} \\ \beta_{21}\beta_{22}\beta_{23}\beta_{24}\beta_{25}\beta_{26} \\ \beta_{31}\beta_{32}\beta_{33}\beta_{34}\beta_{35}\beta_{36} \\ \beta_{41}\beta_{42}\beta_{43}\beta_{44}\beta_{45}\beta_{46} \\ \beta_{51}\beta_{52}\beta_{53}\beta_{54}\beta_{55}\beta_{56} \\ \beta_{61}\beta_{62}\beta_{63}\beta_{64}\beta_{65}\beta_{66} \end{bmatrix} \begin{bmatrix} \Delta LARP_{t-1} \\ \Delta LPRC_{t-1} \\ \Delta LPCF_{t-1} \\ \Delta LCUL_{t-1} \\ \Delta LPGP_{t-1} \\ \Delta LARP_{t-1} \end{bmatrix} + \begin{bmatrix} \gamma_{11}\gamma_{12}\gamma_{13}\gamma_{14}\gamma_{15}\gamma_{16} \\ \gamma_{21}\gamma_{22}\gamma_{23}\gamma_{24}\gamma_{25}\gamma_{26} \\ \gamma_{31}\gamma_{32}\gamma_{33}\gamma_{34}\gamma_{35}\gamma_{36} \\ \gamma_{41}\gamma_{42}\gamma_{43}\gamma_{44}\gamma_{45}\gamma_{46} \\ \gamma_{51}\gamma_{52}\gamma_{53}\gamma_{54}\gamma_{55}\gamma_{56} \\ \gamma_{61}\gamma_{62}\gamma_{63}\gamma_{64}\gamma_{65}\gamma_{66} \end{bmatrix} \begin{bmatrix} \Delta LARP_{t-\rho} \\ \Delta LPRC_{t-\rho} \\ \Delta LPCF_{t-\rho} \\ \Delta LCUL_{t-\rho} \\ \Delta LPGP_{t-\rho} \\ \Delta LARP_{t-\rho} \end{bmatrix} + \begin{bmatrix} \pi_1 \\ \pi_2 \\ \pi_3 \\ \pi_4 \\ \pi_5 \\ \pi_6 \end{bmatrix} [ect_t] + \begin{bmatrix} \mu_1 \\ \mu_2 \\ \mu_3 \\ \mu_4 \\ \mu_5 \\ \mu_6 \end{bmatrix} \quad (3)$$

per kg at time t.

$LPCF_t$ is annual average price of catfish in MYR per kg at time t.

$LCUL_t$ is annual total number of culturists in the industry at time t.

$LPGY_t$ is annual average price of goby in MYR per kg at time t, and $LPGP_t$ is annual average price of giant prawn in MYR per kg at time t.

All variables are measured in natural logarithmic values, ρ > 1 represents lag length greater than one, and Δ is the difference notation. US$1 = MYR3.1.

α_i are intercepts of the multivariate equations, γ_{ij} β_{ij} for i = j = 1, 2, ..., n $(\beta_{ij} \neq \beta_{ji}, \gamma_{ij} \neq \gamma_{ji})$ represent the coefficients of endogenous and exogenous variables of the multivariate equations to be estimated in the model. The terms ect_i refer to the error correction terms, whose coefficients measure speeds of adjustment and are derived from the long-run cointegrating relationships, μ_i represents the disturbance terms, and ρ is the lag lengths for the respective multivariable equations. Causalities tests between these endogenous variables in multivariate analyses using vector error correction model requires that each of these endogenous variables will be used as right hand side (RHS) equation as shown in Equation (3).

4.2. Number of Cointegrating Variables and Order of Lags

Tables 2(a) and (b) show the number of cointegrating variables, r, for the case of restricted intercepts and no trends in vector autoregressive (VAR) model. Both maximal eigenvalue and trace of the stochastic matrix suggest that at least the number of cointegrating variables, r = 2 for the order of lag, VAR = 2 at either 95% or 90% significant level respectively. Pesaran & Pesaran [5] caution that the figures reported by Microfit differ from those obtained by Johansen and Juselius [8] and also by Osterwald-Lenum [9]. However, results from the model selection Akaike Information Criterion (AIC), Schwarz Bayesian Criterion (SBC) and Hannan-Quinn Criterion (HQC) in **Table 2**(c) suggest the cointegrating variables of r = 3, r = 2 and r = 3 respectively. The cointegrating vectors based on the theoretical formulation of the endogenous variables that was originally intended for this investigation and those suggested using econometric criteria will be tested in order to get the most appropriate results. The lag order is limited to two after several trials were carried out for higher degree of lags had proven to yield no significant results. With higher degree of lags such as three and above the loss in the degree of freedom is somewhat serious given the limitation on the number of observations for this study.

Table 1. Unit root tests for the related variables using dickey-fuller.

		Without Trend		With Trend	
		DF	ADF (1)	DF	ADF (1)
At level form	*LCUL*	−4.2311	−3.0532	−4.1422	−2.9732
	LPRC	−3.4457	−3.1533	−2.7205	−1.9461
	LPCF	−5.2088	−3.7823	−5.5283	−3.7478
	LPGY	−1.8526	−1.4778	−3.0975	−2.6430
	LPGP	−0.9844	−0.6692	−4.1355	−3.5040
	LARP	−7.0818	−4.2542	−6.9190	−4.2648
At 1st difference	*LCUL*	−7.4110	−4.7610	−7.5820	−4.7988
	LPRC	−5.9698	−3.9135	−6.3416	−4.5241
	LPCF	−9.4382	−7.3577	−9.2954	−7.5165
	LPGY	−6.5456	−4.7248	−6.4142	−4.6090
	LPGP	−6.2512	−6.0686	−6.0686	−6.7573

Note: Dickey-Fuller statistics at 95 percent critical value are −2.9798 *I*(0) and −3.6027 *I*(1) with trend and 2.9850 *I*(0) and 3.0532 *I*(1) without trend.

5. AQUACULTURE ERROR CORRECTION MODEL

Freshwater species which is widely raised by the majority of culturists is the catfish. Catfish species are relatively easy to breed and they are available in local markets. Furthermore, the species are believed by majority of Malaysians to have the specialty of healing capacity for bleeding. The price of catfish is relatively low compared to other freshwater delicacies like giant prawn. Hence, there is sufficient reason to believe that catfish prices may have a significance influence in attracting culturists as well as the cultivated area for the long run cointegration analysis. These three policy variables are grouped as endogenous variables in the current study while the other prices of river carps, giant prawn and goby are included as the exogenous variables as they are being influenced by unobservable factors outside the currently available data.

The log-linear error correction models (ECM) for the three endogenous variables namely the area per pond in hectares (dLARP), number of culturist (dLCUL) and average price of catfish (dLPCF) are shown in Tables 3 (a)-(c). These regression equations were first estimated using the ordinary least square (OLS) omitting their dynamic characteristics; that is, the differences and lags of the estimated endogenous variables are of order one I(1). The residuals have satisfied the necessary condition for the existence of long-run relationships.

The short-run characteristics of the ECM coefficients need to be substantiated from the difference equations of **Table 3**(a) thru **Table 3**(c). The Granger representation theorem (GRT) requires that the ECM coefficients are negatives and statistically significant.

As apparent ecm1(−1) =−0.81661 (**Table 3**(a)), ecm2(−1) = −1.0587 (**Table 3**(b)) and ecm3(−1) = −0.58028 (**Table 3**(c)) demonstrate that they are statistically significant at 99 percent probability level for the ecm1(−1) and ecm2(−1) respectively. The ecm3(−1) is also significant but at a probability level of 95 percent. Thus, the endogenous vectors used in current aquaculture study are cointegrated in the long-run. This is illustrated in Figures 1 thru 3 showing the impact of generalized impulse responses to one standard deviation shock on the cointegrated variables.

Table 2. (a) Cointegration with restricted intercepts and no trends in the VAR maximal eigenvalue of the stochastic matrix 1978-2007; (b) Cointegration with restricted intercepts and no trends in the VAR trace of the stochastic matrix, 1978-2007; (c) Cointegration with restricted intercepts and no trends in VAR using model selection criteria 1978 to 2007.

(a)

Null	Alternative	Statistic	95% Critical Value	90% Critical Value
$r = 0$	$r = 1$	21.7735	22.0400	19.8600
$r <= 1$	$r = 2$	18.6515	15.8700	13.8100
$r <= 2$	$r = 3$	6.4055	9.1600	7.5300

(Order of $VAR = 2$).

(b)

Null	Alternative	Statistic	95% Critical Value	90% Critical Value
$r = 0$	$r >= 1$	46.8305	22.0400	31.9300
$r <= 1$	$r >= 2$	25.0570	20.1800	17.8800
$r <= 2$	$r = 3$	6.4055	9.1600	7.5300

(Order of $VAR = 2$).

(c)

Rank	Maximized LL	AIC	SBC	HQC
$r = 0$	4.1437	−13.8563	−26.4671	−17.8906
$r = 1$	15.0305	−8.9695	−25.7839	−14.3486
$r = 2$	24.3562	3.6438	−23.2606	−9.9194
$r = 3$	27.5589	−2.4411	−23.4590	−9.1649

(Order of $VAR = 2$) AIC = Akaike Information Criterion; SBC = Schwarz Bayesian Criterion; HQC = Hannan-Quinn Criterion.

Further investigation of interest to the current study that utilizes the information given in **Table 3**(a) through **Table 3**(c) is to derive Granger's causality test. According to Rosilawati Amiruddin et al. [10] the works of Granger [11], Hendry [12], Engle and Granger [6] and Johansen and Juselius [13] on the extension of cointegration technique had opened up new avenues for evidence of causalities in one and bi-directional which were once uncertain due to spurious regression results.

With cointegration between two or more variables and either one or bi-directional causalities could be carried out on appropriate model specification and this is done with the Wald test using Microfit. The section is discussed in next section under Granger's causalities.

6. GRANGER'S CAUSALITY

Causality between variables in time series multivariate analysis was initially investigated by Granger [11,14]. In general it provides richer information than the single or univariate equation model because several endogenous and exogenous variables could be tested simultaneously. These causalities are obviously essential for identifying national strategies such as that on the current policy of promoting aquaculture to supplement the existing depleting marine fisheries. It looks reasonable to augment the present shortage of marine supply with aquaculture undertakings but it would be wise to further investigate and to ask; are the firms responsive to this strategy? They may not be responsive if the non-aquaculture investments outside the industry such as manufacturing, services and agricultural sector itself are more profitable and suitable to the investors' skills and interests.

Table 3. (a) Log-linear ECM for dependent variable dLARP estimated by OLS based on cointegrating VAR (2), 1978-2007; (b) Log-linear ECM for dependent variable dLCUL estimated by OLS based on cointegrating VAR(2), 1978-2007; (c) Log-linear ECM for dependent variable dLPCF estimated by OLS based on co-integrating VAR(2), 1978 to 2007.

(a)

Regressor	Coefficient	Standard Error	T-Ratio (Prob)
dLARP1	0.32130	0.17621	1.8234 (0.083)
dLCUL1	0.06382	0.19527	0.3269 (0.747)
dLPCF1	0.28802	0.16021	1.7978 (0.087)
ecm1(−1)	−0.81661	0.19549	−4.1774 (0.000)
ecm2(−1)	−0.18977	0.28099	−0.6754 (0.507)
ecm3(−1)	−0.62160	0.22416	−2.7731 (0.011)
LPRC	−0.14416	0.08403	−1.7157 (0.101)
LPGP	−1.07080	0.29160	−3.6721 (0.001)
LPGY	0.56681	0.17696	3.2030 (0.004)
R-Squared	0.58779	R-Bar-Squared	0.43076
S.E. of Regression	0.20525	F-stat.F (8,21)	3.74320 (0.007)
Mean of Dependent Variable	−0.06535	S.D. of Dependent Variable	0.27205
Residual Sum of Squares	0.88470	Equation Log-likelihood	10.28740
Akaike Info. Criterion	1.28740	Schwarz Bayesian Criterion	−5.01800
DW-statistic	2.37310	System Log-likelihood	27.55890

(b)

Regressor	Coefficient	Standard Error	T-Ratio (Prob)
dLARP1	−0.10651	0.19277	−0.552530 (0.586)
dLCUL1	0.06630	0.21363	0.310360 (0.759)
dLPCF1	−0.19197	0.17527	−1.095300 (0.286)
ecm1(−1)	0.16165	0.21386	0.755850 (0.458)
ecm2(−1)	−1.05870	0.30741	−3.444000 (0.002)
ecm3(−1)	0.38988	0.24523	1.589900 (0.127)
LPRC	−0.14773	0.09193	−1.607000 (0.123)
LPGP	−0.01180	0.31901	−0.036998 (0.971)
LPGY	0.06295	0.19360	0.325130 (0.748)

R-Squared	0.50601	R-Bar-Squared	0.31782
S.E. of Regression	0.22455	F-stat.F (8.21)	2.6888 (0.033)
Mean of Dependent Variable	0.01944	S.D. of Dependent Variable	0.27187
Residual Sum of Squares	1.05890	Equation Log-likelihood	7.59180
Akaike Info. Criterion	−1.40820	Schwarz Bayesian Criterion	−7.71360
DW-statistic	1.82090	System Log-likelihood	27.55890

(c)

Regressor	Coefficient	Standard Error	T-Ratio (Prob)
dLARP1	−0.01249	0.18242	−0.06848 (0.946)
dLCUL1	−0.11646	0.20215	−0.57613 (0.571)
dLPCF1	−0.28045	0.16585	−1.69090 (0.106)
ecm1(−1)	−0.24628	0.20237	−1.21700 (0.237)
ecm2(−1)	0.35571	0.29089	1.22280 (0.235)
ecm3(−1)	−0.58028	0.23206	−2.50060 (0.021)
LPRC	0.26741	0.08699	3.07410 (0.006)
LPGP	0.44522	0.30187	1.47480 (0.155)
LPGY	−0.29847	0.18320	−1.62920 (0.118)

R-Squared	0.63170	R-Bar-Squared	0.49139
S.E. of Regression	0.21248	F-stat.F (8,21)	4.50230 (0.003)
Mean of Dependent Variable	0.04446	S.D. of Dependent Variable	0.29795
Residual Sum of Squares	0.94814	Equation Log-likelihood	9.24860
Akaike Info. Criterion	0.24860	Schwarz Bayesian Criterion	−5.01800
DW-statistic	2.21240	System Log-likelihood	27.55890

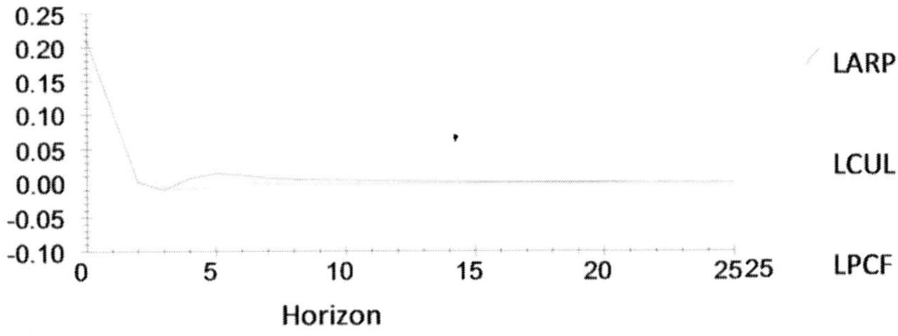

Figure 1. Generalized impulse response(s) to one S.E. shock for LARP.

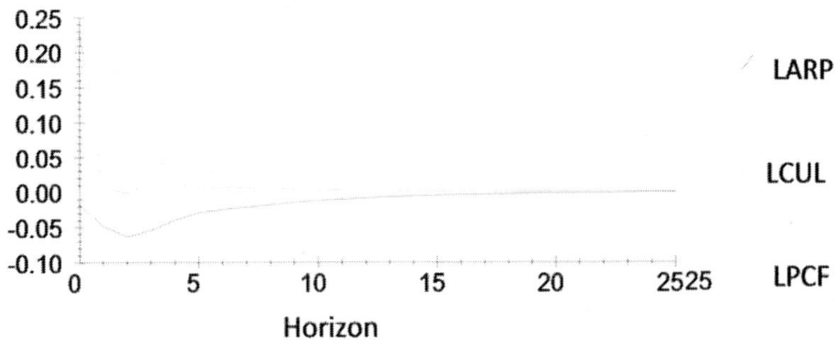

Figure 2. Generalized impulse response(s) to one S.E. shock for LCUL.

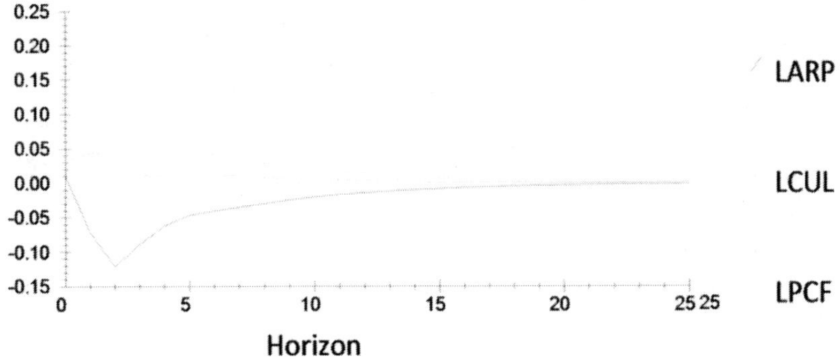

Figure 3. Generalized impulse response(s) to one S.E. shock for LPCF.

Table 4 illustrates the existence of causality among the investigated economic variables in the Malaysian aquaculture industry. There should be either one, bi-directional or nonexistence of causality between these policy variables. The null hypothesis indicated that the size of aquaculture pond area (in ha.) had no causality whatsoever with the number of culturists in the industry and vice-versa. An increase in size of the aquaculture area had not contributed significantly to the course for raising employment. Alternatively, an increase in the number of culturists will not have the potential to increase the size of aquaculture area. However, changes in the one-period lag price of catfish (dLPCF), current prices of river carps (LPRC), giant prawn (LPGP) and current price of goby (LPGY) would draw a direct change in the size of aquaculture pond area. On the other hand, a change in price of catfish will have no effect on inducing one-period lag of size of aquaculture area (dLARP1).

With the exception of direct and current change in price of river carps that have some influence on changing the number of culturists, all other variables in the EVCM model apparently show no significant causality on culturists. Price of river carps may attract those anglers who fish for recreational purpose. Sometimes river carps can fetch a good price for local markets and restaurants and at other times organized competition in sea and river fishing are held. Private individuals also provide special fishing grounds for river carps and selected local freshwater fish species for anglers at reasonable fees.

The one-period lag catfish price (dLPCF1) has shown some degree of causality with its own current price (dLPCF) at about 90 percent probability level. Similarly, the price of exogenous variable goby (LPGY) has exhibited some degree of causality on the current price of catfish. The degree of causality between the current price of river carps and the difference price of catfish are stipulated to be high at 99 percent probability level. In all of these cases the null hypotheses that support no causality between the prices mentioned above are rejected. It is true that the price of giant prawn has no association with the price of catfish. These species have their own market specialties.

Table 4. Wald statistical tests for multivariate vector error correction model (VECM), 1978-2007.

	ΔLARP	ΔLCUL	ΔLPCF
ΔLARP1	3.3249	0.3053	0.00469
	(0.068)*	(0.581)NS	(0.945)NS
ΔLCUL1	0.10683	0.09632	0.33192
	(0.744)NS	(0.756)NS	(0.565)NS
ΔLPCF1	3.2322	1.1996	2.8593
	(0.072)*	(0.273)NS	(0.091)*
ecm1(−1)	17.4503	0.5713	1.481
	(0.000)***	(0.450)NS	(0.224)NS
ecm2(−1)	0.45612	11.8611	1.4953
	(0.499)NS	(0.001)***	(0.221)NS
ecm3(−1)	7.6899	2.5277	6.2531
	(0.006)***	(0.112)NS	(0.012)**
LPRC	2.9437	2.5825	9.4502
	(0.086)*	(0.108)*	(0.002)***
LPGP	13.484	0.00137	2.1752
	(0.000)***	(0.970)NS	(0.140)NS
LPGY	10.2592	0.10571	2.6544
	(0.001)***	(0.745)NS	(0.103)*

Notes: Figures in parentheses represent the probability values (p-values) of significant levels. ***denotes significant at 0.01 probability level. **denotes significant at 0.05 probability level. *denotes significant at 0.10 probability level.

In general, the finding revealed that price of catfish Granger causes the aquaculture area which is only onedirectional since the reverse influence of area on price is not significantly different from zero. Furthermore, the one-period lag aquaculture area Granger causes the current area and the one-period lag price of catfish Granger causes the current price of catfish. This trend is not observable in the case of number of culturists. The exogenous variables of the current prices of giant freshwater prawn, goby and river carps have greatly influenced the increase in size of the aquaculture area. The price of river carps and to a lesser extent the price of goby have a similar influence on the price of catfish. River carps and catfish are substitutes because their prices move in the same direction while goby and catfish are complementary products since their price movements are in the opposite direction.

7. CONCLUSIONS

The multivariate econometric vector error correction model was estimated using Microfit software to investigate the relationship between endogenous and exogenous variables namely the size of pond area, the number of culturists and the prices of the major fresh fish species. The findings revealed that price movements, specifically the price of catfish has caused changes in the size of the aquaculture pond area but the reverse is not true. This indicates that the direction of causality is one way. Furthermore, the potential of the variation in aquaculture area to draw larger investors as indicated by the number of culturists in the industry is most likely not going to happen. The causality between the determining variables is unfortunately insignificant, thus employment wise the promotion for aquaculture projects has not been able to place big hope in terms of employment generation. Possible explanations could be that aquaculture activities require a large amount of initial capital to start, high technical knowhow and it is a risky investment. This is without considering problems relating to marketing of fish harvests, getting supply of fries and the ever-increasing prices of land for fishponds compete with other economically potential investments. In the case of cage culture although it requires high initial outlays competition for water use is less profound; however, water quality needs to be assured from pollution.

The idea of utilizing freshwater and brackish water fisheries to supplement marine catch is nevertheless a worthy strategy that needs to be supported. However, special care for degradation of the environment due to aquaculture development and activities should be taken into consideration by the policy makers. Research and development should be conducted from time to time to find the best way to invest in aquaculture. It would be possible to allow individual households to start a small aquaculture projects to reduce costs and to be self-sufficient in food supply with fishery-based programs in addition to a several large privately owned aquaculture projects with government assistance.

ACKNOWLEDGEMENTS

The authors would like to express their grateful thanks to the Ministry of Higher Education for providing the financial support under the fundamental research grant scheme (FRGS) vote number 59189 for the duration of April 2009-March 2012. They also wish to thanks the unanimous reviewers of this article for their suggestions and recommendations.

REFERENCES

1. Department of Fisheries Malaysia, "Annual Fisheries Statistics (1987-2007)," 2010. http://www.dof.gov.my/59

2. Government of Malaysia, "Rancangan Malaysia Kesembilan 2006-2010," Unit Perancang Ekonomi Malaysia, 2010.

3. EarthTrends AQUASTAT, "Information System on Water and Agriculture Country Profiles," 2003. http://www.fao.org/waicent/faoinfo/ agricult/agl/aglw/aquastat/countries/index.stm

4. P. J. Pradeep, T. C. Srijaya, H. Anuar, S. Faizah and C. Anil, "The Aquatic Chicken Tilapia and its Future Prospects in Malaysia," Institute of

Tropical Aquaculture, University Malaysia Terengganu, Terengganu. Prospect, Malaysian Premier Higher Education Magazine Times Guides Sdn Bhd, Pelangi Damansara, 2010, pp. 52-56.

5. W. Enders, "Applied Econometrics Time Series. New Jersey," John Wiley & Sons Inc., Hoboken, 2003.

6. M. H. Pesaran and B. Pesaran, "Working with Microfit 4.0 Interactive Econometric Analysis," Oxford University Press, Oxford, 1997.

7. R. F. Engle and W. J. Granger, "Cointegration and Error Correction: Representation, Estimation and Testing," Econometrica, Vol. 55, No. 2, 1987, pp. 251-276.

8. S. Johansen and K. Juselius, "Testing Structural Hypotheses in a Multivariate Cointegration Analysis of the PPP and UIP for UK," Journal of Econometrica, Vol. 53, No. 1-3, 1992, pp. 211-244.

9. Osterwald-Lenum, "A Note with Quantiles of the Asymptotic Distribution of the Maximum Likelihood Cointegration Rank Test Statistics," Oxford Bulletin of Economics and Statistics, Vol. 54, No. 3, 1992, pp. 461-472.

10. Rosilawati, A. H. Shaari, and I. Ismail, "Test on Dynamic Relationship between Financial Development and Growth in Malaysia," Gadjah Mada International Journal of Business, Vol. 9, No. 1, 2007, pp. 1-17.

11. W. Granger, "Developments in the Study of Cointegrated Variables," Oxford Bulletin of Economics and Statistics, Vol. 48, No. 3, 1986, pp. 213-227.

12. F. Hendry, "Econometric Modeling with Cointegrated variables: An Overview," Oxford Bulletin of Economics and Statistics, Vol. 48, No. 3, 1986, pp. 201-212.

13. S. Johansen and K. Juselius, "Maximum Likelihood Estimation and Inference on Cointegration with Application to the Demand for Money," Oxford Bulletin of Economics and Statistics, Vol. 52, No. 2, 1990, pp. 169-210.

14. C. W. Granger, "Investigating Causal Relationships by Econometric Models and Cross-Spectral Methods," Econometrica, Vol. 37, No. 3, 1969, pp. 424-438.

Index